11 C

Trenchless Technology Piping

About the Author

Dr. Mohammad Najafi, P.E., is an assistant professor and the director of the Center for Underground Infrastructure Research and Education (CUIRE) in the Department of Civil Engineering at The University of Texas at Arlington. He received his B.S. from Texas  Tech University, M.S. from Purdue University, and Ph.D. from Louisiana Tech University. He is the founder and editor-in-chief of the new *ASCE Journal of Pipeline Systems Engineering and Practice*. Dr. Najafi's research interests are in the area of pipelines and underground infrastructure systems construction, renewal, and asset management, with emphasis on the utilization of innovative methods and trenchless technology. His research has been widely published in technical journals and professional and trade conference proceedings, and he has been a consultant and expert witness on many pipelines and trenchless technology projects. He can be reached at najafi@uta.edu.

Trenchless Technology Piping

Installation and Inspection

Mohammad Najafi, Ph.D., P.E.
The University of Texas at Arlington

WEF Press
Water Environment Federation
Alexandria, Virginia

ASCE PRESS

New York   Chicago   San Francisco
Lisbon   London   Madrid   Mexico City
Milan   New Delhi   San Juan
Seoul   Singapore   Sydney   Toronto

Cataloging-in-Publication Data is on file with the Library of Congress

621 .8672 NAJAFI   2010

Najafi, Mohammad.

Trenchless technology piping

1 2 3 4 5 6 7 8 9 0   DOC/DOC   1 6 5 4 3 2 1 0

ISBN 978-0-07-148928-7
MHID 0-07-148928-2

Sponsoring Editor	Project Managers	Indexer
Larry S. Hager	Somya Rustagi and Vastavikta Sharma, Glyph International	Arc Films, Inc.
Editing Supervisor		**Art Director, Cover**
Stephen M. Smith		Jeff Weeks
	Copy Editor	
Production Supervisor	Ragini Pandey,	**Composition**
Pamela A. Pelton	Glyph International	Glyph International
Acquisitions Coordinator	**Proofreader**	
Alexis Richard	C&W Shields, Inc.	

Printed and bound by RR Donnelley.

McGraw-Hill books are available at special quantity discounts to use as premiums and sales promotions, or for use in corporate training programs. To contact a representative, please e-mail us at bulksales@mcgraw-hill.com.

This book is printed on acid-free paper.

It took more than two years to plan and complete this book. During this time, my colleagues, students, friends, and family have supported my efforts. I would like to thank Dr. Tom Iseley, my dissertation advisor at Louisiana Tech University, who introduced me to the exciting field of trenchless technology. I would like to acknowledge the support of my family and dedicate this book to my wife Homa, my daughters Maryam and Zinat, my son-in-law Robert Bernacki, and my grandson Ashton.

Contents

Preface

This book complements another title, *Trenchless Technology: Pipeline and Utility Design, Construction, and Renewal*, published by McGraw-Hill and WEF Press in 2005. The main idea for this book came from a series of trenchless technology operator and inspector training schools the author has offered since 1998. The need for proper planning and partnering among owners, engineers, and contractors in design, contract and project execution is especially critical in trenchless technology projects where many uncertainties are involved. However, development of new equipment and methods and increased level of equipment sophistication and capabilities is not a substitute for experience and training of operators and field personnel, even it has increased this need. A safe and successful trenchless project depends on skills, training, and experience of operators, field personnel, and project inspectors. The main objective of this book is to highlight some critical aspects of trenchless installation and inspection, and provide conceptual guidelines for proper execution and quality control of these projects. This book will be of use for all parties in a trenchless project as it provides planning, design and construction, inspection, and project management concepts.

Chapter 1 begins with an overview of buried pipe history, pipe-soil interaction, and continues with comparison of open-cut and trenchless method of pipe installation. A description of trenchless installation methods, including conventional pipe jacking, utility tunneling, horizontal earth boring (horizontal auger boring, horizontal directional drilling, microtunneling, pilot-tube microtunneling, and pipe ramming), are included. An important part of this chapter includes characteristics and applications, including capabilities and limitations for trenchless installation methods. This information will be helpful to utility owners and design and consulting engineers to properly select appropriate trenchless installation methods based on their project specifics and site conditions. This chapter concludes with a summary of safety considerations and potential risks/impacts to pavements and adjacent utilities.

Chapter 2 focuses on existing pipeline renewal and replacement methods. It includes planning and design process, method applicability, and descriptions of different renewal and replacement methods. To help with planning and method selection process, this chapter includes decision support systems for both gravity and pressure pipeline systems. This chapter concludes with an overview of emerging design concepts, and presents detailed design calculations for partially deteriorated and fully deteriorated pipes and provides sample technical specifications.

Due to emerging importance of water conservation and quality of service to customers, Chap. 3 is dedicated to spray-on coatings and linings for potable water pipe renewals. Such topics as method selection criteria, site investigation requirements, pipe inspection requirements, installation considerations, quality control, and project closeout requirements are covered in this chapter. In conclusion, descriptions of four common spray-on linings and coating materials, namely, cement mortar, epoxy, polyurea, and polyurethane are described and compared.

Chapter 4 covers one of the most important topics in pipeline installation and inspection, namely, the pipe-soil interactions. To present a better understanding of the concept, a comparison of open-cut and trenchless installation is provided in this chapter. Other important topics, such as the concept of flexible and rigid pipe as well as common methods of pipe failures are included in detail. This chapter continues with pipe selection considerations for cement-based pipes, asbestos-cement pipes, vitrified clay, plastic pipes, glass-reinforced pipes, and metallic pipes. For each pipe type, manufacturing process, applicable standards, joint types, and advantages and limitations are presented.

Due to the importance of the horizontal directional drilling (HDD) method of pipe installation, Chap. 5 is devoted to this method. The topics included are method descriptions for large, medium, and small (Maxi, Midi, and Mini) methods of HDD, site investigation requirements, drilling operations, drilled path design, drilling fluids, installation loads, contractual considerations, inspection and construction monitoring. To present an understanding of concepts and parameters involved in design of HDD projects, at the conclusion of the chapter, a simplified bore planning and pipe load calculation for use of high-density polyethylene pipe (HDPE) for Mini-HDD projects is provided. Similar concepts can be developed for larger size HDD projects and different pipe materials.

Chapter 6 presents a full coverage of pipe replacement method which is another growing application for trenchless technology. The main branch of pipe replacement is pipe bursting, which is covered in great detail. The topics presented include design considerations, investigation of existing pipe and site conditions, geotechnical investigation requirements, risk assessment plan, ground movements, plans and specifications, submittals, quality control/quality assurance issues, construction considerations, and safety considerations. The last section of this chapter is devoted to pipe load calculations, to provide a simplified approach for better understanding of forces encountered in the field. While this chapter is mainly focused on HDPE, which is product of choice for pipe bursting, the same concept can be applied to other pipe materials.

Chapter 7 concentrates on Cured-in-Place Pipe (CIPP) technology, which is the most common method for renewal of deteriorated existing pipes. The chapter covers full description of CIPP technology, and continues with site compatibility and applications, effects of existing pipe defects, and installation methods. Due to a large number of factors influencing quality installation of CIPP, there have been a number of issues regarding final inspection and acceptance by owners. Therefore, this chapter provides a section on inspecting installation of CIPP, and covers possible CIPP defects and causes, accepted tolerances, repair types, and possible monetary penalties. While throughout this book, partnering among owners, design and consulting engineers and contractors/installers are emphasized, there might be situations that owners may have to correct a defective CIPP installation, and contractors may offer a deduction in order for owners to be able to accept the installation and pay the balance of payment to the contractor. Therefore, these suggested guidelines can be used as bases for considerations that are mutually accepted to both parties. It should be noted that including harsh language in the bid documents and/or unacceptable and unreasonable tolerances in the contract and specifications will result in contractors submitting higher bids (with more contingencies) or not submitting bids at all.

Chapter 8 is dedicated on quality assurance/quality control considerations for trenchless installation and replacement methods. The idea for this chapter came around the concept that most trenchless installations are conducted in urban settings and "under a road, street, or pavement." So basically, the trenchless installation must be safe for the general public, road embankment and the operational use. This chapter provides suggested guidelines for the "road authority" or the "transportation agency" to evaluate requests for trenchless installations (such as road crossings), issue permit, and inspect the installation. It provides simple charts and checklists for the inspectors to observe compliance with the main parameters of permit, to ensure a safe and quality installation is completed. This chapter is written in such a way that provides a standalone presentation of trenchless installation methods (microtunneling, horizontal directional drilling, horizontal auger boring, pipe ramming, conventional pipe jacking, pilot-tube microtunneling, and pipe replacement) with introduction, pipe material, and construction considerations. For each method, it presents a simple inspection guide, which includes preinspection plan review and construction inspection.

Chapter 9 covers two main areas of any construction and specifically trenchless technology project: planning and safety. Without a good plan, there will not be a safe and productive project. Obviously, planning starts at the project inception and continues with project installation and completion. So all parties in a construction project have a role in planning, but the initial planning conducted by the owner and the design/consulting engineer has the greatest impact over successful completion of the project. Among decisions made and information collected during the initial planning include, project objectives, surface and subsurface investigations, jobsite logistic requirements, screening of alternatives and method selection, reasonable and realistic expectations and tolerances, risk and hazard assessment, contingency plans, and safety requirements. The owner of the project has the main responsibility in the safe and productive completion of the project, as he or she set all the rules and expectations for the contractor/installer to follow. Simply said, without a quality plan and quality contract, there will not be a quality and safe trenchless installation.

Concluding our thoughts will require an emphasis upon planning ahead, proper surface and subsurface investigations, proper technology selection, appropriate design and specifications, quality control and inspections, team play and partnering among all parties. The best time to begin avoiding and resolving disputes is at the very beginning by making absolutely sure that each party understands the conditions on which the bid proposal is based. Contracts, whether written or oral, should be carried out in good faith to their full intent.

This book has addressed important aspects of typical trenchless technology installation and inspections. It is intended as a reference for design and consulting engineers, utility owners, pipeline professional, government agencies, municipalities, contractors, manufacturers, professionals who are involved with planning, design, construction, operation, and maintenance of underground pipelines and utility systems. This book is also intended as a reference for formal and classroom training of both young and experienced engineers and pipeline professionals to complement knowledge gained from field operations.

Mohammad Najafi, Ph.D., P.E.

New Developments

Since 1990, when, by the establishment of the North American Society for Trenchless Technology (NASTT), the trenchless technology industry started and organized in the United States, many major accomplishments have been made. The organizational activities have been led by many developments in trenchless equipment and methods. These developments include manufacturing more powerful and versatile horizontal directional drilling equipment, including new locating and tracking tools by Digital Control, manufacturing of first microtunneling boring machine (MTBM) in the United States in 1997 by Akkerman, Inc., and development of more capable pipe bursting/replacement and pipe ramming equipment. On the area of renewing deteriorated and old pipelines, we have seen developments in pipeline inspection technologies, and new gravity and pressure pipelines (such as new potable water distribution spray-in-place pipe by 3M water infrastructure) and manhole renewal methods. Other developments include new methods and tools in existing pipeline inspections (such as use of laser and pipe scanning technologies, and use of ground penetrating radar*), underground mapping and utility locating, see ahead technologies, curved pipe jacking and microtunneling, new MTBM cutterheads (such as high pressure flushing systems), new steering and tracking tools for horizontal auger boring machines, and new on-grade installations for gravity pipelines. The examples of on-grade installation are Vacuum Boring System (AXIS®) by Vermeer (see Fig. 1), ArrowBore® by Trenchless Flowline (see Fig. 2), Ditch Witch Grade Drilling Method® (see Fig. 3), and new restraint joint PVC pipe for trenchless applications by S & B Technical Products (see Fig. 4).

Many times, contractors may choose more than one method to address project challenges. Examples of these "hybrid" technologies

---

*Ground penetrating radar (GPR) is a real-time, nondestructive testing technique that uses radio waves to image pipes. Referred to as Pipe Penetrating Radar® (PPR) by SewerVUE, this emerging method can be used to detect pipe wall cracks, changes in material, reinforcing location and placement, and pipe wall thickness for concrete, polyethylene, vitrified clay, and similar pipe materials.

Figure 1   AXIS® Vacuum Boring System. (*Source: Ariaratnam et al., 2010.*)

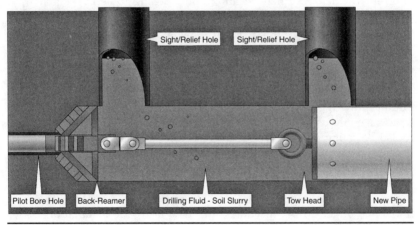

Figure 2   ArrowBore®. (*Source: Trenchless Flowline.*)

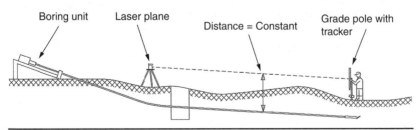

Figure 3   Ditch Witch Grade Drilling Method®. (*Source: Gunsaulis and Levings, 2008.*)

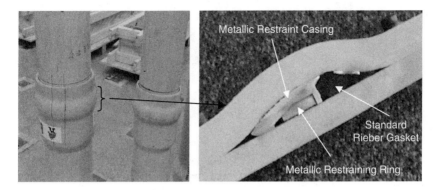

FIGURE 4   New Bulldog Restraint Joint System for trenchless applications of PVC pressure pipes. (*Source: S & B Technical Products.*)

are horizontal directional drilling (HDD) with pipe ramming, such as Conductor Barrel®, Drill Stem Recovery, Bore Salvage®, Pullback Assist®, and Pipe Ram Rescue® (see Fig. 5) for retrieving a disabled MTBM, all offered by TT Technologies.

When HDD procedure reaches its limits—for example, on extremely long crossings with large pipe diameters—the Herrenknecht Pipe Thruster offers the necessary extra power. Installed at the target pit, it pushes the pipe string with a thrust force of up to 750 tons towards the entry point. Furthermore, it distributes the forces acting on the pipe, since the pipe string is pulled (by the HDD rig) and pushed (by the Pipe Thruster; see Fig. 6) at the same time. The pipe string is handled with the specially developed clamping segments which are suitable for pipe coatings and diameters of 20 to 60 in. A further advantage of the Pipe Thruster is its mobility. It can be installed in densely built-up areas and in inaccessible terrain.

Design engineers, utility owners, and contractors should consider the existing and proven methods, and those viable methods that will

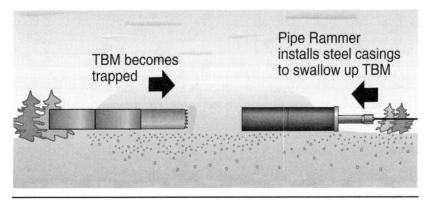

FIGURE 5   Pipe Ram Rescue. (*Source: TT Technologies.*)

Figure 6   Pipe Thruster. (*Source: Herrenknecht AG.*)

be available or under development. Further, it should be noted that underground pipeline construction and renewal projects present many risks. There is no guarantee that the methods presented in this book will be successful at all times and at all project and site (surface and subsurface) conditions. Moreover, there is no endorsement and/or recommendation of the proprietary methods, pipe materials, and brand names mentioned in this book. The specified method characteristics, diameter range, maximum installation, typical application, and accuracy is based on project and site conditions, type of equipment used, and experience and training of the crews and the operators. Design engineers, project owners, contractors, government agencies, and all other parties involved in trenchless technology projects should consider the risks involved and assume appropriate contingencies in the contract documents. Methods successfully used in some applications may not be applicable in other conditions due to change in project, site, and/or soil conditions. Design engineers and pipeline owners need to be involved in the selection of appropriate trenchless methods for their specific project conditions and do not leave the trenchless method selection entirely to the contractor.

As a final word, trenchless technology is developing at a fast rate. Therefore, everyone involved in a trenchless technology construction and renewal should keep abreast of latest developments. Some sources of further information are provided at the end of this book. Other sources are professional and trade conferences, industry magazines, technical journals, professional and trade associations, and companies and organizations involved in trenchless equipment manufacturing, engineering, and contracting. Many information and case studies of past projects are accessible through Web sites of these organizations.

Acknowledgments

This book would not be possible without the help, support, and patience of Larry Hager, senior editor, McGraw-Hill Professional. The author would like to thank McGraw-Hill Professional and the editorial staff for their support of this publication. The cooperation and hard work of Somya Rustagi and Vastavikta Sharma, project managers, Glyph International, and their colleagues are acknowledged. The author would like to acknowledge the contributions of his current and former students at Michigan State University and The University of Texas at Arlington. The efforts of Abhay Jain, CUIRE research assistant, who spent many long hours coordinating the review process and preparing the manuscript, are appreciated. The author would like to acknowledge the support and contributions of the board members of the Center for Underground Infrastructure Research and Education (www.cuire.org) and his colleagues in the Department of Civil Engineering at The University of Texas at Arlington.

The author is indebted to many contributors and reviewers, whose names follow, for helping to complete this book.

Contributors

Alan Atalah, Bowling Green State University *(portions of Chap. 5)*

Grady Bell, Laney Directional Drilling Co. *(Fig. 5.1)*

Ralph Carpenter, American Ductile Iron Pipe Co. *(portions of Chap. 4)*

Mark Dionise, Michigan Department of Transportation *(portions of Chap. 8)*

Mustafa Kanchwala, CUIRE *(portions of Chap. 3)*

Trupti Kulkarni, CUIRE *(portions of Chap. 3)*

Rich Maxwell, StraightLine *(portions of Chap. 9)*

Terry McArthur, HDR Engineering, Inc. *(portions of Chap. 9)*

Lynn Osborn, Insituform Technologies *(Secs. 2.7 and 2.8)*

Alhad Panwalkar, Jacobs Associates *(portions of Chap. 4)*

Plastics Pipe Institute (PPI) *(portions of Chap. 6)*

Jeff Pocket, J. D. Hair and Associates *(portions of Chap. 5)*

Shah Rahman, Northwest Pipe Company *(portions of Chap. 4)*

Larry Slavin, Outside Plant Consulting Services, Inc. *(Secs. 5.4 and 6.10)*

Reynolds Watkins, Utah State University *(Sec. 1.1)*

Reviewers

Bill Adams, WL Plastics *(Chap. 4)*

Dan Akkerman, Akkerman Inc. *(Chap. 8)*

Sam Arnaout, Hanson Pipe and Precast *(Chap. 4)*

Chris Brahler, TT Technologies *(Chap. 8)*

Craig Fisher, S&B Technical Products *(Chap. 4)*

Ahmad Habibian, Black & Veatch *(Chaps. 1 and 9)*

Joanne Hughes, RS Lining Systems *(Chap. 3)*

Daniel Liotti, Midwest Mole, Inc. *(Chap. 8)*

Joe Lundy, Hanson Pipe and Precast *(Chap. 4)*

Gary Natwig, 3M Water Infrastructure *(Chap. 3)*

Paul Nicolas, Akkerman Inc. *(Chap. 8)*

Lynn Osborn, Insituform Technologies *(Chaps. 2 and 7)*

Mario Perez, 3M Water Infrastructure *(Chap. 3)*

Larry Petroff, Performance Pipe *(Chap. 4)*

Camille Rubeiz, Plastics Pipe Institute *(Chaps. 4 and 6)*

Mike Schwager, TT Technologies *(Chap. 8)*

Eric Skonberg, Trenchless Engineering Corp. *(Chap. 5)*

Rick Turkopp, Hobas Pipe Inc. *(Chap. 4)*

Mike Vandine, National Clay Pipe Institute *(Chap. 4)*

CHAPTER 1

New Pipeline Installations

1.1 Buried Pipe History

The history of buried pipes started about 2500 B.C., when the Chinese delivered water through bamboo pipes. In some Mediterranean countries, clay pipes supplied water to villagers at a central well. Buried pipes in Persia, called "khanats," were rock-lined tunnels, dug by hand back under the mountains, to collect clean water and pipe it as much as 30 mi to parched cities on the plains. In ancient Greece, pipelines and tunnels were constructed to distribute water in urban areas. From A.D. 100 to 300 in Rome, with plenty of low-cost slave labor, pipes became the guts of the infrastructure for the emperor and elite. Water was delivered to Rome in aqueducts and distributed in lead pipes to the mansions of the elite and to their luxurious Roman baths. The "fall of Rome" may have been brought about, in part, by those lead pipes. The acidic water of Rome dissolved lead from the pipes. The elite were lead-poisoned. Lead caused impotence, and the few successful pregnancies produced heirs who were imbeciles.

In North America, European settlers fashioned pipes by boring logs. Later they made wooden pipes from carefully sawed staves held together by steel hoops. The concept was adapted from coopers who made wooden barrels and tubs. Some old wooden stave pipes are still in service. Iron had been known since 1000 B.C., but before the Renaissance, iron was used mostly to make spears, swords, and shields. In A.D. 1346, iron was used to make guns. Those guns became the inspiration for iron pipe—the dream of "ingeniators" (the ingenious ones)—because of the demand for water in the burgeoning cities and because iron was stronger than bamboo or clay. Iron pipes became reality in England in 1824 when James Russell invented a device for welding iron tubes (gun barrels) together into pipes. Costly, handmade, iron pipes supplied gas for the gas lamps in the streets and the dwellings of the elite. In 1825, Cornelius Whitehouse made long iron pipes by drawing flat strips of hot iron through a bell-shaped die. Then came the Bessemer process for making steel, and the open-hearth furnace for production of large quantities of steel. Steel pipes became reality. The urban way of life changed. Community expanded into metropolis. The "guts of the city" became steel pipes for water and clay pipes for sewage.

1.1.1   The Pipe-Soil Interaction

The evolution of pipes was by trial and error. How many lives were lost in cave-ins when the Persians excavated tunnels ahead of rock lining? How could Romans predict that lead pipes would poison them? Evolution of pipes proceeded empirically—not by design. Design of buried pipes started in 1913, when Anson Marston became the dean of engineering at Iowa State College. He noted, with concern, how transportation bogged down during every rainstorm and every spring thaw when "dirt" roads became quagmires of mud. The dean sought a remedy. He called for action with publicized resolve: "Let's get Iowa out of the mud." He recognized that to get Iowa out of the mud, he had to get the water out of the roads. His remedy was buried drainpipes along the roads. It worked. The nation was impressed. A Federal Highway Research Board was formed with Marston as the first director. Marston came out with the first engineered design of buried pipe, an equation for soil load on pipe. In those days, drainpipes were clay or concrete—both rigid. He left design of the pipe up to the manufacturers to make pipe strong enough to withstand his *Marston load*. The idea was radical—performance specification, not the traditional procedure specification. The test for pipe strength was a three-edge-bearing (TEB) load. A pipe was supported on two wooden 2 × 4's close together and one 2 × 4 on top onto which load was applied. The load at failure, divided by a safety factor, was to be greater than the Marston load. Clearly, design was conservative.

The Armco Company had developed flexible pipe of corrugated steel. Steel was available in coils from which corrugated pipe could be fabricated. But the flexible pipe could not support a Marston load under a three-edge-bearing test in the yard. When buried, however, flexible pipe worked. Armco sponsored a research project at Iowa State College. The project was assigned to a young faculty member, M. G. Spangler. Using a soil box, Spangler showed that flexible pipe deflects under soil load and develops horizontal soil support on the sides of the pipe. Corrugated steel pipe could be used as culverts and drain pipes. Spangler derived the *Iowa formula* for predicting the horizontal expansion of buried, flexible pipe based on a horizontal soil modulus of elasticity, E'. The derivation was elegant and correct, except for the definition of E'. The soil modulus was corrected by Spangler's student Watkins and the modified Iowa formula was published (Watkins and Spangler, 1955). The Iowa formula is not a basis for design of pipe, but rather it predicts ring deflection of flexible pipe when buried and shows the importance of the soil embedment. Spangler is considered as "the father of pipe-soil interaction."

1.2   Open-Cut Method of Pipe Installation

The conventional method for construction, replacement, and renewal of underground utilities has been trenching or open cut (OC), an *indirect method of pipeline installation*. Open-cut methods involve digging a trench along the alignment of the proposed pipeline, supporting trench walls (or sloping sides of trench), constructing bedding or foundation, placing the pipe sections in the trench, embedding the pipe sections, and backfilling the trench and compacting operations (see Fig. 1.1). In open-cut construction, the construction effort is concentrated on such activities as managing the traffic flow through detour roads, trench excavation and shoring, shielding or sloping, dewatering (if needed), backfilling and compaction operations, and reinstatement of the surface. This leads to a small part of the construction effort actually being focused on the final product, which is the pipe installation itself. In some cases, the double handling of the soil, including trenching, stockpiling, hauling, backfilling, compacting, and reinstating of the ground and pavement alone may amount to *70 percent of the total cost of the project.*

Additionally, open-cut methods include higher "social costs." Social costs include costs to general public (such as traffic disruptions, noise, and dust), negative impacts to the environment (higher

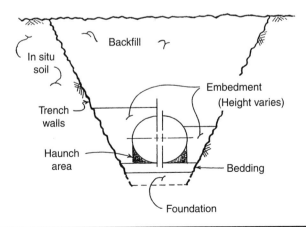

FIGURE 1.1   Open-cut installation. (*Howard, 2002.*)

carbon emissions due to higher fuel consumptions), safety hazards of trenching (cave-in hazards, traffic accidents, falls into trench, and so on), damage to trees and tree roots, reduction in service life of pavement,* and possible damage to existing utilities and structures due to trenching operations. Najafi (2005) presents a breakdown of the social costs including an analysis of the life-cycle cost of the project. As such, and considering all the project parameters, the open-cut method in most cases could be a time-consuming and inefficient method of pipe installation.

Trenchless technology is a process for construction, renewal, and replacement of underground pipelines and utilities with minimal surface and subsurface disruptions. Trenchless technology methods provide more opportunities for direct installation of pipelines and ducts with welded and fused or bell and spigot and retrained joints, thereby, reducing possibilities of leaks and joint misalignments and settlements. Jacking pipes, commonly used in trenchless installation, have higher degree of accuracy, which requires the installation by specialized and qualified contractors. Due to the larger wall thickness requirements, higher strength materials, and lower tolerances requirements, jacking pipes have a higher resistance to stresses due to shear loads. The pushing-in of the jacking pipes guarantees a higher positional accuracy and a lower disturbance of the natural soil in the embedment area so that optimum bedding is achieved.

Another important benefit of trenchless technology is the fact that in open-cut method, through the soil prism above the pipe, soil backfill and traffic loads act directly on top of the pipe, causing pipe deflections in flexible pipes and more stress in rigid pipes. In comparison, studies have shown that due to arching effects of soil in trenchless technology installations, the loads on the pipe are considerably less and more uniformly distributed around the pipe surface, resulting in minimum pipe deflections. Figure 1.2 illustrates a comparison of pipe loads in trenchless and open-cut methods.

Trenchless installation methods significantly reduce the amount of excavation, backfilling, compaction, and pavement replacement. In the trenchless installation method, the most costly item is the pipe installation, usually estimated on a linear foot basis, while in the open-cut method excavation, backfilling, pavement replacement, and shielding or shoring are the major cost items and are estimated on a cubic yard or square foot basis.

---

*Studies have shown approximately up to 60 percent reduction in the life of pavement due to lateral cracking after pavement cuts.

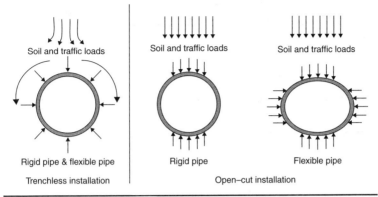

FIGURE 1.2 Comparison of pipe loads in trenchless and open-cut methods.

1.3   Comparison of Construction Operations

Table 1.1 presents a comparison of construction operations for open-cut and trenchless technology methods.

1.4   Trenchless Technology Methods

Since mid-1980s, when trenchless technology as an industry was established, the trenchless technology equipment and methods have grown tremendously. Many developments have occurred in the trenchless equipment and methods. These developments include manufacturing more powerful and versatile horizontal directional drilling (HDD) equipment, utility locating and tracking tools, more sophisticated microtunneling (MT) equipment, and more capable pipe-ramming, horizontal auger-boring (HAB), and tunneling equipment, as well as better pipe materials and proprietary joints. In the area of renewing deteriorated and old pipelines, developments have been in pipeline inspection technologies, pipe bursting, and new pipeline renewal methods in all areas of water, sewer, gas, and oil applications.

Trenchless technology methods are divided into three main areas: (1) construction and installation of new pipelines and utilities (discussed in this chapter); (2) renewal; and (3) replacement of existing, old, and deteriorated pipelines and utilities (discussed in Chaps. 2 and 3). Trenchless technology applications for new installations include pressure pipes, gravity pipes, culverts, and drainage structures under roads, railroads, and river crossings, and installation of cables and telecommunications ducts. Table 1.2 presents primary and alternative applications for trenchless technology methods for new installations, renewals, and replacements. Figure 1.3 illustrates different trenchless new installation method to

Construction Operations	Open-Cut	Trenchless
Route surveying	Yes	Yes
Land and easement acquisitions	Yes	Yes
Mobilizing equipment and personnel	Yes	Yes
Preparing the right-of-way (ROW) (cleaning and grubbing)	Yes	Maybe[b]
Transporting and storing pipe and other materials	Yes	Yes
Topsoil stripping	Yes	No
Grading	Yes	No
Stringing (transporting and laying of pipe on the ROW)	Yes	Maybe[b]
Transporting welding machines and other equipment to site	Yes	Yes
Welding, ultrasonic, and x-ray checking of welds	Yes	Yes
Installing protective coating at pipe joints	Yes	Yes
Testing pipe for external coating integrity	Yes	Yes
Trenching (including shoring, sloping or shielding)	Yes	No
Dewatering	Yes	Yes
Lowering pipe into trench or shaft/pit	Yes	Yes
Installing block valves and terminus equipment	Yes	Yes
Hauling select soil	Yes	No
Backfilling	Yes	No
Compacting backfill soil	Yes	No
Disposing extra soil	Yes	No
Leak testing (hydrostatic testing) and/or internal inspection	Yes	Yes
Reinstatement of ground	Yes	Maybe[b]
Final inspection	Yes	Yes
Demobilizing equipment and personnel	Yes	Yes
Installing cathodic protection facilities	Yes	Yes
Preparation of as-built	Yes	Yes

---

[a]Trenching, backfilling, dewatering and reinstatement of ground make up 70% of cost for open-cut projects.

[b]Required for certain types of horizontal directional drilling, pipe bursting, and continuous sliplining methods.

TABLE 1.1   A Comparison of Construction Operations for Open-cut and Trenchless Technology Methods[a]

Application	Primary Trenchless Technology Methods	Possible Trenchless Technology Methods[a]
Installation of a New Pipe		
Gravity pipe (sanity and storm sewers)	Conventional pipe jacking (CPJ), utility tunneling (UT), microtunneling (MT)	Horizontal directional drilling (HDD)
Pressure pipes (water, gas, and oil)	Horizontal directional drilling (HDD)	Conventional pipe jacking (CPJ), utility tunneling (UT)
Road and railroad crossings	Horizontal auger boring (HAB), pipe ramming (PR)	Horizontal directional drilling (HDD)
Cables and telecommunications	Mini horizontal directional drilling (Mini-HDD)	Piercing method (PM)
Renewing an Existing Pipe		
Gravity pipes (sanity and storm sewers)	Cured-in-place pipe (CIPP), sliplining (SL), modified sliplining (MSL)	Close-fit pipe (CFP)
Pressure pipes (water, gas, and oil)	Close-fit pipe (CFP), cured-in-place pipe (CIPP), spray-in-place pipe (SIPP)	Sliplining, coatings and linings (CL)
Culverts and drainage structures	Cured-in-place pipe (CIPP), sliplining (SL), modified sliplining (MSL)	Coatings and linings (CL)
Replacing an Existing Pipe		
Pressure pipes (water, gas, and oil)	Pipe bursting (PB), pipe reaming (PRM)	Parallel pipeline using horizontal directional drilling (HDD)
Gravity sewer pipes (sanity and storm sewers)	Pipe bursting (PB), pipe reaming (PRM)	Parallel pipeline using microtunneling or pilot tube microtunneling (PTMT)
Service laterals	Pipe bursting (PB)	Parallel pipeline
Culverts and drainage structures	Pipe bursting (PB), pipe ramming (PR)	Horizontal auger boring (HAB), microtunneling method (MTM)

[a]Application of these methods requires special tools or equipment, specialized contractor and operator experience, improved construction practices, and/or may cost more than the main methods. In some cases, the method might be new in the area of this application, and capabilities of the method currently are not fully proven.

TABLE **1.2** Primary and Alternative Applications for Trenchless Technology Methods

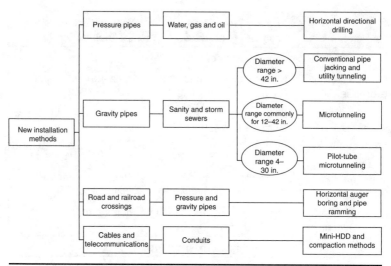

Figure 1.3 Applications of trenchless installation methods.

be used based on the application. Section 1.5 provides more information on specific trenchless installation methods.

1.5 Three Main Divisions of Trenchless Technology Methods

Figure 1.4 divides trenchless technology methods into three main areas: (1) new installation or construction methods, (2) renewal methods, and (3) replacement methods. New installation methods include all the methods for new utility and pipeline installations as shown in Fig. 1.3, where a new pipeline or utility is installed. Trenchless renewal methods include all the *direct* (compared to indirect for open cut) methods of renewing, rehabilitating, and/or renovating an existing, old, or host pipe (in this book collectively called existing pipe or utility). Figure 1.5 also illustrates the subcategories of the new installation methods. Trenchless replacement methods include all the *inline* (compared to installing a new parallel pipeline) where an existing, old, or host pipe is fractured and pushed into the soil or fractured and

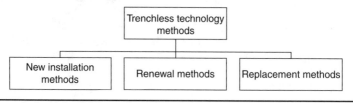

Figure 1.4 Main divisions of trenchless technology methods.

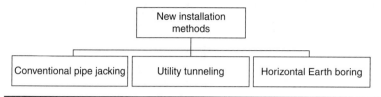

FIGURE 1.5 Trenchless installation methods.

removed and a new pipe is installed in its place. In the inline replacement method, the existing pipe acts like a pilot hole to guide the installation of the new pipe in the same space as the existing pipe occupied originally. It is understood that sometimes point repairs may be required before a renewal or replacement method can be used. Many times, most cost-effective new installation, renewal, or replacement may include utilization of several trenchless methods to adapt to the specific conditions of the project. For a safe, successful, and cost-effective trenchless operation, certain conditions and requirements must be met as explained in this book.

1.6   Trenchless Installation Methods

Trenchless installation methods include all the methods of installing *new pipelines or utility* systems below ground surface without direct installation into an open-cut trench. As Fig. 1.5 illustrates, trenchless installation methods are further divided into two broad categories of worker-entry and nonworker-entry installations. Conventional pipe jacking (CPJ) and utility-tunneling (UT) techniques require workers inside the tunnel during the excavation and pipe installation. However, horizontal Earth-boring methods includes techniques in which the tunnel or borehole excavation is accomplished through mechanical means *without* workers being inside the tunnel or borehole. Conventional pipe jacking is differentiated from utility tunneling by the soil support structure. It can be identified as a *"one-phase"* installation, where pipe sections are installed at the same time when soil excavation is made. In utility tunneling, first the tunnel is excavated and lined with a tunnel liner plate or wood lagging and steel ribs. After completion of the tunnel, the pipe sections are "transported" one by one and installed in the tunnel. After all pipe sections are installed, the annual space between the liner and the outside face of the pipe may be grouted. Both conventional pipe jacking and utility tunneling may utilize the same excavation equipment [such as a tunnel boring machine (TBM)].

1.6.1   Conventional Pipe Jacking

Conventional pipe jacking (PJ), a method of pipe installation where prefabricated pipe sections are jacked or pushed behind the tunnel boring machine or other tunnel excavation methods. In this method,

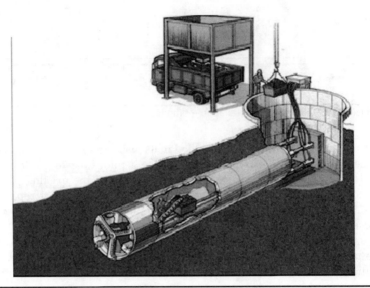

Figure 1.6 Conventional pipe jacking operation.

the operator works behind the tunneling face, so the applicable dimaters are for work-entry pipes (usually more than 42 in.). Pipe jacking is also a concept used in several trenchless technology methods, including microtunneling, horizontal auger boring, and pilot tube microtunneling (see Chap. 8).

Where prefabricated pipe sections are *jacked* behind the cutterhead. Figure 1.6 illustrates a mechanical cutting method. In this the conventional pipe jacking method, new pipe sections are jacked from a drive or entry shaft or pit (also called jacking shaft or pit) toward exit or receiving shaft or pit so that the complete string of pipe is installed *simultaneously* with the excavation of the tunnel. Since the pipe sections are jacked in this method, the pipes are designed and manufactured to take both jacking and permanent loads (soil, hydrostatic, and traffic). However, usually jacking loads (while different from permanent loads) control the pipe design. The main components of a conventional pipe jacking operation include

- Entry and exit shafts
- Tunnel-boring machine (TBM), earth-pressure balance machine (EPBM), open shield excavator, and so on, as described in later sections.
- Spoil removal system (such as a conveyer belt and haul units/moving carts over tracks)
- Jacking frame
- Hydraulic jacks

- Thrust block
- Intermediate jacking stations (if needed)
- Lubrication and pumping equipment
- Ventilation system (for the operator who works at the tunnel face)
- Laser guidance system
- Pipe sections
- Ancillary equipment (crane, backhoe, loader, dump trucks, and so on)

Types of Conventional Pipe Jacking   Pipe jacking concept is divided into two main categories: worker entry and non-worker entry. Figure 1.7 divides worker entry pipe jacking into open shield and closed shield. Non-worker entry is divided into three categories, i.e., microtunneling, guided boring (such as pilot tube microtunneling), and unguided boring (such as horizontal auger boring and pipe ramming).

Open shield pipe jacking can be carried out manually known as open hand shield or mechanically. Open hand shield is an open face shield in which manual excavation takes place as shown in Fig. 1.8.

Open shield wheel tunnel boring machine (TBM) is a shield having a rotating cutting head in which the face may be separated from the rest of the shield by a bulkhead. Various cutting heads are available to suit a broad range of ground conditions. Figure 1.9*a* shows the wheel tunnel boring machine.

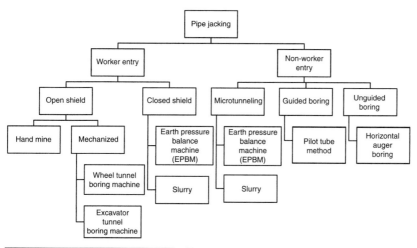

FIGURE **1.7**   Types of pipe jacking.

FIGURE 1.8   Open hand shield. (*Source: Pipe Jacking Association.*)

Open shield excavator tunnel boring machine (TBM) is an open face shield in which a mechanical excavator is mounted for excavation purposes. Figure 1.9*b* shows the excavator tunnel boring machine

Closed shield is divided into two categories, i.e., worker entry operation and microtunneling remote operation. Worker entry operation is further divided into three categories which are defined as follows:

1. *Earth-pressure balance machine (EPBM):* A full face tunnel boring machine in which the excavated material is transported from the face by a balanced screw auger or screw conveyor. The face is supported by excavated material held under pressure behind the cutter head in front of the forward bulkhead. Pressure is controlled by the rate of passage of excavated material through the balanced screw auger or valves on the screw conveyor. Figure 1.10*a* shows the worker entry EPM machine.

2. *Slurry machine:* A soft ground full face tunnel boring machine in which the excavated material is transported from the face in a slurry. Various cutting heads are available to suit broad range of ground conditions. The pressure of slurry is used to balance the ground face pressure. Figure 1.10*b* shows the worker entry slurry machine.

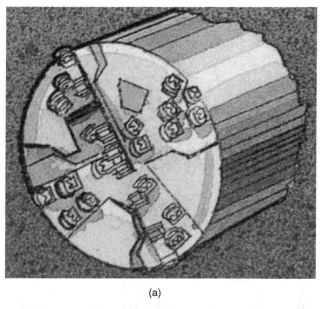

(a)

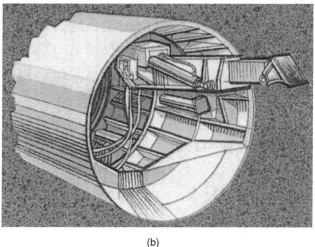

(b)

Figure 1.9 (*a*) Open shield wheel TBM. (*b*) Excavator TBM. (*Source: Pipe Jacking Association.*)

1.6.2 Utility Tunneling

In the utility-tunneling (UT) technique, the same process as conventional pipe jacking is used, except that in the utility-tunneling method, a temporary support structure (called a liner) is simultaneously constructed as the excavation of tunnel proceeds. Normally, the support structure or liner is traditional tunnel liner plates (TLP) or steel ribs (SR) with wooden lagging (WL). After completion of the tunnel, pipe

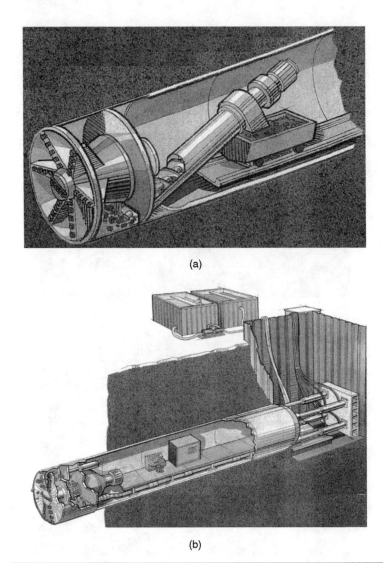

(a)

(b)

FIGURE 1.10 (*a*) Worker entry EPBM. (*b*) Worker entry slurry machine. (*Source: Pipe Jacking Association.*)

sections are transported inside the tunnel, and the annular space between the liner and the pipe is normally grouted. Due to separate phases of liner installation and pipe installation, the utility tunneling is called a two-step or two-phase operation, while conventional pipe jacking is called a one-step or one-phase operation. In general, personnel are required inside the pipe to perform the tunneling and soil-removal operations. The main components of the utility-tunneling method are the same as conventional pipe jacking, except in the utility

FIGURE 1.11 Inside view of a utility tunneling operations.

tunneling, the jacking frame, intermediate jacking stations, thrust block, and lubrication and pumping equipment are not required. Since the pipe sections are not jacked in the utility-tunneling method, the pipes are designed and manufactured to take only permanent loads. Figure 1.11 presents an inside view of a utility-tunneling project.

1.6.3 Horizontal Earth Boring

In the horizontal Earth-boring methods (HEBs), workers may work in the shaft or pit, but usually do not enter the borehole or enter the installed pipe. Therefore, these methods can be used for small-diameter pipe installations (4 in. or more). The horizontal Earth-boring method is further divided into a number of methods as shown in Fig. 1.12. These methods are briefly described in the following sections.

Horizontal Auger-Boring Method

The horizontal auger-boring (HAB) method is a cost-effective pipe-jacking operation for installing a steel casing pipe where the pipe is

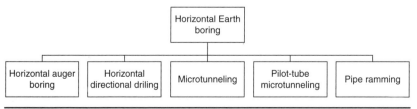

FIGURE 1.12 Horizontal Earth-boring methods.

crossing a road, highway, or railroad track. This process simultaneously jacks a steel casing from a drive pit under the road or railroad, while removing the spoil through the steel pipe by means of a rotating flight auger. The auger is a flighted tube having couplings at each end that transmit torque to the cutting head from the auger-boring machine located in the bore pit and transfers spoil back to the drive pit. The casing supports the soil around it as spoil is being removed. Usually, after completion of casing installation, a product pipe is installed using spacers and the annular space is filled with a grout. The conventional HAB equipment provides a water hose for grade control and has no steering capability for alignment. The line and grade control for HAB equipment are available but not all contractors have this capability.

Figure 1.13 illustrates a schematic of HAB operation. *ASCE Manuals and Reports on Engineering Practice No. 106, Horizontal Auger Boring Projects (ASCE, 2004),* provides full details on HAB methods.

The key components of HAB include

- A cutting head, consisting of a set of cutters mounted on the front face of the boring machine, to cut earth by the rotation of the cutters

- An auger with its front end connected to the cutting head and its tail end connected to the prime mover that drives the system, to convey the spoil to outside the borehole

- A nonrotating casing around the rotating auger, which is the pipe to be installed

- A prime mover that provides the torque to rotate the auger and the cutters, and provides the thrust to advance the pipe (casing) along with the cutting head and the auger

- A system to inject lubrication (bentonite slurry) around the pipe, to reduce friction between the pipe and the surrounding earth in order to facilitate the advancement of the pipe during the action of boring

- Control and monitoring equipment, such as water-level (most common), radio tracking systems, or inertia systems

Horizontal Directional Drilling Method

Horizontal directional drilling (HDD) is a steerable system for the installation of pipes, conduits, and cables along a desired profile using a surface-launched drilling rig. This method requires the drilling of a pilot bore which is then enlarged with the use of a reamer prior to the installation of the product pipe. Depending on the diameter of the product pipe, multiple enlargements may be required. The excavation is performed by fluid-assisted mechanical action of the

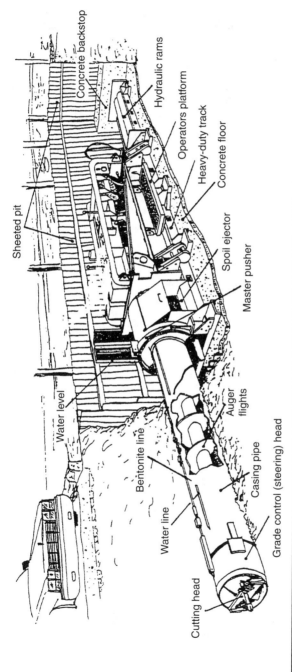

Figure 1.13 Track-type horizontal auger-boring operation. *(Iseley et al., 1999.)*

17

cutterhead. The first stage of installation consists of drilling a small-diameter pilot hole along the desired centerline of a proposed profile. The subsequent stages of installation consist of enlarging the pilot hole to the desired diameter to accommodate the product pipe and the eventual pull of the product pipe through the enlarged borehole. To avoid heaving of the surface and reduce possibility of damaging adjacent utilities, the reaming or enlargement process should be achieved by pulling only and not by pushing the reamer. This can be achieved by adding drill rods from the product pipe side and continuing reaming operation by pulling a slightly bigger-size reamer (*step-up*) until the desired borehole size is achieved.

As the name horizontal *directional* drilling implies, this method has a unique capability to track the location of the cutterhead and steer it during the drilling process. The result is a greater capability in placing the utilities in difficult underground conditions. The directional drilling methods can be classified into three broad categories of small (Mini-HDD), medium (Midi-HDD), and large (Maxi-HDD). Table 1.3 presents comparison and main features of HDD methods.

Drilling Procedures   Trailer-mounted drill rig is brought to one side of the obstacle (river, lake, road, and so on). Sections of the product pipe to be used for crossing, and other ancillary equipment, are brought to the opposite side of the obstacle. Drilling and pipe installation are done in two or three steps. In the first step, a small pilot hole of 2 to 6 in. diameter is drilled along the desired path (i.e., a near-horizontal curved path beneath the obstacle to be crossed) of the pipeline. As drilling proceeds, segments of the drill pipe are added to form the pilot string. Through the pilot string, drilling fluids (also called drilling mud, usually a bentonite and/or polymer slurry) is pumped through the nozzles in the drill bit to lubricate the drill, and to carry the cuttings (spoil) back to the rig side. Step 1 ends when the drill bit has emerged from the ground of the pipe side.

In step 2, called prereaming, the drill bit is removed from the product pipe side, and a reamer assembly is attached to the pilot string to enlarge the borehole. Prior to pulling in large pipes, it is often desirable to pull the larger reamer through the borehole without the pipe attached to swab the borehole. By reversing the direction of the rotation of the pilot string, the rig is now used to pull the reamer into the pilot hole. Segments of the drill pipe are added on the pipe side to the pilot string as they are being pulled back. This step is specifically required to prevent the contractor from pushing the reamer through the borehole, which may cause heaving of ground surface and damage to nearby utilities. Step 2 ends when the reamer covers the entire path, and starts to emerge from ground on the rig side. Several passes of step 2 may be needed for large-diameter pipes which require large boreholes that cannot be created in a single pass.

Type	Diameter (in.)	Depth (ft)	Drive Length (ft)	Torque (ft-lb)	Thrust/Pullback (lb)	Machine Weight (ton)	Typical Applications
Large (Maxi)	24–60	Less than equal to 200	Less than equal to 10,000	More than 100,000	More than 100,000	More than equal to 30	Pressure pipelines for river, shore approaches, and highway crossings
Medium (Midi)	12–24	25–75	Less than equal to 900	900–7000	20,000–100,000	Less than equal to 18	Pressure pipelines for river and highway crossings
Small (Mini)	2–12	Less than equal to 15	Less than equal to 600	Less than equal to 950	Less than equal to 20,000	Less than equal to 9	Telecom and power cables, ducts and gas lines

TABLE 1.3 Comparison of Main Features for Typical HDD Methods

Product Pipe Outside Diameter	Approximate Reamer Diameter
Less than 8 in.	OD[a] + 4 in.
Between 8 and 24 in.	OD × 1.5 in.
More than 24 in.	OD + 12 in.

[a]OD—outside diameter of product pipe

TABLE 1.4   Reamer Diameters

In step 3, a larger reamer is used to achieve the desired size bore-hole (usually 1.5 times the outside diameter of the product pipe) and to pull the product pipe under the obstacle along the borehole. The product pipe is connected to the reamer (usually in the shape of a flywheel and swivel). The swivel, connected between the reamer and the carrier pipe, allows the pipe to be pulled through and not to be rotated with the reamer and the drill pipe. Drilling fluids are pumped through nozzles in the reamer to lubricate the pipe being pulled and to remove the spoil. Step 3 ends when the entire pipeline is pulled into the borehole beneath the obstacle. Table 1.4 presents approximate reamer diameters with respect to outside diameter of the product pipe.

In large-size and some medium-size HDD systems, a downhole survey system is used to pinpoint the location of the drill head from aboveground. Several such systems are available commercially, and all provide good accuracy. For small-size and some medium-size HDD systems, a walkover survey system is used, where the drill head is located from signals emitted by a transmitter, housed behind the drill head. The signals are identified and interpreted on the surface by a receiving instrument, which is usually walked over the drill head location. Figures 1.14 through 1.16 illustrate the three phases of an HDD operation. Due to importance and popularity of the HDD method in the installations of a wide variety (pipe materials, diameters, lengths and applications) of pipes and conduits, Chap. 5 is dedicated to this method.

Microtunneling

The microtunneling (MT) method is mainly used for installation of a gravity pipeline such as a sanitary or storm sewer. Microtunneling boring machines (MTBM) are laser guided and remotely controlled which permit accurate monitoring and adjusting of the alignment and grade as the work proceeds so that the pipe can be installed on precise line and grade. While initially microtunneling methods were developed for pipes 36 in. (900 mm) or less, currently the same technology is used for larger pipes where a remote-controlled technology

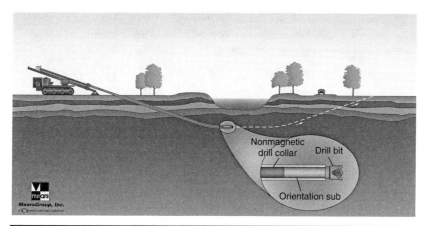

FIGURE 1.14 HDD step 1: Pilot-hole drilling. (*Source: Mears.*)

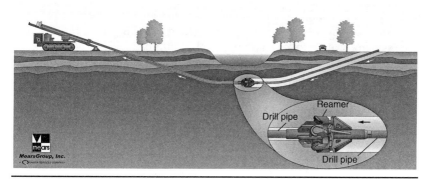

FIGURE 1.15 HDD step 2: Prereaming. (*Source: Mears.*)

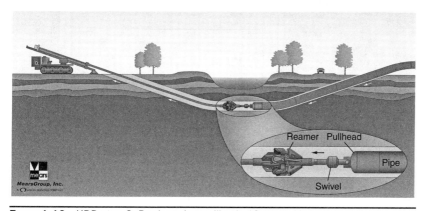

FIGURE 1.16 HDD step 3: Product-pipe pullback. (*Source: Mears.*)

is required. These methods require a drive shaft for jacking the pipe and an exit shaft for retrieving the MTBM. The microtunneling method include main components of:

- *Remote controlled*: The MTBM is remotely operated from a control panel, normally located on the surface. This method excavates and removes the soil and installs the pipe.

- *Laser guided*: The guidance system usually is a laser beam projected onto a target in the MTBM, capable of installing gravity sewers or other types of pipeline which require accurate line and grade.

- *Jacking system*: The process of constructing a pipeline by consecutively pushing the MTBM through the ground using a jacking system.

- *Face support*: Continuous pressure is provided to the face of the excavation to balance groundwater and earth pressure.

The microtunneling method installs pipe with an accuracy of ±1 in. in both the horizontal and vertical alignments. The most common spoil-removal system for microtunneling is a slurry transportation system. With proper planning, soil identification, and MTBM selection, microtunneling methods are applicable to many types of soils, either above or below groundwater table. Figure 1.17 illustrates

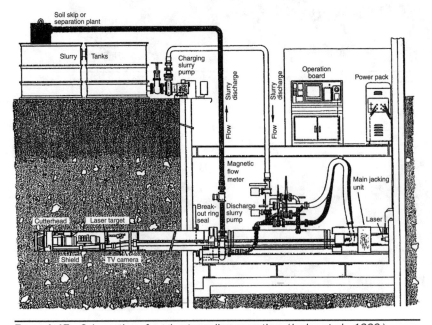

FIGURE 1.17   Schematics of a microtunneling operation. (*Iseley et al., 1999.*)

a schematic of a microtunneling operation. The American Society of Civil Engineers (ASCE) has also published *Standard Construction Guidelines for Microtunneling* (CI/ASCE 36-01, currently under revision) with information on planning, design, pipe materials, and construction aspects of microtunneling (ASCE, 2001). Figure 1.18 presents sequence of microtunneling and pipe-jacking operation.

Pilot-Tube Microtunneling

Pilot-tube microtunneling (PTMT), also called guided auger boring, guided boring, and auger drilling, originated in Europe for installation of 4- to 6-in. service lateral connections and was introduced in 1995 in the United States. Pilot-tube microtunneling is an alternate and cost-effective method to conventional microtunneling. Pilot-tube microtunneling combines the accuracy of microtunneling, the steering mechanism of a directional drill, and the spoil-removal system of an auger-boring machine. PTMT employs auger and a guidance system using a camera-mounted theodolite and a target with electric light emitting diodes (LEDs) to secure high accuracy in line and grade. When conditions are favorable (mainly soft soil conditions, drive distances less than 300 ft, and pipe diameters less than 30 in.), pilot-tube microtunneling can be a cost-effective tool for the installation of gravity pipes. This technique also can be used for house connections direct from existing sanitary sewer manholes. Among the most notable reasons for the popularity of this method are its low initial cost, shallow installations, small workspace requirement, and small jacking pits.

PTMT Operation   The PTMT operation begins with excavation of driving and receiving shafts. These shafts are usually 6.5 to 8.0 ft in diameter. The larger shafts are usually square or rectangular. The PTMT machine is then lowered down into the drive shaft and is set to precise line, grade, and height from a control point established using surveying techniques. The guidance system for PTMT consists of a theodolite with a camera independent of jacking frame, and a monitor screen. The accuracy of completed gravity sewer depends upon setup of a theodolite, which is also adjusted to height, grade, and line. Figures 1.19, 1.20, and 1.21 illustrate the stepwise PTMT installation operation.

The first step involves the installation of a pilot tube precisely over the center line of the prospective sewer or water line as shown in Fig. 1.19. During this process the spoil is displaced by a slant-faced steering head. Pilot tube is directed at precise line and grade during advancement. The hollow stem of the pilot tube provides an optical path for the camera to view the LED target in the steering head, displaying the head position and steering orientation. This establishes the center line for the installation of the new pipe.

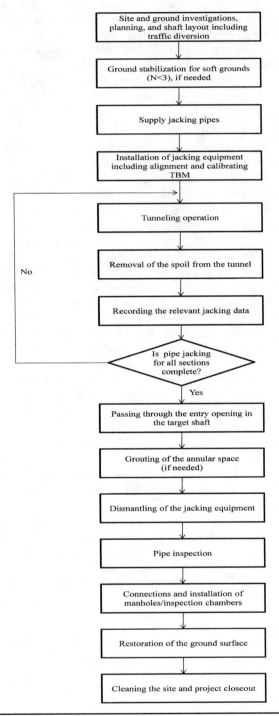

Figure 1.18 Microtunneling and pipe-jacking construction sequence.

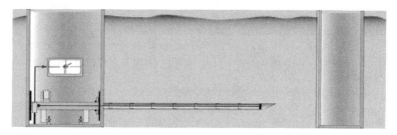

Figure 1.19 PTMT operation, step 1: Installation of pilot tube. (*Source: Bohrtec, 2008.*)

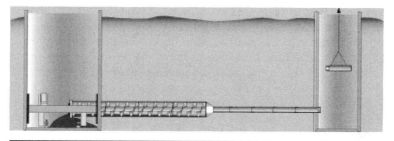

Figure 1.20 PTMT operation, step 2: Insert auger casing. (*Source: Bohrtec, 2008.*)

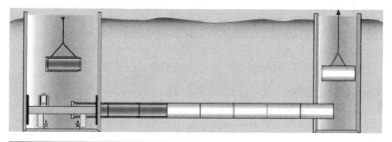

Figure 1.21 PTMT operation, step 3: Replace auger casing with product pipe. (*Source: Bohrtec, 2008.*)

In step 2, the pilot tubes at the drive shaft are connected to a reamer with a diameter slightly larger than the diameter of the product pipe. Following the reamer is the auger casing, which helps in transporting the spoil back to the shaft. The spoil is then removed by a muck bucket or a vacuum truck, depending on the soil type. Step 2 is complete when the reamer and the auger reach the reception shaft. This operation is shown in Fig. 1.20.

The third step is to replace the auger casing with the product pipe which has the same diameter as the auger casing. The product pipes are pushed one by one as the auger casings are removed from the opposite shaft as shown in Fig. 1.21. There is no spoil removal in this step.

Depending on the application, the second and third steps of the three-step PTMT can be combined, resulting in a two-step PTMT. In this case, the reamer funnels the excavated material into the auger casing which is coupled inside the product pipe. Once the product pipes are installed, the auger casings can be removed by jacking through either shaft.

Pipe-Ramming Method

Similar to horizontal auger-boring method, pipe-ramming method is mainly used for installation of pipelines and utilities for road and railroad crossings. Using an air compressor, this method hammers a steel casing pipe inside the ground from a drive pit. The pipe might be hammered closed end (for diameters less than 8 in.), or open end (for diameters of 8 in. or more). When using large diameters, the spoil is pushed out of the steel casing using air pressure, fluid pressure, and mechanical means, such as a bobcat (for large diameters) or combination of these methods.

The dynamic energy of a percussion hammer attached to the end of the casing pipe is used to install the pipe from a drive pit to a reception pit. In this method, the ramming equipment does not create a borehole; rather, it acts as a hammer to drive the pipe through the soil. The product pipe can be used mainly for pressure applications (such as water and gas); however, if the casing is large compared to the product pipe, the product pipe may be installed using spacers on grade for sewer applications. In this method, casing pipe provides a continuous soil support and overexcavation is not required.

Method Description For pipes of up to 8 in. diameter, the pipe can be driven either by having the leading end of the pipe in a wedge or cone shape. For ramming operations with the pipe face closed (also called impact moling or compaction method; see the following section), the soil is compressed around the pipe as it is being rammed and there is no spoil removal.

For pipes larger than 8 in., the leading end is usually left open and a band is installed around the outside edge of the leading section. The band serves a dual purpose: (1) it reinforces the leading edge; and (2) it decreases the friction around the casing. Figure 1.22 illustrates a schematic of open-faced pipe-ramming operation. The leading edge cuts a borehole equal to the diameter of the leading-edge band. The spoil enters the pipe, is compacted, and is forced to the rear of the pipe. For long lengths or in certain soil conditions (such as stiff clays or sands), a steel pipe is installed on the top of the

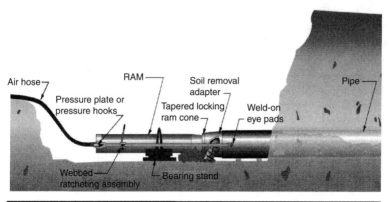

FIGURE 1.22 Pipe ramming. (*Source: TT Technologies, Inc.*)

rammed pipe to supply water, bentonite, and/or other drilling lubricants to facilitate spoil removal and reduce friction. Special adapters connect pipe of different sizes to the ramming tool. After the adapters are in place, the ramming tool is connected with lugs welded to the pipe. These lugs are used to hold straps, chains, or hoists linked to the ramming tool. After each section of the pipe is rammed, another segment is welded or joined by interlocking and the procedure is repeated until the whole length of the pipe is completed. After the completion of ramming, cleaning out of the pipe can be done by a variety of methods, including a pipe cleaning pig or plug and high-pressure air, or auguring.

The size of bore pit required for pipe ramming (PR) and the lengths of the rammed pipes are dependent on project conditions. If sufficient space is available, the bore pit is constructed to enable the pipe to be driven in longer sections. In the case where the area is congested, the pipe is driven in smaller sections, which requires more time for welding these smaller sections.

For situations where the line and grade are not critical, the pipe in the pit can be supported by suspension from construction equipment such as backhoes, cranes, and side-boom tractors; by concrete, wood, or block supports; or directly on the pit floor. In cases where the line and grade are critical, the pipe is supported by adjustable bearing stands, launch cradles or platforms, I-beams, and auger boring machine tracks.

ASCE Manuals and Reports on Engineering Practice No. 115 on Pipe Ramming Projects (ASCE, 2008), provides complete information on the pipe-ramming technique. The following sections describe new pipe ramming innovations:

Using Pipe Ramming in HDD Projects *Bore (product pipe) salvage:* A pneumatic pipe rammer may effectively be used to salvage a stuck product pipe. For this procedure, a pneumatic pipe rammer is attached to the exposed end of the partially installed product pipe in an orientation that tends to pull the pipe from the ground. This can be accomplished through a fabricated sleeve. A winch or other type of pulling device is used to assist the rammer during the pipe removal operation. The percussive power of the pipe rammer is often sufficient to free the stuck pipe and allow it to be readily removed from the ground.

Drill Stem Recovery: There are two possible configurations for applying a pneumatic pipe rammer to assist in drill stem recovery. Depending on the situation, contractors can directly pull the drill stem from the ground using the power of a ramming tool or, if the stem is still attached to the drill rig, they can use the ramming tool power to push on the opposite end of the stem to assist with the drill rig pullback action.

Pullback Assist: The pullback-assist technique helps install the product pipe in problematic situations. For example, when drilling underwater or in loose flowing soil conditions, or when there is loss of drilling fluid circulation, a condition known as hydrolock can occur. Hydrolock results when the pressure at the leading end of the product pipe restricts its forward movement. Alternatively, soil pressure along the side of the pipe due to partial collapse of the borehole may cause additional frictional drag, thus inhibiting the pipe movement. In such cases, the required pull forces may exceed the drill rig's pullback capability or the product pipe's tensile strength. The percussive pushing action of a pipe rammer applied at the tail end of the product pipe may be used to help free the immobilized pipe.

The pullback assist technique has been successfully used on steel pipe as well as high-density polyethylene. The technique can be used initially as a precaution in anticipation of possible problems, such as those described above, or after the pipe has become immobilized. Response time, however, is a key factor. The rate of success greatly improves as the response time decreases. Therefore, many drilling contractors bring ramming equipment to directional drilling sites enabling them to respond quickly to problems that may develop.

Conductor Barrel: The success of a drilling operation can often be determined before initiating the drilling operation. If the soil conditions at the planned entry point are problematic, the success of the entire project may be jeopardized. In such cases, the conductor barrel process may be appropriately used. The conductor barrel technique differs slightly from the preceding methods because it is incorporated into the initial boring plan, rather than being deployed only in the event of a problem that may arise at some stage. In this method, a clear pathway is created through poor soil conditions, allowing the actual drilling operation to begin in more preferable soil conditions. Areas with loose, unsupported soils are prime candidates for the conductor barrel method.

During the conductor barrel process, open-face casings are rammed into the ground at a predetermined angle until desirable soil conditions are met. The spoil is removed from the casing with an auger or core barrel. Drilling then proceeds within the casing, beginning at a point where more desirable soil conditions are encountered. In addition to assisting drilling operations at the start of the installation, the conductor barrel can prevent situations in unstable soils in which drilling fluids under pressure force their way into waterways or wetlands, acting in a similar fashion to containment cells. The conductor barrel may also serve as a low friction section, facilitating pullback.

Impact Moling[*]   Impact moling (also known as compaction method, earth-piercing tools, soil-displacement hammers, impact hammers, percussive moles, or pneumatic moles) is used to install small-diameter pipes, ducts, and cables (less than 10 in.). In this method, percussion or hammering action of a pneumatic piercing tool is used to create the bore by compacting and displacing the soil rather than removing it. The method typically is nonsteerable and should be used with caution. Steerable systems have reached the market in recent years.

When *properly planned and executed*, impact moling can be a simple and cost-effective trenchless installation method. Utility companies widely use this technique for installation of service connections to gas, water, and sewer mains, usually under sidewalks, driveways, and other short crossings under 150 ft. General advantages of impact moling are low operational and investment costs, relative simplicity in operating, minimal or no excavation beyond the necessary connection pits or termination points for the installed product, and minimal public disruption. Support equipment is limited to a small air compressor, and perhaps a small backhoe or trencher, to open and reinstate the connection/termination pits.

Feasibility of the method is restricted by its generic limitations (limited boring diameter and length) and by local ground conditions that can greatly affect performance. Adverse ground conditions may include cobbles, dense dry clays, and other noncompactable soils. Such soils may drastically reduce penetration rates and contribute toward surface upheaval and/or deviation from the desired

---

[*]For more information, visit *Guidelines for Impact Moling*, TTC Technical Report #2001.03, available at: http://www.ttc.latech.edu/publications/guidelines_pb_im_pr/moling.pdf.

straight-line path. It is highly recommended to trace the bore path while boring to detect path deviation so the bore may be aborted for another attempt before damage to pavement or nearby facilities are caused. Steerable moles address the path deviation shortcoming of nonsteerable moles and are poised to expand the usability of the method by virtue of being able to bore the curved path.

Method Description   Impact moles consist of an enclosed steel tube containing an air-powered piston (also referred as the striker) that strikes the nose of the tool driving it forward. A bore is formed by displacing and compacting the soil laterally. The friction between the ground and the mole body prevents the mole from rebounding backward. Repeated impacts of the piston advance the unit through the ground. There is no rigid connection between the mole and the insertion pit, and the progress of the mole relies upon the frictional resistance of the ground for its overall forward movement. There are two main types of impact moling:

1. *Nonsteerable moles* typically involve the excavation of two pits: an insertion pit and a receiving pit. After the careful alignment of the mole in the insertion pit, the tool is expected to advance through the ground in a straight line. A single person can operate the mole. Due to potential steering control difficulties and possibilities of damaging the road structure, some highway departments and municipalities have banned this method.

2. *Steerable moles* may be launched from the surface or from a pit. The operation requires a two-man crew. A walkover tracking system is used, as in directional drilling industry, where one operator walks the bore route with a walkover locator device and monitors the progress of the tool in the ground. The other operator is a tool operator who implements the required course corrections using the guidance controls. A product pipe, cable, or cable duct can either be directly towed into the bore during the boring procedure or subsequently inserted into place after the borehole is completed (providing suitable soil conditions for unsupported borehole). Usually the mole first creates the unsupported bore, and on removing the unit from the receiving pit, the product pipe is attached to the air hose and pulled into the bore (the most popular mode of operation), or the pipe is sometimes pushed into its place.

Impact moling can also be used for dead-end bores, in which case the tool is reversed after the bore is completed and removed from the ground through the insertion pit. Figure 1.23 illustrates a schematic of impact moling operation.

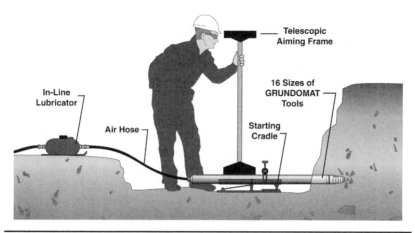

Figure 1.23 Impact moling. (*Source: TT Technologies.*)

1.7   Characteristics and Applications of Trenchless Construction Methods

This section provides information on the various trenchless methods and their applicability to the individual types of pipelines and utility installations. Restriction on the selection of a construction method may include ground conditions, availability of experienced trenchless technology contractors and appropriate equipment, cost, safety, and the technical feasibility of the method desired. The materials presented in this section should only be used as a guide. Each project is unique and for a safe and successful operation, judgment of a professional engineer knowledgeable in trenchless technology methods is required. Standard pipe sizes, bore lengths, and bore depths are considerations in determining the appropriate method. Table 1.5 presents main characteristics of trenchless methods for new installations including range of lengths and diameters. Table 1.6 provides appropriate techniques for specific applications.

Table 1.7 presents minimum safe installation depths for various trenchless installation methods categorized into different pipe diameters and soil conditions. While minimum installation depths depend heavily on subsurface conditions, method used and pipe diameter, other factors such as type of trenchless equipment used, operator's skills, project location, surface conditions, and adjacent utilities and structures are also important. It should be noted that as technology develops, each method may expand its capabilities including range of depth, length, and diameter applications. The installation depth should be determined on case-by-case basis, considering all the risks.

Table 1.8 presents selection of trenchless method based on the type of pipe, pipe installation, and method of excavation and soil removal.

TABLE 1.5 Main Characteristics of Trenchless Installation Methods

Trenchless Method	Diameter Range (in.)	Inst. Length (ft)	Max Depth (ft)	Pipe Material[c]	Typical Application	Accuracy
Conventional pipe jacking & utility tunneling	42 & up (144 in.)	1600–3500	N/A[a]	RCP, GRP, steel	Pressure & gravity pipes	± 1 in.
Auger boring (conventional)	4–60	600	N/A[a]	Steel	Road & railroad crossings	± 1% of the bore length
Auger boring steered on grade[a]	4–60	600	N/A[a]	Steel	Pressure & gravity pipes	± 1 in.
Auger boring steered on line + grade[a]	4–60	600	N/A[a]	Steel	Pressure & gravity pipes	± 12 in.
Mini-HDD	2–12	600	60	HDPE, steel, PVC, ductile iron	Pressure pipes/conduits/cables	Varies
Midi-HDD	12–24	600–2000	100	HDPE, fusible PVC, steel, ductile iron	Pressure pipes	Varies
Maxi-HDD	24–60	2000–6000	200	HDPE, steel	Pressure pipes	Varies
Microtunneling	12–136	1500	N/A[b]	RCP, GRP, VCP, DIP, steel, PCP	Gravity pipes	± 1 in.
Pilot tube microtunneling	6–30	300	N/A[b]	RCP, GRP, VCP, DIP, steel, PCP	Small-diameter gravity pipes	± 1 in.
Pipe ramming	4–140	400	N/A[b]	Steel	Road & railroad crossings	Dependent on setup

[a]Not all the contractors have the necessary equipment and required experience to use this method.
[b]Maximum depths are dependent on practical capability of using a shaft or pit to remove spoil and provide the necessary tools and materials.
[c]Steel—Steel Casing Pipe; RCP—Reinforced Concrete Pipe; GRP—Glass-Fiber Reinforced Plastic Pipe; PCP—Polymer Concrete Pipe; VCP—Vitrified Clay Pipe; DIP—Ductile Iron Pipe; PVC—Polyvinyl Chloride Pipe; HDPE—High Density Polyethylene Pipe; MDPE—Medium Density Polyethylene Pipe.

Trenchless Method	Water	Sanitary and Storm Sewers	Gas	Electricity	Telecommunications
PJ and UT	N/A	Yes	No	No	N/A
HAB	Yes	Marginal	Yes	Yes	Yes
HDD	Yes	Marginal	Yes	Yes	Yes
MT	No	Yes	No	No	No
PTMT	Yes	Yes	No	No	No
PR	Yes	Marginal	Yes	Yes	Yes

Yes—Method is suitable to this application, No—Method is *not* suitable to this application, Marginal—Special tools or considerations/contractor experiences are required, N/A—*not* applicable or *not* usually applied.

TABLE 1.6   Appropriate Techniques for Specific Applications

1.8   Capabilities and Limitations of New Installation Methods

Trenchless technology methods have many benefits over traditional open-cut and trenching techniques such as reducing or eliminating traffic disruptions, damage to pavement and road structure, noise and dust, and safety hazards. While trenchless technology methods provide many benefits, it should be recognized that there are some conditions where trenchless applications are not appropriate, and open-trench excavation may be necessary. Adverse subsurface conditions such as presence of boulders and cobbles, abandoned manmade objects and structures, specific project conditions, and/or uncertain location of existing utilities may preclude the use of trenchless technology. Refer to Chap. 10 for more information on project management and safety considerations for trenchless construction methods.

1.8.1   Conventional Pipe Jacking and Utility Tunneling

Advantages

Conventional pipe jacking and utility tunneling can be accomplished through almost all types of soils with a high degree of accuracy. Since the operator is located at the excavation face, he can see what is taking place and take immediate corrective action for changing subsurface conditions. The tunnel face can be readily inspected visually or by using a video camera. When unforeseen obstacles are encountered,

Pipe Diameters (in.)	Soil Conditions	PJ/UT	HAB (ft)	HDD (ft)	MT	PTMT	PR (ft)
Small (<12)	Clayey	6 ft of cover or 3 times outside diameter whichever is more	4	4	6 ft of cover or 3 times outside diameter whichever is more	4 ft of cover or 3 times outside diameter whichever is more	2
	Silty		4				
	Sandy		6				
	Gravely		6				
Medium (12~24)	Clayey		6	8			3
	Silty		8				
	Sandy		12				
	Gravely		20				
Large (>24)	Clayey		10	25			4
	Silty		14				
	Sandy		20				
	Gravely		25				

[a]Among other factors, minimum depth depends primarily on the type of soil and diameter of the pipe. Unstable soil conditions (such as running sand) require deeper installations, while more stable soils such as hard clays would allow for shallower installations. Also, small diameters (less than 12 in.) would allow for shallower installations.

TABLE 1.7 Minimum Depth Guidelines Based on Soil Conditions and Diameters for Trenchless Installation Methods[a]

Method[a]	Pipe/Casing Installation Mode	Suitable[b] Pipe/Casing	Soil Excavation Mode	Soil Removal Mode
CPJ	Jacking	Steel, RCP, GFRP, PCP	Manual or mechanical	Augers, conveyors, manual carts, power carts, or hydraulic
UT	Manual	Steel, RCP, GFRP, PCP	Manual or mechanical	Augers, conveyors, manual carts, power carts, or hydraulic
HAB	Jacking	Steel	Mechanical	Augers
HDD	Pulling	Steel, PVC, HDPE, DI	Mechanical and hydraulic	Hydraulic
MT	Jacking	Steel, RCP, GFRP, PCP, VCP, DIP	Mechanical	Hydraulic (slurry)
PTMT	Jacking	RCP, GFRP, VCP, DIP, Steel, PCP	Mechanical	Augers
PR	Hammering/driving	Steel	Mechanical	Augers, hydraulic, compressed air, mechanical skid
PB[c]	Pulling/pushing	(Steel, clay, HDPE, PVC, GFRP)[d] (Clay, concrete, cast iron, steel, DIP)[e]	Mechanical	Compaction

[a]HAB—horizontal auger boring; PR—pipe ramming; CPJ—Conventional pipe jacking; HDD—horizontal directional drilling; MT—microtunneling; PB—pipe bursting.

[b]Steel-steel casing pipe, RCP—reinforced concrete pipe, GFRP—glass-fiber reinforced plastic pipe, PCP—polymer concrete pipe, VCP—vitrified clay pipe, DIP—ductile iron pipe, PVC—polyvinyl chloride pipe, HDPE—high-density polyethylene pipe, MDPE—medium density-polyethylene pipe.

[c]See Chaps. 2 and 6 for more information on pipe bursting.

[d]Material suitable for use as a new pipe during pipe bursting.

[e]Material suitable for existing pipe to be replaced using pipe bursting.

TABLE 1.8   Selection of Trenchless Method Based on Type of Pipe, Pipe Installation, and Method of Excavation and Soil Removal

they can be identified and removed. Many options are available for handling the soil conditions.

In utility tunneling, only a small jacking force sufficient to drive only the shield or TBM has to be developed. Also, jacking pipes are not required, so potentially pipes with less wall thickness than jacking pipes can be used.

Limitations

Pipe jacking and utility tunneling are specialized operations. As with any other trenchless technology methods, these methods require a lot of planning and coordination. While these operations can potentially be conducted on a radius, it is recommended that all directional changes be made at the shafts. In the pipe-jacking method, the pipe should be strong enough to resist jacking forces. Hence not all types of pipes can be used for this operation.

In the utility-tunneling method, the liner is classified as a temporary structure. Therefore, a product pipe (also called carrier pipe) must be transported and installed inside the tunnel and the annular space between the product pipe and the tunnel liner filled with grout to provide adequate support.

1.8.2 Horizontal Auger Boring

Advantages

The major advantage of auger boring is that the casing is installed as the borehole excavation takes place. Hence, there is no uncased borehole that substantially reduces the probability of a cave-in, which could result in surface subsidence along the bore path. Also, with use of proper cutterhead and construction practice, this method can be used in a wide variety of soil types, which makes it a very versatile method.

Limitations

Horizontal auger boring is generally unsteerable; however, some basic steering systems are available and some contractors may use innovative methods (walk-over tools, inertia systems, etc.) to achieve accurate line and grade installations. This method requires entry that must accommodate pipe section lengths (usually 20 ft and up) and reception pits to retrieve the cutterhead. To identify obstacles such as large boulders, running sands, or very soft ground conditions, a thorough site investigation is required. Horizontal auger boring can accommodate rocks up to one-third of the diameter of the casing. The casing must be made of steel, to accommodate the steel augers turning inside the casing. With larger casing pipe installations (worker entry), it is possible to use smaller casing tubes with augers inside to remove spoils. If excessive thrust force is applied at the excavation face and not enough soil materials excavated and removed, there is a risk of heaving. Unless a special push-on jointing system is used, the welding of two pipe sections may take several hours. After completion of casing installation, the product pipe is installed using spacers or other means (a two-phase process).

1.8.3   Horizontal Directional Drilling

Advantages
The major advantage of HDD is its steering capability. In case of hitting obstacles, the drill head can be pulled back and guided around the obstacle. As the HDD system can launch from the ground surface, only small entry and reception pits (to collect drilling fluids) are required. Therefore, the setup time is relatively shorter than other trenchless installation methods. The single HDD drive length that can be achieved (up to 6000 ft) is longer than any other non-worker-entry trenchless method.

Limitations
The disposal of the slurry mixed with soil cuttings needs to be considered in advance, especially when no fluid recycling method is to be used. Although the U.S. Environmental Protection Agency (EPA) does not consider bentonite and certain polymers toxic materials, the acceptability of such spoil material varies among local agencies as well as landfill owners.

A thorough site and subsurface investigation is very important, as corrective measures applied midway in the drilling or back-reaming operation can be very time consuming and costly. When boring under roadways or other environmentally sensitive areas, the use of pressured fluid may cause serious concerns regarding the possible deleterious effect of slurry migration laterally and vertically (also called frac-outs). Care should also be taken to prevent possible ground movement and loss of slurry to the pavement for shallow soil cove installations.

While Mini-HDD equipment is portable, self-contained, and designed to work in small, congested areas, Maxi-HDD operations may require a relatively large area for the drilling rig and associated equipment at the rig side. A comparable large area is generally required at the product pipe side. Other limitations include the possibility that the bore may collapse in some granular (gravelly and sandy) soils. In very soft soils, the steering ability may become difficult. Not maintaining enough fluid flow to remove cuttings may cause a "hydrolock" situation, potentially causing heaving of soil or pavement above.

1.8.4   Microtunneling

Advantages
The microtunneling method is capable of installing pipes to accurate line and grade tolerances. It has the capability of performing in difficult ground conditions without expensive dewatering systems and/or compressed air systems to pressurize the tunnel face. Pipe can be installed at a great depth without a major cost increase. The depth factor becomes increasingly important as underground congestion is increased, or a high water table and difficult ground conditions are encountered; where open-cut method becomes very costly. In this method, safety is enhanced, as workers are not required to enter the tunnel. The carrier (product) pipe, with sufficient axial load capacity, can be jacked directly without the need of a separate casing pipe or lining for soil support.

Limitations
The capital cost of equipment is high. However, on projects where this method has been competitively bid against other tunneling methods, the unit price costs have been competitive. Some MTBM systems have difficulty in soils with boulders more than one-third of the machine diameter.

1.8.5   Pilot-Tube Microtunneling

Advantages
The change-over and setup times for this method are short. The machine technology and operation is simple. The investment costs for the equipment is reasonable. The space needed to set up the equipment is small.

Limitations
The use of the soil displacement principle is only possible in soil that can be displaced [approximately with a standard penetration test (SPT) of less than 50]. Steerability becomes difficult with increasing jacking distance. Because of the absence of position monitoring and steering capabilities during the reaming process, obstacles or collapsing soil layers may cause directional deviations. This may lead to constraining forces at the coupling point of the pilot drill-rod string with the reamer and, in extreme cases, fracture the joint.

1.8.6   Pipe Ramming

Advantages
The pipe-ramming method is an effective method for installing pipes and utilities under roads and railroad tracks. The versatile pit sizes,

maximum drive length, and ability to handle different soil conditions makes this method a practical and economical technique for installing steel pipe casing. This method does not require any thrust block as the ramming action impulses the pipe by the percussion tool. A single size of pipe-ramming tool and the air compressor can be used to install a wide variety of pipe lengths and sizes. Ramming tool can be used for vertical pile driving, angular ramming, or pipe replacement (bursting). It can also be used for assisting in HDD and MTBM rescue operations.

Limitations

The major limitation of the pipe-ramming method is the minimal amount of control over line and grade. Therefore, the initial setup is of major importance. Also, in the case of obstructions, like boulders or cobbles, especially for pipes with small diameter, the pipe may be deflected. Therefore, sufficient information on the existing soil conditions must be available to determine the proper size of casing to be used. Other drawbacks include high noise levels that are typical for pipe ramming (if no noise suppression is used). Unless special push-on jointing system is used, the welding of two pipe sections may take several hours. After completion of casing installation, the product pipe is installed using spacers or other means (a two-phase process).

1.9　Planning and Safety Considerations

Although trenchless installation methods significantly reduce the negative impacts of open-cut trenching to quality of life and reduce damage to pavements, and surrounding infrastructure, there are some potential effects that should be understood. Many of the trenchless methods described in this book may have several potential risks. Some of these risks are common to several methods, while others apply to a particular method. These potential risks should be considered in the planning, design, and preconstruction phase of the project, and when deciding on a specific trenchless technology method. Some risks can be reduced or mitigated by selection of experienced, qualified contractors and trained/certified machine operators. While generally jobsite safety is the responsibility of the contractors, project owners and consulting/design engineers also have a responsibility for ensuring proper method selection, and eliciting information on safety and operational plans of the contractor by including requirements for submittals in the bid documents.

A safe job starts with a good planning and design, proper geotechnical and site investigations, and decisions made during the inception of the project, including the profile and alignment of the installation. It is important that the owner and the engineer convey all the data, including geotechnical reports and soil bore information to the contractor. This will help the contractor to make adjustments in machine selection and installation methods.

The following are *examples* of these risks with more information provided in Chap. 10.

- Borehole collapse/subsidence.
- Ground displacement/upheaval.
- Ground vibrations damaging nearby utilities and structures.
- Excessive fluid pressures (such as in HDD operations) damaging/breaking nearby utilities.
- Excessive use of torque/thrust.
- Use of forward thrust power in HDD operations to ream the borehole (this practice may heave the ground, damage the pavement, and damage nearby utilities).
- Striking underground utilities during reaming and pullback operations (such as HDD method). It should be noted that in HDD operations, the back-reaming and installation of product pipe may take a shallower profile than original planned borehole.
- Lack of keeping a safe distance from parallel and crossing utilities.
- Lack of proper entry/access/exit pit excavation and support systems.
- Risks involved in operating boring and jacking machines and use of hydraulic and air hoses, and power tools.
- Confined space entries.
- Work zone traffic accidents.
- Fall into shafts and pits.
- Overhead power lines.
- Striking natural obstacles (tree roots, boulders, etc.) and manmade obstacles (mass of concrete, abandoned underground tanks, and other structures).
- Ground variability and mix-face conditions.
- Exceeding jacking capacity or torque capacity of boring machines (such as machine upset in horizontal auger boring).
- Exceeding jacking capacity of the thrust block or backstop.
- Exceeding jacking capacity or pulling capacity of the product pipe.
- Uncovered or unmarked surface potholes.

- Lack of proper work gear [personal protection equipment (PPE)], such as lack of insulated boots and gloves (in HDD) and eye protection glasses.

- Use of machine power to change back-reamers, cutterhead or other tools.

- Eye injuries from high-pressure drilling, other fluids, or flying objects.

- Caught-in/crushed-by accidents from rotating/moving components.

- Fire and/or explosion from breached gas lines.

- Lack of safety training of crew and/or lack of "tool box" or safety meetings.

- Improperly cutting unrestrained or spooled plastic pipe.

- Lack of good communication (hand signals, two-way radios, etc.) among machine operators and crew members.

- Lack of good planning and job site organization, such as assigning responsibilities and tasks to subcontractors, parties, or organizations that do not have the type of expertise and authority required to conduct those tasks.

Loose, cohesionless, and granular soils are more susceptible to borehole collapse if a casing or a pipe is not advanced properly while the soil is being excavated. Pipe-jacking and horizontal auger-boring are most affected by this type of soil with respect to tunnel or borehole collapse or ground subsidence.

The main concern with pipe bursting (see also Chap. 6) is surface heaving which potentially can cause outward ground displacement along the pipe alignment. The displacement is typically localized, and their effects dissipate rapidly from the bursting operation. Some causes for displacement or upheaval during pipe bursting include

- Shallow depth of the existing pipe

- The ground displacement is directed upward

- The new pipe diameter is significantly larger than that of the existing pipe

Ground displacements in pipe bursting can cause damage to nearby utilities if they are within 2 to 3 times the diameter of the new pipe. Ground vibrations can affect the surrounding soil and adjacent utilities and structures within a few feet of pipe alignment. This can be caused by pneumatic pipe bursting, as well as pipe ramming.

As a final word, a safe and productive construction project starts with a good plan and follows with implementing that plan (like the saying "plan the work and work the plan"). Underground construction in general and pipelines and utility construction specifically are risky operations, therefore, for a safe and successful project, cooperation of all parties in the construction process (owner, engineer, contractor, subcontractor, regulatory agency, and so on) is required with a well-defined plan and organization to complete the work. Figure 1.24 presents a summary of planning and safety considerations for planning and design phase and construction phase of the project.

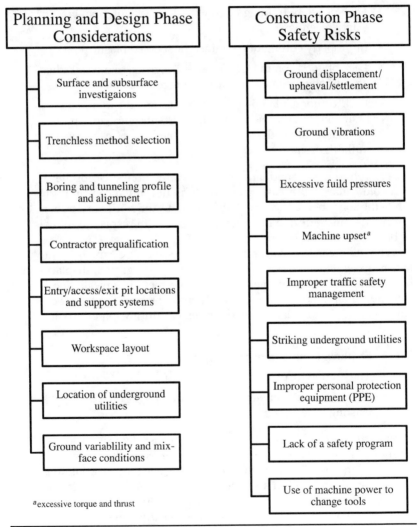

Planning and Design Phase Considerations	Construction Phase Safety Risks
Surface and subsurface investigaions	Ground displacement/ upheaval/settlement
Trenchless method selection	Ground vibrations
Boring and tunneling profile and alignment	Excessive fuild pressures
Contractor prequalification	Machine upset[a]
Entry/access/exit pit locations and support systems	Improper traffic safety management
Workspace layout	Striking underground utilities
Location of underground utilities	Improper personal protection equipment (PPE)
Ground variablility and mix-face conditions	Lack of a safety program
	Use of machine power to change tools

[a]excessive torque and thrust

FIGURE 1.24   Sample planning and safety considerations.

1.10   Cost Estimating and Bidding

Unit price contracting is the most common method for bidding and cost estimating of trenchless projects. Other methods of bidding trenchless installations include lump sum and cost-plus or time and materials. Table 1.9 presents a sample unit price bid for a HDD water pipe installation project.

Item No.	Item Description	Quantity	Unit	Unit Cost	Total Price
1	Mobilization, bonds, insurance	1	ls	$_____	$_____
2	Stormwater Pollution Prevention Plan	1	ls	$_____	$_____
3	Trench and Excavation Safety Plan	13,936	lf	$_____	$_____
4	12" DR-18 C900 PVC water line	5,010	lf	$_____	$_____
5	12" Gate valve and valve box	2	ea	$_____	$_____
6	Service line pressure reducing valve	19	ea	$_____	$_____
7	12" × 12" tapping saddle	1	ea	$_____	$_____
8	Cut and plug existing 6" water line	52	ea	$_____	$_____
9	Fire hydrant assembly	13	ea	$_____	$_____
10	Reconnect existing fire hydrant	11	ea	$_____	$_____
11	Remove existing fire hydrant	11	ea	$_____	$_____
12	Bore without encasement – 12" DR18 C900	861	lf	$_____	$_____
13	16" bore without encasement – 8" DR18 C900	156	lf	$_____	$_____
14	New service line and meter stop-recon to ex. meter	133	ea	$_____	$_____
15	Asphalt pavement replacement	2,855	lf	$_____	$_____
16	Gravel pavement replacement	2,360	lf	$_____	$_____
17	Traffic control plan	1	ls	$_____	$_____
18	Disinfection and testing	1	ls	$_____	$_____
		TOTAL ORIGINAL BID:			$_____

TABLE 1.9   A Sample Unit Price Bid for a HDD Water Pipe Installation Project

1.11 Summary

This chapter presented an overview of different trenchless installation methods with their range of applications, capabilities, advantages and limitations. Owing to its many benefits, trenchless construction has become the method of choice to replace traditional open-cut pipeline installations. Through trenchless construction a better-quality pipe is installed, permanent soil and traffic loads on the pipe are reduced, which results in extending the pipe's service life and lowering life cycle cost of the project.

CHAPTER 2

Existing Pipeline Renewal and Replacement Methods

2.1 Introduction

Trenchless renewal and replacement methods can be used to renew both gravity and pressure pipelines. Range of applications include sanitary sewers, storm sewers, culverts and drainage structures, potable water pipes, natural gas and oil pipelines, sewer manhole structures, and so on. Each method has certain capabilities and limitations that make them more suitable for specific applications and specific conditions. The decision process to select a specific method should consider many factors, including nature and extent of existing pipeline deterioration and problems, type of application, pipe geometry (see Fig. 2.1), as well as plans for future pipe applications, costs and availability of contractors and technology providers. Although numerous cross-sectional shapes are available, the circular shape is the most common shape for a pipe because it is hydraulically and structurally efficient under most conditions. A pipe-renewal selection should also consider construction cost, potential for clogging by debris, limitations on headwater elevation, pipe depth, and hydraulic performance of the new pipe. The site (soil conditions, surface conditions, and availability of space for installation), and project-specific conditions (length of pipe, bends, alignment conditions, bypassing requirements, future and current land use, and so on) would also influence selection of a specific method.

2.1.1 Existing Pipe Underperformance

There are a wide variety of factors that affect performance of existing pipes. Among these factors, structural loads (soil and live loads), corrosion, excessive fluid pressure, inadequate flow capacity, scour, and erosion of streambed and embankments are most common. Existing pipe performance is closely related to the rate of deterioration and the service life. Noninspected and nonmaintained pipes deteriorate faster than expected due to various service,

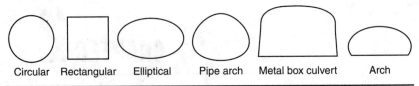

Circular   Rectangular   Elliptical   Pipe arch   Metal box culvert   Arch

Figure 2.1 Common existing pipe shapes.

environmental, hydraulic, and social conditions, which often lead to emergency repairs. Table 2.1 provides a summary of factors to consider, grouped under surface, subsurface, and existing pipeline conditions, pipe service requirements, constructability. and strengths and limitations, when a project is considered for trenchless renewal and/or replacement.

The pipeline renewal and replacement method selection process is indeed a complicated one. As mentioned above, many parameters and factors must be considered to reach an optimum solution. The decision must be made during the planning phase and reevaluated during the design and preconstruction phase of the project. The project owner and design engineer must identify the appropriate solutions and consult with contractors and technology providers during the design and preconstruction phase. However, the decision-making process should not be left to contractors to be decided in the bidding phase.

2.2   Planning Trenchless Renewal/Replacement Project

The total process of a construction project includes project definition, preliminary planning, project design, procurement of major items, project construction, and project startup. During the project definition, preliminary planning, and project design stages, major decisions are made concerning overall project size and complexity, project location, time constraints, level of quality, and costs.

Generally speaking, planning is done by all the parties in different stages of a construction project, but the initial planning is typically done by the project owner using the services of an engineer. The project background may also have a great impact on the planning process itself. The planning phase for a conventional project can proceed in a more regulated and ordered fashion than that for an emergency pipe repair or replacement due to a collapse in a roadway (see Fig. 2.2). In an accelerated schedule such as for an emergency, the priority of speed of completion may outweigh all other priorities. This may lead to selection of a project delivery system and contracting method that allows a greater focus on speed, such as design-build, cost plus, and the like, with a much higher cost to accelerate operations.

• Surface Conditions

• What is the topography?
   • What are the surface features?
   • What and where are the existing utilities located?
   • What and where are the sensitive areas within the project site?
   • What historical data is available for the project site?
   • Is the existing pipeline under a road, lake, or river? What are accessibility issues?

• Subsurface Conditions

• What are the general soil types, locations, and in-situ conditions?
   • What is the potential for the presence of rocks, cobbles, or boulders?
   • Is groundwater present? If yes, what is its depth?
   • What is the soil's corrosion potential?
   • What is the soil's settlement potential?

• Existing Pipeline Conditions

• What is the external condition of the existing pipe?
   • What is the potential for the presence of external voids around the existing pipe?
   • What are the internal conditions of the pipe?
   • What is the condition of existing manholes/on-line structures?
   • Where are bends, fittings, valves, service connections, concrete encasements, casing pipes, and other factors specific to the existing pipe located? Does existing pipe have repair sleeves installed?

• Determination of the New Pipe Service Requirements

• What flow capacity is required?
   • What length of pipe is under consideration?
   • What is the corrosion potential?
   • What are the structural requirements?

• Constructability and Site Limitations

• What safety issues need to be considered?
   • What type of access into the existing pipe is available?
   • What are the surface impacts of the construction techniques?
   • What are the easement needs of the construction techniques?
   • What impacts will groundwater have?
   • What are the scheduling limitations and constraints?
   • Will bypass pumping and flow control be required during the construction?
   • What is the impact of other utilities?

• Strengths and Limitations of Potential Renewal/Replacement Methods

• Are the proposed material and method a proven technology with available competent contractors?
   • What is the availability of the technology?
   • What is the anticipated service life?
   • What are the potential method's maintenance requirements?
   • What are the initial and long-term costs?
   • How well does the potential solution satisfy the identified service requirements?
   • What level of quality assurance/quality control is available?

TABLE 2.1 Pipeline Renewal/Replacement Parameters

FIGURE 2.2 High emergency cost expected for the pipe failures in urban areas (*Source: City of Fort Lauderdale Public Works Department.*)

It is important in the planning process for the project owner to clearly identify, define, and communicate the project priorities to all those working on the project. The project owner should also perform a self-audit of their project management to help ensure that conflicting priorities are not inadvertently introduced into the project. The following sections present detailed activities for the various renewal and replacement methods.

2.2.1 Planning Activities

Proper planning helps to ensure that the project will meet the needs and priorities of the owner. The major activities that are typically performed during the planning phase include establishing project requirements and objectives, assessing background, such as identifying screening risks, constraints, and alternatives, collecting data, and evaluating and selecting alternatives for design. Project planning involves thorough considerations of all the project parameters such as soil conditions, productivity, surface access, waste disposal, available surface area, and so on. The trenchless contractor should also be creative for any feasible approaches to enhance the project efficiency.

Background Assessment

This activity identifies and evaluates the background factors for the project. The background factors are usually nontechnical factors that establish the conditions and limitations of the project along with setting the project priorities. Many of these factors may be available through existing pipe records, such as original installation data such as, age of the existing pipe, type and class of existing pipe material, diameter, profile and alignment, size of trench, and backfill materials, bedding, and compaction, history of pipe problems with solutions provided (such as leaks or collapse, overflow, and types of repairs and/or actions taken in the past).

Screening of Alternatives

This activity will provide a general screening of potential solutions based upon the technical conditions and needs of the project. The purpose of this activity is to further eliminate candidate solutions, and identify the type of design data that must be collected for the project to proceed.

Data Collection

This activity includes investigations that provide the information required to evaluate and screen candidate alternatives for consideration in the design phase. Data collection may provide information relating to flow conditions, physical conditions, subsurface conditions, and existing pipe conditions. Surface survey, geotechnical investigations, and internal cleaning and subsequent inspection of existing pipe (with use of CCTV or other methods) will accomplish this goal.

Evaluation and Selection of Alternatives

This activity includes the final screening of alternatives and the selection of alternatives for proceeding into design. The final screening is based upon the preliminary known data in addition to the data collected during this activity.

2.2.2   Design Process

Pipe-renewal design involves a set of equations considering factors such as groundwater, soil, traffic loads, and other loadings condition. As for any structural design, there are many variables and parameters used in the design of the pipe wall thickness, including material properties, pipe geometry, and loading conditions, including assumptions for a safety factor and long-term material properties, which provide the required performance requirements to resist the loads. There is no single design equation that can be used for all different conditions that must be taken into account for the proper design of a pipeline renewal and replacement method. For proper design, it is necessary to divide these conditions into different groups.

For both gravity flow and internal pressure, design equations have been categorized into partially and fully deteriorated on the basis of ASTM F1216-09, as described in the following sections. See Secs. 2.7 and 2.8 for sample design calculations and sample specifications, respectively.

Partially Deteriorated Pipe

A partially deteriorated existing pipe is defined as one where the pipe is cracked or corroded, but has no missing pipe sections. The soil-pipe system in this case is capable of supporting all the soil, and surface loads. In this case, the new pipe is designed to support external hydrostatic loads due to groundwater as well as withstand the internal pressure in spanning across specific cracks and holes in the original pipe wall.

Fully Deteriorated Pipe

In this condition, the existing pipe is not structurally sound and cannot support soil and live loads or is expected to reach this condition over the design life of the renewed pipe. This condition is evident when sections of the original pipe are missing, the pipe has lost its original shape (possibly collapsed), or the pipe has corroded or failing due to the effects of the fluid, atmosphere, soil, or applied loads. The two basic design concepts are summarized in Table 2.2.

The specific design process for potable water pipelines (see Chap. 3) will also include the following considerations:

- *Internal corrosion:* Lining must provide a highly effective, corrosion-resistant barrier between the inside diameter (ID) of the existing pipe and the conveyed product.

- *Water quality problems associated with pipeline internal corrosion and deposits:* Lining must remove the pipe internal tuberculation and corrosion which contributes to conveyed water-quality problems.

Existing Pipe Conditions	Design Objectives
Partially deteriorated	Liner design guarantees long-term hole/ gap spanning capability under design operating conditions. Structural support is necessary from the existing pipe to withstand general operating pressures/ external loads.
Fully deteriorated	No structural support available from existing pipe against internal/external pressures/loads.

TABLE 2.2   Current Structural Design for Pipeline Renewal Systems

- *Flow capacity problems arising from pipe internal tuberculation and deposits:* Lining must present a smooth surface to the conveyed liquid, which helps maximize the flow capacity of the lined pipe. Due to extremely low friction coefficient of the new liner, it is possible to increase the hydraulic capacity of heavily tuberculated pipes.

- *Leakage from corrosion holes and failed pipeline:* Lining must provide a continuous and pressure-tight envelope inside the existing pipeline which must be designed to effectively span corrosion holes and joint gaps.

- *Vacuum Collapse:* Ensure that the liner will not collapse if the lined pipe is subject to an internal vacuum during the service.

- *Groundwater level is above the invert level of the existing pipe:* The groundwater may enter the annulus of existing pipe and liner through leaking joints/corrosion holes, and others in the existing pipeline. The resulting external hydrostatic pressure of water on the liner has the potential to cause the liner to collapse (buckle).

Sample design calculations are presented at the end of this chapter.

2.3 Applicability of Trenchless Renewal and Replacement Methods

Trenchless renewal and replacement methods can be used to *line, rehabilitate, upgrade, or renovate* (collectively called *renewal methods*) existing pipelines. There are also methods that can *replace* and/or *enlarge* existing pipelines in situ. These methods are collectively called *in-line replacement methods*. Throughout this book, the term "renewal" is used when lining methods are applied to extend the design life of pipelines. When trenchless methods are used to repair pipelines without extending their design life, the term "repair" is used. The basic trenchless renewal and replacement methods are presented in Fig. 2.3 and can be categorized according to their application for *gravity* and *pressure* pipes, as presented in Table 2.3.

Factors to be considered while implementing any pipe renewal and replacement method and how the method is advantageous over other competing techniques include:

- Social cost
- Life span (durability)
- Reliability
- Environmental impacts

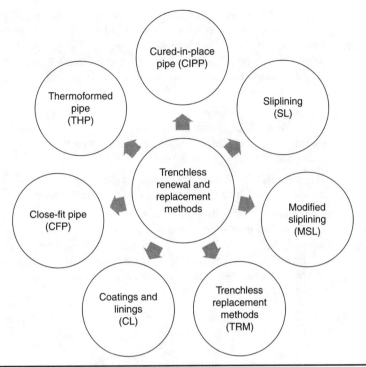

FIGURE **2.3**   Basic trenchless renewal methods.

Basic Method	Gravity Pipe	Pressure Pipe
Pipe-Lining Methods		
Cured-in-place pipe	Yes	Yes
Sliplining	Yes	Yes
Close-fit pipe	Yes	Yes
Coatings and linings	Yes	Yes
Modified sliplining	Yes	No
Thermoformed pipe	Yes	Yes
Trenchless Replacement Methods		
Pipe bursting	Yes	Yes[a]
Pipe removal	Yes	No
Pipe extraction	Yes	No

[a]Only pipe splitting method (a variation of static pipe bursting), is suitable for pressure water or natural gas pipes (such as existing cast iron or ductile iron pipes).

TABLE **2.3**   Application of Trenchless Renewal and Replacement Methods

- Productivity and schedule
- Quality of installation
- Project cost-effectiveness
- Applicability and constructability

Table 2.4 present specific applications of renewal and replacement methods for common pipeline problems. Table 2.5 presents the main characteristics of specific renewal and replacement methods. More information on the main methods of lining and replacement systems are provided in the following sections with additional installation considerations for pipe replacement and CIPP are presented in Chap. 6 and Chap. 7 respectively. Figure 2.4 illustrates a guide to select a trenchless renewal, replacement or repair method. Due to the importance of coatings and linings (CL) in potable water pipe applications, Chap. 3 is dedicated to these methods.

2.3.1   Cured-in-Place Pipe

The cured-in-place pipe (CIPP) process involves the insertion of a resin-impregnated fabric tube into an existing pipe by use of water or air inversion or winching. Usually the fabric is a polyester material, reinforced fiberglass, or similar. Usually water or air is used for inversion process. The pliable nature of the resin-saturated fabric prior to curing allows installation around curves, filling of cracks, bridging of gaps, and maneuvering through pipe defects. CIPP can be applied for structural or nonstructural purposes. Figure 2.5 shows a CIPP installation process.

2.3.2   Sliplining

Sliplining (SL) is mainly used for structural applications when the existing pipe does not have joint settlements or misalignments. In this method, a new pipeline of smaller diameter is inserted into the existing pipe and usually the annular space between the existing pipe and new pipe is grouted. This installation method has the merit of simplicity and is relatively inexpensive. However, there can be a loss of hydraulic capacity.

Sliplining can be categorized into two main categories: *continuous* and *segmental*. The continuous sliplining method involves accessing the deteriorated pipe at strategic points and inserting high-density polyethylene (HDPE) or polyvinyl chloride (PVC) pipe, joined into a continuous pipe string. The segmental sliplining method involves the use of short sections of pipe that incorporate a flush sleeve joint commonly used in microtunneling and pipe-jacking processes. Figure 2.6 illustrates a schematic diagram for the conventional sliplining process.

TABLE 2.4 Applications of Renewal/Replacement Methods[a]

Renewal Method	Joint Problems	Corrosion	Cracks/ Holes	Inflow, Infiltration, & Exfiltration	Structural Problems	Inadequate Hydraulic Capacity	Extension of Service Life of Existing Pipe
Cured-in-place pipe	Marginal	Yes	Yes	Yes	Yes	Marginal	Yes
Sliplining	Marginal	Yes	Yes	Yes	Yes	No	Yes
Close-fit pipe	Marginal	Yes	Yes	Yes	No	Marginal	Yes
Thermoformed pipe	Marginal	Yes	Yes	Yes	Yes	Marginal	Yes
Coatings and linings	No	Yes	Marginal	Yes	Marginal	No	Marginal
Modified sliplining	Marginal	Yes	Yes	Yes	Yes	No	Yes
Trenchless replacement methods	Yes	Yes	Yes	Yes	Yes	Yes	Yes

[a] Marginal means special preparation (such as a point repair) may be required before installation of the renewal/replacement method and/or the renewal/replacement solution may be effective dependent on the specific product (manufacturer), quality of application, and project conditions.

Method	Diameter Range (in.)	Approx Inst. Length (ft)	Liner Material	Applications
Cured-In-Place Pipe				
Inverted in place	4–108	3000	Thermoset resin/fabric composite	Gravity and pressure pipelines
Winched in place	4–108	1000	Thermoset resin/fabric composite	Gravity and pressure pipelines
Sliplining				
Segmental	4–100	1000–2000	HDPE, PVC, GRP	Gravity and pressure pipelines
Continuous	4–63	1000	HDPE, PVC	Gravity and pressure pipelines
Modified Sliplining				
Panel lining	More than 48 in.	Varies	GRP	Gravity pipelines
Spiral wound	4–100 (Worker entry application can go larger)	1000	HDPE, PVC	Gravity pipelines
Coatings and Linings				
Coatings and linings	4–180 and larger	1500	Cement mortar/shotcrete	Gravity and pressure pipeline
	3–180 and larger	1500	Epoxy	Gravity and pressure pipeline
	4–98 and larger	500	Polyurethane	Gravity and pressure pipeline
	Dependent on the type of product and manufacturer	500	Polyurea	Gravity and pressure pipeline

TABLE 2.5   Main Characteristics of Trenchless Renewal and Replacement Methods

Method	Diameter Range (in.)	Approx Inst. Length (ft)	Liner Material	Applications
Close-Fit Pipe				
Close-fit pipe structural	3–63	1000	HDPE, MDPE	Gravity pipelines
Close-fit pipe AWWA class III or class IV (depends on floor pressure)	3–63	1000	HDPE, MDPE	Pressure pipelines
Thermoformed Pipe				
Thermoformed pipe	4–24	500–1500	HDPE, PVC	Gravity and pressure pipelines
Lateral Renewal				
Lateral renewal	4–8	100	Any	Gravity pipelines
Point Source Repair or Localized Repair				
Robotics	8–30	N/A	Epoxy resins/cement mortar	Gravity
Grouting	N/A	N/A	Chemical gel grouts, cement-based grouts	Gravity
Internal seal	4–24	N/A	Special sleeves	Gravity
Point CIPP	4–48	50	Fiberglass, polyester, etc.	Gravity
Trenchless Replacement Methods				
Pipe bursting	4–48	1500	HDPE, PVC, DI, GRP	Gravity and pressure pipelines
Pipe removal	Up to 36	300	HDPE, PVC, DI, GRP	Gravity and pressure pipelines
Pipe extraction	Up to 24	500	Clay, Ductile Iron	Gravity and pressure pipelines

Table 2-5 Main Characteristics of Trenchless Renewal and Replacement Methods (Continued)

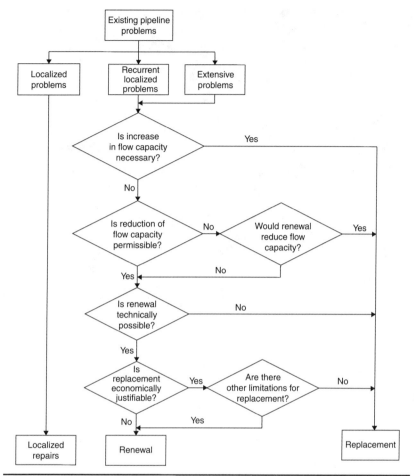

Figure 2.4 Decision support system for selection of a trenchless renewal, replacement or repair methods.

2.3.3 Modified Sliplining

Modified sliplining (MSL) includes methods in which pipe sections or plastic strips are installed in close-fit inside existing pipe and the annular space is grouted. There are two variations of modified sliplining method: panel lining (PL), and spiral wound process (SWP).

Panel linings can be used to structurally renew large-diameter (more than 48 in. or worker-entry) pipes. This method can accommodate different shapes, such as noncircular pipelines. The main type of material for this method is fiberglass. Figure 2.7 presents a panel lining method.

FIGURE 2.5 CIPP installation process. (*Source: Insituform Technologies.*)

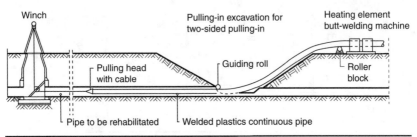

FIGURE 2.6 Schematic diagram for conventional sliplining process with annular space. (*Stein, 2001.*)

The spiral wound process uses a layered composite PVC liner and cementitious grout to renew both worker or non-worker entry into the existing pipes. The combination of the ribbed profile on the PVC liner and the highly fluid nature of the grout produce a highly integrated structure with the PVC liner "tied" to the existing pipe through the grout. The structural strength of the renewed pipe is determined by the grout characteristics. Figure 2.8 presents a schematic of a spiral wound process.

2.3.4   Coatings and Linings

The spraying of a thin cemnet-mortar or a polymer coating on the internal surface of existing pipe is another method of pipeline renewal.

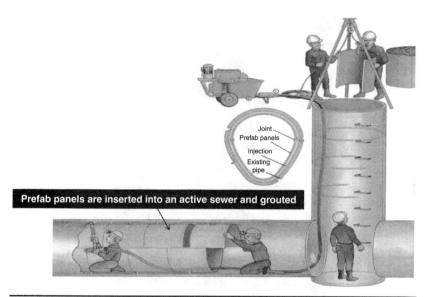

FIGURE 2.7   Panel lining method. (*Source: Channel Line.*)

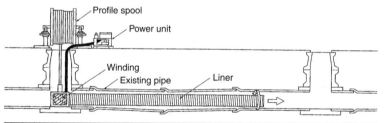

FIGURE 2.8   Schematic diagram for spiral wound process. (*Source: Sekisui SPR.*)

For non-worker-entry pipes (usually for existing pipe less than 48 in. diameter), coatings and linings methods can provide improved hydraulic characteristics and corrosion protection. Additionally, specific polymer materials may enhance the structural integrity of the pipeline and seal joints or leaks. The lining materials may include cementitious materials, epoxy, polyester, silicone, vinyl ester, polyurea and polyurethane. They are sprayed directly onto pipe walls using remote-controlled traveling sprayers. For worker-entry pipes, sprayed cement-mortars (shotcrete or gunite) are effective and widely used for renewing pressure pipes and gravity sewers and can be used for structural purposes. Figure 2.9 illustrates a coating process. Chapter 3 presents a detailed analysis of coatings and linings methods for potable water pipe applications.

FIGURE 2.9   A resin lining process. (*Source: Raven Lining Systems.*)

2.3.5   Close-Fit Pipe

This type of trenchless pipeline renewal temporarily reduces the cross-sectional area of the new pipe before it is installed. After placement, liner expands to its original size and shape at the jobsite, just to provide a close fit with the existing pipe and for pressure (more common) and gravity applications. This method can be used for both structural and nonstructural purposes. Lining pipe can be reduced on-site and reformed by heat and/or pressure or in case of thin polyethylene pipe. There are two versions of this approach: structural and semi-structural. Figure 2.10 presents a semi-structural close-fit pipe (CFP) process.

2.3.6   Thermoformed Pipe

Thermoformed is a terminology in North America for pipes that are reduced in cross-section in factory, by folding. After insertion, the liner is heated (thermoform) to conform to the existing pipe dimensions with a close fit. Both PVC and PE can be used for this method, but PVC is more common. A sample thermoformed pipe (THP) is illustrated in Fig. 2.11.

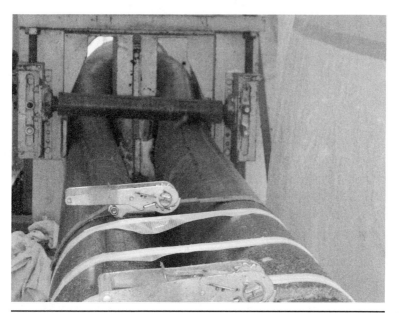

Figure 2.10   A semi-structural close-fit pipe going through diameter reduction process.

Figure 2.11   Thermoformed pipe.

2.3.7   Lateral Renewal

Sewer service laterals can be renewed using any of the methods used for renewal of main pipelines such as chemical grouting, cured-in-place pipe, close-fit pipe, pipe bursting, and spray-on coatings and linings.

2.3.8   Point Source Repair or Localized Repairs

Point source repairs are considered when local defects are found in a structurally sound pipeline. Remote-controlled resin injection to seal localized defects in the range of 4 to 24 in. (100 to 600 mm) in diameter are available. Four specific applications can be addressed with these methods:

1. To maintain the loose and separated pieces of unreinforced existing pipe in alignment to ensure load-bearing equivalent of the masonry arch.

2. To provide added localized structural capacity or support to assist the damaged pipes to sustain structural loads.

3. To provide a seal against infiltration and exfiltration.

4. To replace missing pipe sections.

Figure 2.12 presents a localized repair technique.

Chemical grouting can be used in cases where compression rings are used, to stop the leakage when grout is "injected" or "forced" into the defective joint. Chemical grouting can also be done from the surface above the repair area via probe grouting. This process not only seals the leak, but also stabilizes the surrounding soils behind the pipe. It should be noted that a new liner in a pipe cannot be considered complete if the soil surrounding the pipe still contains voids potentially creating an unstable and shifting environment.

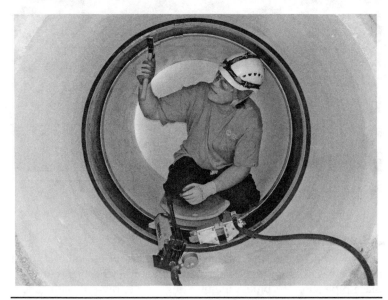

Figure 2.12   Internal seal point repair. (*Source: Miller Pipeline Corporation.*)

2.3.9   Trenchless Replacement Methods

When capacity of pipelines is found to be inadequate, the pipe can be replaced with a trenchless replacement method. There are two main types of trenchless replacement methods: pipe bursting and pipe removal (also called pipe eating, see Table 2.5). See Chap. 6 for more information on pipe replacemnt methods.

Pipe Bursting

Pipe bursting, as the name implies, uses a hammer to break the existing pipe and force broken fragments bursting head into the surrounding soil while a new pipe is pulled and/or pushed in its place simultaneously. There are different variations of pipe bursting method:

- *Pneumatic pipe bursting:* a pneumatic hammer is used to break the existing pipe.

- *Static pipe bursting:* the energy to break the existing pipe is in the pulling with no percussion action. Compared to pneumatic method, this is a quiet operation and action preferable in clayey soils or when there is need to cut (split) cast iron or ductile iron pipe.

- *Hydraulic pipe bursting:* the bursting head articulates to create the bursting action, without the noise of the pneumatic systems but pulled along with a cable like pneumatic systems.

- *Insertion method (also called pipe expansion):* this method jacks a new rigid pipe (such as clay) into the existing pipe. Clay and ductile iron are the two most widely used segmental pipes.

Figure 2.13 illustrates a schematic of pipe bursting operation. This method can be used to replace natural gas, water, and sewer pipes. This technique is useful in size-for-size replacement and up-sizing of pipeline sections. A pit is excavated to make new insertion possible based on the required bending radius of the new pipe. Pipe bursting is applied for pipes ranging from 4 to 48 in. The length of installation is based on the project and site conditions and can be in the 400 ft range.

The pipe-bursting technique may not be applicable when replacement occurs in hard soil conditions, such as "expansive" clays, densely compacted soils and backfills, or soils below the water table. In addition, pipe bursting projects could be complicated further by close proximity to other underground utilities (less than 10 times outside diameter of new pipe), past point repairs that reinforce the existing pipe with ductile materials, and a collapsed section of existing pipe. See Chap. 6 for a complete discussion of pipe bursting method.

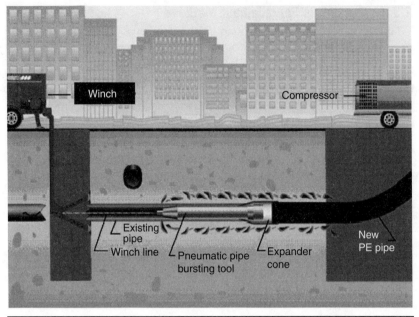

Winch

Compressor

Existing
pipe

Winch line

Pneumatic pipe
bursting tool

Expander
cone

New
PE pipe

Figure 2.13   Pneumatic pipe bursting method. (*Source: TT Technologies.*)

ASCE Manual of Practice on Pipe Bursting Projects (MOP No. 112) provides a full description of pipe bursting method and different existing and new pipe applications.

Pipe Removal

Pipe removal, also known as pipe eating, can be performed by use of a horizontal directional drilling (HDD) rig, a horizontal auger boring (HAB) machine or a *modified* microtunnel boring machine (MTBM). In this method the existing pipe is broken into small pieces and taken out of ground by means of slurry (in HDD or MTBM method) or auger (in HAB method).

2.4   Sample Decision Support Systems for Gravity and Pressure Pipes

Considering the project requirements and capabilities and limitations of available trenchless renewal and replacement methods, Figs. 2.14 and 2.15 illustrate sample decision support systems that can be used for gravity and pressure pipes, respectively. The reason to use a decision support system is to optimize the method-selection process based on the requirements of the project and priorities set by the project owner. No textbook or software program can make a proper

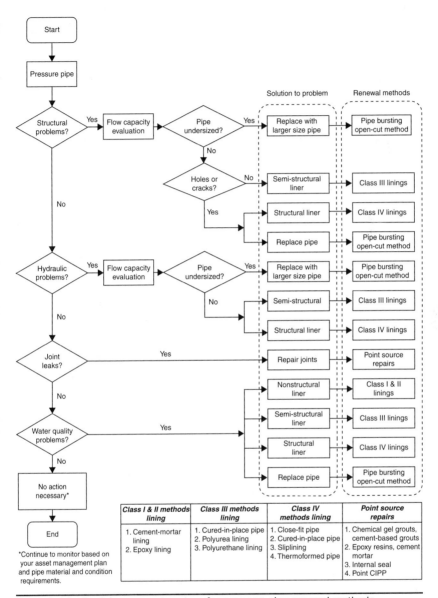

FIGURE 2.14 Decision support system for pressure pipe renewal methods.

recommendation for specific conditions and requirements of a project, so it is recommended that consulting and design engineers work with project owners to rank and weigh project priorities and then make recommendations for specific methods. The selected methods can then be advertised for bids.

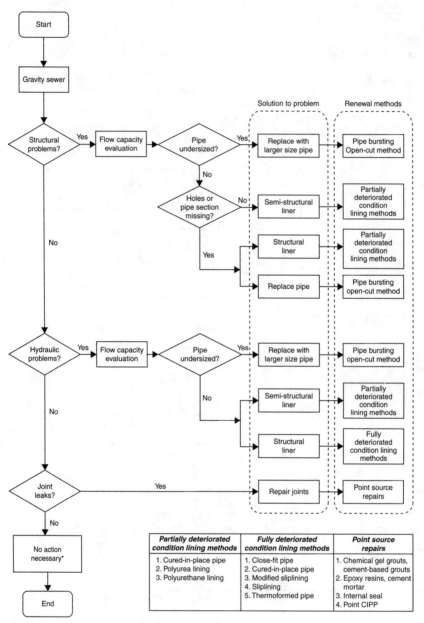

FIGURE 2.15   Decision support system for gravity pipe renewal methods.

2.5 Emerging Design Concepts for Pipeline Renewal Systems*

A conventionally buried pipe is designed to "carry" or "transfer" all the loads, which includes the weight of soil backfill placed over it, hydrostatic pressures, vacuum, internal working pressures, and loads applied at the ground surface. As mentioned earlier, current methods for structural design of flexible gravity pipe liners by the Appendix X1 of ASTM F1216-09 consider two external load cases for dimensioning of the liner pipe. The first is sustained hydrostatic pressure due to groundwater acting in the annular space between the liner pipe and the existing pipe. The second load case assumes that overtime earth and traffic loads will be transferred from the existing pipe-soil structure to the new liner pipe. This concept is particularly true in the treatment of the "fully deteriorated pipe condition" by ASTM F1216-09, for two reasons described below:

1. The soil load reaching the liner pipe is overestimated by treating the liner as it had been directly buried in a trench. The more correct liner pipe condition, even in circumstances where the existing pipe may continue to deteriorate after renewal, is a tunnel lining situation (Schrock and Gumbel, 1997).

2. The formula used to describe liner pipe response to transferred soil load, and hence calculate the required wall thickness, has been incorrectly modified from an already conservative theory for open-cut applications, which entail further irrational safety factors (Gumbel, 1998).

The basic design concept proposed at the ASCE Pipeline Division Web site (www.pipelinedivision.org) is to develop buckling pressure or safe water head charts for each renewal technique. These charts incorporate existing pipe imperfections, but allow variations of the existing pipe imperfections. The most appropriate form of chart may vary according to the type of renewal technique and/or liner material (ASCE, Pipeline Division). Reader is encouraged to refer to the publications by ASCE pipeline division for more information.

2.5.1 Long-Term Testing

The design life of pipeline systems has always been a major concern due to early deterioration of some pipe materials as well as unexpected and excessive repair and maintenance costs. For example, there are limited established design methodologies for all pipe lining systems. The existing ASTM Standard F1216-09,

---

*For more information see www.pipelinedivision.org.

provides information on design methods and procedures for the renewal of gravity and pressure pipes, for cured-in-place pipe (CIPP) method. The data obtained from short-term tests cannot be used for forecasting design life of plastic liner pipes. For prediction of structural failure of flexible pipes, when subjected to an external hydrostatic pressure, there are various methods available. The most widely used approach involves the modification of the Timoshenko equation covering the buckling of unconfined pipe (Timoshenko and Gere, 1961).

Additionally, ASTM D2990-01, "Test Method for Tensile, Compressive and Flexural Creep and Creep-Rupture of Plastics" provides a test of simply supported beam samples of plastic where "creep modulus" is measured over 10,000 hours with extrapolating long-term values. These data are used to define an apparent long-term flexural modulus of elasticity for use in the buckling equations. ASTM D1598-02, ASTM D2992-06, and ASTM D2837-08 are used to test pipe samples that are subjected to a varying internal pressure to fail the lining pipes in 10,000 hours. Graph of hoop stress with respect to time is plotted on the log scale and extrapolated to determine long-term hoop stress on the existing pipe and the lining. Table 2.6 presents the expected useful life of various renewal methods in service life extension of the pipe based on various publications indicated.

2.6 Summary

With the continued aging of existing pipelines and underground infrastructure, the need for renewal of these pipelines is increasing. The needs for improving quality of life, reduction of inconvenience to the society make trenchless renewal methods more attractive for pipeline owners and public agencies. The main benefits of trenchless renewal methods are not only being cost-effective, but also these methods provide environmentally friendly and sustainable construction operations. Extensive project planning and background assessment is required for proper execution of pipeline renewal and replacement projects.

This chapter presented a summary of different trenchless renewal and replacement methods with factors to consider when selecting a specific method. These methods have become method of choice to traditional open-cut pipeline installations and replacements. Trenchless renewal and replacement methods stop leaks, resist corrosion and abrasion, and install a new pipe in place of the existing and deteriorated pipe and provide a new design life. Pipeline replacement systems can break or remove the existing pipe and install a new pipe with the same or larger diameter without digging a trench. This chapter

Method	Material Used	Expected Useful Life	Reference
Spray-in-place-pipe	Cement mortar	> 50 years	Deb et al. (AWWA, 1990)
	Epoxy resin	> 75 years	Watson, 1998
	Polyurea	> 50 years	3M water
Cured-in-place-pipe	Thermoset resin/ fabric composite	> 50 years	TTC report, 1994
Sliplining	Polyethylene	50 years	Silbert et al. (AWWA, 2002)
Thermoformed ThP	PE & PVC	100 years	Najafi & Gokhale, 2005
Close-fit pipe	PE & PVC	> 50 years	Selvakumar et al. (EPA 2002)
Modified sliplining	HDPE, PE & PVC	> 50 years	Silbert et al. (AWWA, 2002)
Pipe bursting	PE, PVC, HDPE & GRP[b]	> 50 years	Silbert et al. (AWWA, 2002)
Pipe removal	PE, PVC, HDPE & GRP[b]	> 50 years	Silbert et al. (AWWA, 2002)

[a] Expected useful life is a loose term that depends on many factors, such as quality of design and installation, liner pipe material, thickness, and its properties, pipe loadings and pipe environmental conditions, type of application, fluid properties, and existing pipe conditions and level of its deterioration.
[b] Glass reinforced pipe (GRP) has a useful life in excess of 100 Years [Silbert et al. (AWWA, 2002)].

TABLE 2.6 Expected Useful Life of Various Pipe Renewal Methods[a]

also presented applications, characteristics, capabilities, and limitations of various trenchless renewal and replacement methods.

2.7 Sample Design Calculations for CIPP

This section presents CIPP design calculations for fully and partially deteriorated pipe. Tables 2.7 to 2.14 represent the minimum CIPP cured polyester composite physical properties of cured liner composites, ovality factors, standard trench widths, ku' values, load influence coefficients for live loads, Manning coefficient 'n' for typical pipeline materials, and gravity pipe flow comparison.

Property	Test Method	Min. per ASTM F1216-09	Enhanced Resin
Modulus of elasticity	ASTM D790-07	250,000 psi	400,000 psi
Flexural stress	ASTM D790-07	4500 psi	4500 psi

TABLE 2.7   Minimum CIPP Cured Polyester Composite Physical Properties

Physical Property	Polyester/ Vinyl Ester	Premium Poly/ Vinyl Ester[b]	Epoxy	ASTM Test Method
Flexural strength, 6 (psi)	4500– 5000	4000– 5000	5000	D 790-07
Flexural modulus of elasticity, E (psi)	250,000– 350,000	350,000– 500,000	300,000	D 790-07
Flexural modulus of elasticity reduced to account for long-term effects, E_L (psi)	125,000– 175,000	175,000– 300,000	75,000– 180,000	—
For pressure pipes only: tensile strength, 6_t (psi)	2500– 3500	N. R.[c]	4000	D 638

[a] Values shown are for typical resins used in the CIPP process. Specific physical properties should be obtained from an CIPP representative for the particular resin system used and local field conditions encountered.

[b] Premium vinyl ester systems will affect the chemical resistance of the final product; contact an CIPP representative for assistance.

[c] N. R. = not recommended.

TABLE 2.8   Typical Physical Properties of Cured Liner Composites[a]

Ovality, % (q)	0	1.0	2.0	4.0	5.0	6.0	8.0	10.0
Factor C	1.00	0.91	0.84	0.70	0.64	0.59	0.49	0.41

TABLE 2.9   Typical Ovality Factors

Liner Pipe Diameter (in.)	Trench Width, B_d (ft)
4	1.75
6	2.00
8	2.25
10	2.50
12	2.50
15	3.50
18	3.75
21	4.00
24	4.50
30	5.00
36	5.50
42	6.00
48	6.50
54	7.00
60	7.50
72	8.50
84	9.50
96	10.50

TABLE 2.10   Standard Trench Widths

Soil Type	w (lb/ft³)	ku′ (dimensionless)
Sand & gravel	110	0.165
Saturated topsoil	115	0.150
Ordinary clay	120	0.130
Saturated clay	130	0.110

TABLE 2.11   Typical ku′ Values

$(B_c/2H_s)^c$ \ $(X/2H_s)^a$		$(Z/2H_s)^b$ or $(1.5/H_s)^d$												
	0.1	0.2	0.3	0.4	0.5	0.6	0.7	0.8	0.9	1.0	1.2	1.5	2.0	5.0
0.1	0.019	0.037	0.053	0.067	0.079	0.089	0.097	0.103	0.108	0.112	0.117	0.121	0.124	0.128
0.2	0.037	0.072	0.103	0.131	0.155	0.174	0.189	0.202	0.211	0.219	0.229	0.238	0.244	0.248
0.3	0.053	0.103	0.149	0.190	0.224	0.252	0.274	0.292	0.306	0.318	0.333	0.345	0.355	0.360
0.4	0.067	0.131	0.190	0.241	0.284	0.320	0.349	0.373	0.391	0.405	0.425	0.440	0.454	0.460
0.5	0.079	0.155	0.224	0.284	0.336	0.379	0.414	0.441	0.463	0.481	0.505	0.525	0.540	0.548
0.6	0.089	0.174	0.252	0.320	0.379	0.428	0.467	0.499	0.524	0.544	0.572	0.596	0.613	0.624
0.7	0.097	0.189	0.274	0.349	0.414	0.467	0.511	0.546	0.584	0.597	0.628	0.650	0.674	0.688
0.8	0.103	0.202	0.292	0.373	0.441	0.499	0.546	0.584	0.615	0.639	0.674	0.703	0.725	0.740
0.9	0.108	0.211	0.306	0.391	0.463	0.524	0.574	0.615	0.647	0.673	0.711	0.742	0.766	0.784
1.0	0.112	0.219	0.318	0.405	0.481	0.544	0.597	0.639	0.673	0.701	0.740	0.774	0.800	0.816
1.2	0.117	0.229	0.333	0.425	0.505	0.572	0.628	0.674	0.711	0.740	0.783	0.820	0.849	0.868
1.5	0.121	0.238	0.345	0.440	0.525	0.596	0.650	0.703	0.742	0.774	0.820	0.861	0.894	0.916
2.0	0.124	0.244	0.355	0.454	0.540	0.613	0.674	0.725	0.766	0.800	0.849	0.894	0.930	0.958

Distribution load

[a] $X/2H_s$ = Distributed load width (ft) divided by twice the depth to the top of liner pipe (ft)

[b] $Z/2H_s$ = Distributed load width (ft) divided by twice the depth to the top of liner pipe (ft)

Concentrated load

[c] $B_c/2H_s$ = Outside diameter of liner pipe (ft) divided by twice the depth to the top of liner pipe (ft)

[d] $1.5/H_s$ = 1.5 divided by depth to the top of liner pipe (ft)

Source: "WPCF Manual of Practice No. FD-5, Gravity Sanitary Sewer Design and Construction," 1982, by ASCE and WPCF.

TABLE 2.12 Load Influence Coefficients for Live Loads

Pipe Material	'n'
Liner pipe[a]	0.009–0.012
Vitrified clay[b]	0.013–0.017
Concrete	0.013–0.017
Corrugated metal	0.019–0.030
Brick	0.015–0.017

[a] Selection of the liner pipe 'n' is based on the condition of the underlying pipe. For example: If a low range 'n' is selected for the existing pipe, select a low range liner pipe 'n'.

[b] The values shown are for pipes in fair to poor condition from "Applied Hydraulics for Technology" by J.D. Kanen.

TABLE 2.13   Manning Coefficient, 'n', for Typical Pipeline Materials

Existing Pipe Roughness Coefficient, 'n'	Insitupipe Dimension Ration (D/t)			
	30	40	50	60
0.014	117%	122%	126%	128%
0.015	125%	131%	135%	137%
0.016	133%	140%	144%	146%
0.017	141%	148%	153%	155%

[a] Assumes liner pipe 'n' = 0.010
Existing pipe flow = 100 %

TABLE 2.14   Gravity Pipe Flow Comparison Percent Existing Full Flow Capacity after Liner Installation[a]

Fully Deteriorated Pipe DR

Parameters: Groundwater 50 percent of soil cover
120 lb/ft^3—soil density
1000 psi soil modulus
50 percent flexural modulus reduction
H20 traffic loading

Maximum Existing Pipe Ovality Percentage	Depth of Pipe (Soil Cover) (ft)	Required DR (d/t)	
		E_i = 250,000 psi	E_i= 400,000 psi
2%	4'–10'	DR57	DR66
	11'–15'	DR48	DR56
	16'–20'	DR42	DR49
	21'–24'	DR39	DR46
5%	4'–10'	DR52	DR61
	11'–15'	DR44	DR52
	16'–20'	DR39	DR45
	21'–24'	DR36	DR42
10%	4'–10'	DR45	DR52
	11'–15'	DR38	DR45
	16'–20'	DR33	DR39
	21'–24'	DR31	DR36

Gravity Pipe, Fully Deteriorated Condition

Designed by:	
Checked by:	
Date:	09/29/2009

Project name	Design example—48 in. diameter
Location	Anytown, U.S.
Client	

Design Criteria

Pipe mean inside diameter (in.)	48.00 in.
Ovality (q) %	2 %
Pipe minimum inside diameter (in.)	47.04 in.
External water (H_w, ft above invert)	12.00 ft*
External water (H_w, ft above crown)	8.00 ft*
Depth of soil (H, ft above top of pipe)	16.00 ft
Soil modulus (E_s) psi	1000 psi
Liner pipe flexural modulus (E) psi	250,000 psi
Liner pipe flexural strength (S) psi	4500 psi
Reduction to account for long-term effects (%)	50%
Soil density (w) lb/ft³ (Table 2.11)	120 pcf
Live load (concentrated) lb (p)	16,000 lb
Live load (distributed), (p) psf	0 psf
Distributed load width, (X) ft	8 ft
Distributed load length, (Y) ft	20 ft
Factor of safety, N	2.00
*Plus internal vacuum if applicable	

Design Calculations for CIPP

I. Buckling analysis (modified AWWA formula)

$$q_t = \frac{1}{N}\left[32 \times R_w \times B' \times E'_s \times C \times \left(\frac{E_L \times I}{D^3}\right)\right]^{\frac{1}{2}}$$

Solve for t,

$$t = 0.721 \times D \times \left[\frac{\left(\dfrac{N \times q_t}{1}\right)^2}{E_L \times R_w \times B' \times C \times E'_s}\right]^{\frac{1}{3}}$$

where q_t = total external pressure on pipe, psi

$$= 0.433 \times H_w + \frac{w \times H_s \times R_w}{144} + W_s$$

H_w = height of water above top of pipe, ft
w = soil density, lb/ft³
H_s = height of soil above top of pipe, ft
R_w = water buoyancy factor

$$= 1 - 0.33 \times \left(\frac{H_w}{H_s}\right), \text{ minimum value} = 0.67$$

W_s = live load, psi (W_{sc} and/or W_{sd})
D = mean inside diameter of existing pipe (or liner pipe
    outside diameter), in.
C = ovality factor

$$= \left[\frac{\left(1 - \dfrac{q}{100}\right)}{\left(1 + \dfrac{q}{100}\right)^2}\right]^3$$

q = ovality

$$q = 100 \times \left(\frac{\text{maximum diameter} - \text{mean diameter}}{\text{mean diameter}}\right)$$

N = factor of safety, typical value = 1.5 to 2
B' = coefficient of elastic support

$$= \frac{1}{(1 + 4e^{-0.65H_s})}$$

E_L = flexural modulus of elasticity of liner pipe, psi, reduced
    to account for long-term effects
I = moment of inertia, in.(4/in.) = $t^{3}/12$
t = liner pipe thickness, in.

a. Determine load

$$q_t = 0.433 \times H_w + \frac{w \times H_s \times R_w}{144} + W_s$$

$$R_w = 1 - 0.33 \times \left(\frac{H_w}{H_s}\right)$$

$$R_w = 0.835$$

$$w = 120 \text{ pcf}$$

1. Live load
   a. Concentrated

$$W_{sc} = \frac{0.33 \times C_{sc} \times P \times F_{sc}}{12 \times D}$$

where W_{sc} = concentrated live load, psi
P = concentrated load, lb
F_{sc} = impact factor (concentrated)
C_{sc} = load influence coefficient (concentrated)
D = inside diameter of existing pipe (or liner pipe outside diameter), in.

   b. Distributed

$$W_{sd} = \frac{C_{sd} \times p \times F_{sd}}{144}$$

where W_{sd} = distributed live load, psi
p = applied surface load, psf
F_{sd} = impact factor (distributed)
C_{sd} = load influence coefficient (distributed)
D = inside diameter of existing pipe (or outside diameter), in.

Concentrated L.L P = 16,000 lb
C_{sc} = 0.03826 (Table 2.12)
F_{sc} = 1.00
Distributed L.L P = 0 psf
C_{sd} = 0.000 (Table 2.12)
F_{sd} = 0.00

$$W_{sc} = \frac{0.33 \times 0.03826 \times 16,000 \times 1}{12 \times 48} = 0.35 \text{ psi}$$

$$W_{sd} = \frac{0.000 \times 0 \times 0}{144} = 0.00 \text{ psi}$$

$$W_s = W_{sc} + W_{sd} = 0.35 \text{ psi}$$

2. Total external load

$$q_t = 0.433 \times 8.00 + \frac{120 \times 16 \times 0.835}{144} + 0.35$$

$$q_t = 14.90 \text{ psi}$$

a. Calculate coefficient of elastic support

$$B' = \left[\frac{1}{1 + (4 \times 2.178^{0.065 \times 16})}\right] = 0.414$$

b. Calculate ovality reduction factor

$$C = \left[\frac{\left(1 - \dfrac{2.0}{100}\right)}{\left(1 - \dfrac{2.0}{100}\right)^2}\right]^3 = 0.836$$

c. Calculate E_L

$E_L = E - (E \times$ reduction factor to account for long-term effects)
$E_L = 250{,}000 - (250{,}000 \times 0.5) = 125{,}000$ psi

d. Determine minimum CIPP thickness in buckling

$$t = 0.721 \times D \times \left[\frac{\left(\dfrac{N \times q_t}{1}\right)^2}{E_L \times R_w \times B' \times C \times E_s'}\right]^{\frac{1}{3}}$$

$$t = 0.721 \times 48.00 \times \left[\frac{\left(\dfrac{2.0 \times 14.90}{1}\right)^2}{125{,}000 \times 0.835 \times 0.414 \times 0.836 \times 1000}\right]^{\frac{1}{3}}$$

$t = 1.01$ in. (external pressure, deteriorated conduit)

e. Check minimum stiffness requirement

$$\frac{E \times I}{D^3} = \frac{E}{12 \times (DR)^3} \geq 0.093$$

where DR = Dimension Ratio, $DR = D/t$
Solve for t,

$$t = \left[\frac{48^3}{\left(\dfrac{250{,}000}{12 \times 0.093}\right)}\right]^{\frac{1}{3}}$$

$t = 0.790$ in. (min. stiff. limitation, deteriorated conduit)

II. Deflection (spangler formula)

$$\frac{Y}{D} = \frac{K \times (L \times W + W_s)}{\dfrac{E}{1.5 \times (DR - 1)^3} + 0.061 \times E'_s} \times 100$$

where Y = vertical deflection of Liner pipe, in

$\dfrac{Y}{D}$ = deflection ratio, expressed as a percentage

L = empirical lag factor, taken as 1.25

K = bedding constant, taken as 0.083

W = earth load, psi

W_s = live load, psi (W_{sc} and/or W_{sd})

DR = liner pipe dimension ratio $= \dfrac{D}{t}$

E'_s = modulus of soil reaction, psi (typical value 700–1500 psi)

a. Determine loads
  1. Earth load (modified Marston Formula)

$$W = \frac{C \times w \times B_d}{144}$$

where W = earth load, psi
    B_d = trench width, ft
    w = soil density, pcf

$$C = \frac{1 - e^{\left(\frac{-2ku'H_s}{B_d}\right)}}{2ku'}$$

e = 2.718
k = a ratio of horizontal to vertical pressures
u' = coefficient of sliding friction between the backfill and trench walls (Table 2.11)
H_s = soil depth to top of pipe, ft
B_d = 6.5 (Table 2.10)
ku' = 0.13 (Table 2.11)

$$C = \frac{1 - 2.718^{\left(\frac{-2 \times 0.130 \times 16}{6.5}\right)}}{2 \times 0.13} = 1.82$$

$$W = \frac{1.82 \times 120 \times 6.5}{144} = 9.85 \, \text{psi}$$

2. Live load

$$W_s = 0.35 \text{ psi (from I1)}$$

a. Calculate deflection

$$\text{Dimension ration (DR)} = \frac{D}{t} = \frac{48.00}{1.01} = 47.6$$

$$\frac{Y}{D} = \frac{0.083 \times [(1.25 \times 9.85) + 0.35]}{\dfrac{250,000}{1.5 \times (47.6 - 1)^3} + (0.061 \times 1000)} \times 100$$

$$\frac{Y}{D} = 1.68\ \% < 5\ \% \text{ allowable?}\quad \text{O.K.}$$

III. Buckling due to external water pressure (modified Timoshenko) restrained buckling analysis

$$P = \frac{2 \times K \times E_L}{(1 - v^2)} \times \frac{1}{(DR - 1)^3} \times \frac{C}{N}$$

solve for t,

$$t = \frac{D}{\left[\dfrac{2 \times K \times E_L \times C}{P \times N \times (1 - v^2)}\right]^{\frac{1}{3}} + 1}$$

where D = mean inside dia. of existing pipe (or liner pipe outside dia.), in.

t = liner pipe thickness, in.

DR = dimension ratio, $\dfrac{D}{t}$

P = allowable restrained buckling pressure, psi (or allowable external pressure measured above the pipe invert)

K = enhancement factor (typically 7)

E_L = flexural modulus of elasticity of liner pipe, psi, reduced to account for long-term effects

C = ovality reduction factor

$$= \left[\frac{\left(1 - \dfrac{q}{100}\right)}{\left(1 + \dfrac{q}{100}\right)^2}\right]^3$$

$$q = 100 \times \left(\frac{\text{maximum diameter} - \text{mean diameter}}{\text{mean diameter}} \right)$$

v = Poisson's ratio (0.30 typical for liner pipe)
N = safety factor (typically 1.5 to 2)

a. Determine load

$$P = \frac{h_w \times 62.4}{144} = \frac{12 \times 62.4}{144} = 5.20 \, \text{psi}$$

b. Calculate thickness

K = enhancement factor = 7.00
v = Poisson's ratio = 0.30

$$t = \frac{48}{\left[\dfrac{2 \times 7 \times 125,000 \times 0.84}{5.20 \times 2 \times (1 - 0.3^2)} \right]^{\frac{1}{3}} + 1}$$

t = 0.88 in. (external pressure buckling)

IV. Pressure limited due to stress

$$1.5 \times \frac{q}{100} \times \left(1 + \frac{q}{100}\right) \times DR^2 - 0.5 \times \left(1 + \frac{q}{100}\right) \times DR = \frac{S}{PN}$$

where S = flexural strength of liner pipe, psi
Solve for t,

$$1.5 \times \left(\frac{2.0}{100}\right) \times \left(1 + \frac{2.0}{100}\right) \times \left(\frac{48}{t}\right)^2 - 0.5 \times \left(1 + \frac{2.0}{100}\right) \times \left(\frac{48}{t}\right) = \frac{4500}{2 \times 5.20}$$

$$\frac{70.5024}{t^2} - \frac{24.48}{t} = 432.7$$

$$432.69t^2 + 24.48t - 70.5 = 0$$

t = 0.376 in. (maximum compressive hoop stress)

V. Select minimum CIPP thickness
  a. Thickness limitations
    1.01 in. : External pressure, deteriorated conduit (AWWA)
    0.790 in. : Minimum stiffness limitation, deteriorated conduit
    0.878 in. : External pressure buckling
    0.376 in. : Maximum compressive hoop stress

  b. Minimum design thickness = 1.01 in. (25.6 mm) DR = 47.6
    Design case: external pressure, deteriorated conduit (AWWA)

Partially Deteriorated Pipe DR

Parameters: Groundwater 50 percent of soil cover

Maximum Existing Pipe Ovality Percentage	Depth of Pipe (Soil Cover) (ft)	Required DR (d/t)	
		E_l = 250,000 psi	E_l = 400,000 psi
2%	4'–10'	DR69	DR80
	11'–15'	DR61	DR71
	16'–20'	DR56	DR66
	21'–24'	DR42	DR62
5%	4'–10'	DR63	DR73
	11'–15'	DR56	DR65
	16'–20'	DR52	DR60
	21'–24'	DR49	DR57
10%	4'–10'	DR54	DR63
	11'–15'	DR49	DR57
	16'–20'	DR45	DR52
	21'–24'	DR42	DR49

Gravity Pipe, Partially Deteriorated Condition

Designed by:	
Checked by:	
Date:	09/29/2009

Project name	Design example—48 in. diameter
Location	Anytown, U.S.
Client	

Design Criteria

Pipe mean inside diameter (in.)	48.00 in.
Ovality (q) %	2%
Pipe minimum inside diameter (in.)	47.04 in.
External water (H_w, ft above invert)	12.00 ft
Depth of soil (H, ft above top of pipe)	16.00 ft
Soil modulus (E_s) psi	1000 psi
Liner pipe flexural modulus (E) psi	250,000 psi
Liner pipe flexural strength (S) psi	4500 psi
Reduction to account for long-term effects (%)	50%
Factor of safety, N	2.00

Design Calculations for CIPP

I. Buckling due to external water pressure (modified Timoshenko)
Restrained buckling analysis

$$P = \frac{2 \times K \times E_L}{(1 - v^2)} \times \frac{1}{(DR - 1)^3} \times \frac{C}{N}$$

Solve for t,

$$t = \frac{D}{\left[\dfrac{2 \times K \times E_L \times C}{P \times N \times (1 - v^2)}\right]^{\frac{1}{3}} + 1}$$

where D = mean I.D. of existing pipe (or Liner pipe outside
diameter), in.

t = liner thickness, in.

DR = dimension ratio, $\left(\dfrac{D}{t}\right)$

P = allowable restrained buckling pressure, psi (or
allowable external pressure measured above the pipe
invert)

K = enhancement factor (typically 7)

E_L = flexural modulus of elasticity of liner, psi, reduced to
account for long-term effects

C = ovality reduction factor

$$= \left[\frac{\left(1 - \dfrac{q}{100}\right)}{\left(1 + \dfrac{q}{100}\right)^2}\right]^3$$

q = ovality

$$q = 100 \times \left(\frac{\text{maximum diameter} - \text{mean diameter}}{\text{mean diameter}}\right)$$

N = factor of safety, typical value = 1.5 to 2

v = Poisson's ratio = 0.3

a. Calculate external pressure

$$P = \frac{H_w \times 62.4}{144} = 5.20 \text{ psi}$$

b. Calculate thickness

$$t = 0.88 \text{ in.}$$

II. Pressure limited due to stress

$$1.5 \times \frac{q}{100} \times \left(1 + \frac{q}{100}\right) \times DR^2 - 0.5 \times \left(1 + \frac{q}{100}\right) \times DR = \frac{S}{P \times N}$$

where S = flexural strength of liner, psi = 4,500 psi
Solve for t,

$$\frac{70.5}{t^2} - \frac{24.48}{t} = 432.7$$

$$432.7t^2 + 24.48t - 70.5 = 0$$

$t = 0.376$ in. (maximum compressive hoop stress)

III. Select minimum CIPP thickness

Thickness Limitations
0.878 in. : External pressure buckling
0.376 in. : Maximum compressive hoop stress
Minimum design thickness = 0.88 in. (22.3 mm)
Design case: External pressure buckling

2.8   Sample Specifications for CIPP
This section presents sample specifications for CIPP.
1. Intent
   1.1 It is the intent of this specification to provide for the reconstruction of pipelines and conduits by the installation of a
   resin-impregnated flexible tube, which is tightly formed
   to the original conduit. The resin is cured using either hot
   water under hydrostatic pressure or steam pressure
   within the tube. The cured-in-place pipe (CIPP) will be
   continuous and tight fitting. The work shall be completed
   within (to be determined) calendar days from the "Notice
   to Proceed."

2. Referenced documents
   2.1 This specification references standards from the American
   Society for Testing and Materials, such as: ASTM F1216-09
   (Rehabilitation of Existing Pipelines and Conduits by the
   Inversion and Curing of a Resin-Impregnated Tube), ASTM
   F1743-08 (Rehabilitation of Existing Pipelines and Conduits
   by Pulled-in-Place Installation of Cured-in-Place Thermoset-
   ting Resin Pipe [CIPP]), ASTM D5813-04 (Cured-in-Place
   Thermosetting Resin Sewer Pipe), ASTM D790-07 (Test
   Methods for Flexural Properties of Un-reinforced and

Reinforced Plastics and Electrical Insulating Materials), and D2990-09 (Tensile, Compressive, and Flexural Creep and Creep-Rupture of Plastics) which are made a part hereof by such reference and shall be the latest edition and revision thereof. In case of conflicting requirements between this specification and these referenced documents, this specification will govern.

3. Product, manufacturer/installer qualification requirements

   3.1 Since sewer products are intended to have a 50-year design life, and in order to minimize the owner's risk, only proven products with substantial successful long-term track records will be approved. All trenchless rehabilitation products and installers must be preapproved prior to the formal opening of proposals.

   Products and installers seeking approval must meet all of the following criteria to be deemed commercially acceptable:

   3.1.1 For a product to be considered commercially proven, a minimum of five successful wastewater collection system projects of a similar size and scope of work shall be performed in the United States and documented to the satisfaction of the owner to assure commercial viability.

   3.1.2 For an installer to be considered as commercially proven, the installer must satisfy all insurance, financial, and bonding requirements of the owner, and must have had at least 5 (five) years active experience in the commercial installation.

   3.1.3 Sewer rehabilitation products submitted for approval must provide third party test results supporting the structural performance (short-term and long-term) of the product and such data shall be satisfactory to the owner. No product will be approved without independent third party testing verification.

   3.1.4 Both the rehabilitation manufacturing and installation processes shall operate under a quality management system which is third-party certified to ISO 9000 or other recognized organization standards. Proof of certification shall be required for approval.

   3.1.5 Proposals must be labeled clearly on the outside of the proposal envelope, listing the product name and installer being proposed. Only proposals using preapproved products and installers will be opened and read. Proposals submitted on products and/or from installers that have not been preapproved will be returned unopened.

3.1.6 The owner authorizes the use of proven materials that serve to enhance the pipe performance specified herein. Proven materials have passed independent laboratory testing, not excluding long-term (10,000 hours) structural behavior testing, and have been successfully installed to repair failing existing pipes in the United States for at least 4 years.

Documentation for products and installers seeking preapproved status must be submitted no less than 2 weeks prior to proposal due date to allow time for adequate consideration. The owner will advise of acceptance or rejection a minimum of three days prior to the due date. All required submittals must be satisfactory to the owner.

4. Materials
4.1 Tube—The sewn tube shall consist of one or more layers of absorbent nonwoven felt fabric and meet the requirements of ASTM F1216-09, Sec. 5.1 or ASTM F1743-08, Sec. 5.2.1 or ASTM D 5813-04, Secs. 5 and 6. The tube shall be constructed to withstand installation pressures, have sufficient strength to bridge missing pipe, and stretch to fit irregular pipe sections.

4.1.1 The wet out tube shall have a relatively uniform thickness that when compressed at installation pressures will equal or exceed the calculated minimum design CIPP wall thickness.

4.1.2 The tube shall be manufactured to a size that when installed will tightly fit the internal circumference and length of the original pipe. Allowance should be made for circumferential stretching during installation.

4.1.3 The outside layer of the tube shall be coated with an impermeable, flexible membrane that will contain the resin and allow the resin impregnation (wet out) procedure to be monitored.

4.1.4 The tube shall contain no intermediate or encapsulated elastomeric layers. No material shall be included in the tube that may cause delamination in the cured CIPP. No dry or unsaturated layers shall be evident.

4.1.5 The wall color of the interior pipe surface of CIPP after installation shall be a relatively light reflective color so that a clear detailed examination with closed circuit television inspection equipment may be made.

4.1.6 Seams in the tube shall be stronger than the non-seamed felt material.

4.1.7 The tube shall be marked for distance at regular intervals along its entire length, not to exceed 5 ft. Such markings shall include the manufacturers name or identifying symbol. The tubes must be manufactured in the United States.

4.2 Resin—The resin system shall be a corrosion resistant polyester or vinyl ester system including all required catalysts, initiators that when cured within the tube create a composite that satisfies the requirements of ASTM F1216-09, and ASTM F1743-08, the physical properties herein, and those which are to be utilized in the submitted and approved design of the CIPP for this project. The resin shall produce a CIPP that will comply with the structural and chemical resistance requirements of this specification.

5. Structural requirements

5.1 The CIPP shall be designed as per ASTM F1216-09, App. X1. The CIPP design shall assume no bonding to the original pipe wall.

5.2 The contractor must have performed long-term testing for flexural creep of the CIPP pipe material installed by his company. Such testing results are to be used to determine the long-term, time dependent flexural modulus to be utilized in the product design. This is a performance test of the materials (tube and resin) as defined within the relevant ASTM standard. A percentage of the instantaneous flexural modulus value (as measured by ASTM D790-07 testing) will be used in design calculations for external buckling. The percentage, or the long-term creep retention value utilized, will be verified by this testing. Retention values exceeding 50 percent of the short-term test results shall not be applied unless substantiated by qualified third party test data to the owner's satisfaction. The materials utilized for the contracted project shall be of a quality equal to or better than the materials used in the long-term test with respect to the initial flexural modulus used in the CIPP design.

5.3 The enhancement factor "K" to be used in "Partially Deteriorated" design conditions shall be assigned a value of seven.

5.4 The layers of the cured CIPP shall be uniformly bonded. It shall not be possible to separate any two layers with a probe or point of a knife blade so that the layers separate cleanly or the probe or knife blade moves freely between the layers. If the layers separate during field sample testing, new samples will be required to be obtained from the installed pipe. Any reoccurrence may cause rejection of the work.

5.5 The cured pipe material (CIPP) shall conform to the structural properties, as listed in Table 2.7*.

5.6 The required structural CIPP wall thickness shall be based as minimum, on the physical properties in Sec. 5.5 or greater values if substantiated by independent lab testing and in accordance with the design equations in the App. X1. Design considerations of ASTM F1216-09, and the following design parameters:

Design safety factor (typically used value)	=	2.0
Retention factor for long-term flexural modulus to be used in design (As determined by long-term tests described in Sec. 5.2 and approved by the owner)	=	50–75%
Ovality[a] (calculated from X1.1 of ASTM F1216-09)	=	%
Enhancement factor, K	= See Sec. 5.3	
Groundwater depth (above invert of existing pipe)[a]	=	ft
Soil depth (above crown of existing pipe)[a]	=	ft
Soil modulus[b]	=	psi
Soil density[b]	=	pcf
Live load[b]	=	H20 Highway
Design condition (partially or fully deteriorated)[c]	=	c

[a] Denotes information, which can be provided here or in inspection videotapes or project construction plans. Multiple lines segments may require a table of values.

[b] Denotes information required only for fully deteriorated design conditions.

[c] Based on review of video logs, conditions of pipeline can be fully or partially deteriorated.

(See ASTM F1216-09 Appendix). The owner will be sole judge as to pipe conditions and parameters utilized in design

5.7 Any layers of the tube that are not saturated with resin prior to insertion into the existing pipe shall not be included in the structural CIPP wall thickness computation.

6. Testing requirements

6.1 Chemical resistance—The CIPP shall meet the chemical resistance requirements of ASTM F1216-09, App. X2. CIPP

---

*Please see "Table 2.7. Minimum CIPP Cured Polyester Composite Physical Properties"

samples for testing shall be of tube and resin system similar to that proposed for actual construction. It is required that CIPP samples with and without plastic coating meet these chemical-testing requirements.

6.2 Hydraulic capacity—Overall, the hydraulic cross-section shall be maintained as large as possible. The CIPP shall have a minimum of the full flow capacity of the original pipe before rehabilitation. Calculated capacities may be derived using a commonly accepted roughness coefficient for the existing pipe material taking into consideration its age and condition.

6.3 CIPP field samples—When requested by the owner, the contractor shall submit test results from field installations of the same resin system and tube materials as proposed for the actual installation. These test results must verify that the CIPP physical properties specified in Sec. 5.5 have been achieved in previous field applications. Samples for this project shall be made and tested as described in Sec. 10.1.

7. Installation responsibilities for incidental items

7.1 It shall be the responsibility of the owner to locate and designate all manhole access points open and accessible for the work, and provide rights-of-access to these locations. If a street must be closed to traffic because of the orientation of the sewer, the owner shall institute the actions necessary to provide access during this for the mutually agreed time period. The owner shall also provide free access to water hydrants for cleaning, installation and other process related work items requiring water.

7.2 Cleaning of sewer lines—The contractor, when required, shall remove all internal debris out of the sewer line that will interfere with the installation of CIPP. The owner shall also provide a dumpsite for all debris removed from the sewers during the cleaning operation. Unless stated otherwise, it is assumed this site will be at or near the sewage treatment facility to which the debris would have arrived in absence of the cleaning operation. Any hazardous waste material encountered during this project will be considered as a changed condition.

7.3 Bypassing sewage—The contractor, when required, shall provide for the flow of sewage around the section or sections of pipe designated for repair. Plugging the line at an existing upstream manhole and pumping the flow into a downstream manhole or adjacent system shall make the bypass. The pump(s) and bypass line(s) shall be of adequate capacity to accommodate the sewage flow. The owner may require a detail of the bypass plan to be submitted.

7.4 Inspection of pipelines—Inspection of pipelines shall be performed by experienced personnel trained in locating breaks, obstacles, and service connections using close circuit television (CCTV) inspection techniques. The pipeline interior shall be carefully inspected to determine the location of any conditions that may prevent proper installation of CIPP. These shall be noted and corrected. A video record and suitable written log for each line section shall be produced for later reference by the owner.

7.5 Line obstructions—It shall be the responsibility of the contractor to clear the line of obstructions such as solids and roots that will prevent the insertion of CIPP. If preinstallation inspection reveals an obstruction such as a protruding service connection, dropped joint, or a collapse that will prevent the installation process, that was not evident on the pre-bid video and it cannot be removed by conventional sewer cleaning equipment, then the contractor shall make a point repair excavation to uncover and remove or repair the obstruction. Such excavation shall be approved in writing by the owner's representative prior to the commencement of the work and shall be considered as a separate pay item.

7.6 Public notification—The contractor shall make every effort to maintain sewer service usage throughout the duration of the project. In the event that a connection will be out of service, the longest period of no service shall be 8 hours. A public notification program shall be implemented, and shall as a minimum, require the contractor to be responsible for contacting each home or business connected to the sanitary sewer and informing them of the work to be conducted, and when the sewer will be off-line. The contractor shall also provide the following:

a. Written notice to be delivered to each home or business the day prior to the beginning of work being conducted on the section, and a local telephone number of the contractor they can call to discuss the project or any potential problems.

b. Personal contact with any home or business, which cannot be reconnected within the time stated in the written notice.

7.7 The contractor shall be responsible for confirming the locations of all branch service connections prior to installing the CIPP.

8. Installation

8.1 CIPP installation shall be in accordance with ASTM F1216-09, Sec. 7, or ASTM F1743-08, Sec. 6, with the following modifications:

8.1.1   Resin impregnation—The quantity of resin used for tube impregnation shall be sufficient to fill the volume of air voids in the tube with additional allowances for polymerization shrinkage and the potential loss of resin during installation through cracks and irregularities in the original pipe wall, as applicable.

8.1.2   Tube insertion—The wet out tube shall be positioned in the pipeline using either inversion or a pull-in method as defined within relevant ASTM standards previously stipulated. If pulled into place, a power winch or its equivalent should be utilized and care should be exercised not to damage the tube as a result of pull-in friction. The tube should be pulled-in or inverted through an existing manhole or approved access point and fully extend to the next designated manhole or termination point.

8.1.3   Temperature gauges shall be placed between the tube and the existing pipe's invert position to monitor the temperatures during the cure cycle.

8.1.4   Curing shall be accomplished by utilizing hot water under hydrostatic pressure or steam pressure in accordance with the manufacturer's recommended cure schedule.

9. Reinstatement of branch connections

9.1   It is the intent of these specifications that branch connections to buildings be reopened without excavation, utilizing a remotely controlled cutting device, monitored by a CCTV. The contractor shall certify a minimum of two complete functional cutters plus key spare components are on the job site before each installation or are in the immediate area of the jobsite and can be quickly obtained. Unless otherwise directed by the owner or his authorized representative, all laterals will be reinstated. No additional payment will be made for excavations for the purpose of reopening connections and the contractor will be responsible for all costs and liability associated with such excavation and restoration work.

10. Inspection

10.1   CIPP samples shall be prepared for each installation designated by the owner/engineer in accordance with ASTM F1216-09, Sec. 8.1. Pipe physical properties will be tested in accordance with ASTM F1216-09 or ASTM F1743-08, Sec. 8. The flexural properties must meet or exceed the values listed in the Table 2.7* of this specification, Table 1 of ASTM

---

*Please see "Table 2.7. Minimum CIPP Cured Polyester Composite Physical Properties"

F1216-09, or the values submitted to the owner/engineer by the contractor for this project's CIPP wall design, whichever is greater.

10.2   Wall thickness of samples shall be determined as described in paragraph 8.1.6 of ASTM F1743-08. The minimum wall thickness at any point shall not be less than 87 ½ percent of the submitted minimum design wall thickness as calculated from the parameters in paragraph 5.6 of this document.

10.3   Visual inspection of the CIPP shall be in accordance with ASTM F1743-08, Sec. 8.6.

11. Clean-up

11.1   Upon acceptance of the installation work and testing, the contractor shall restore the project area affected by the operations to a condition at least equal to that existing prior to the work.

12. Payment

12.1   Payment for the work included in this section will be in accordance with the prices set forth in the proposal for the quantity of work performed. Progress payments will be made monthly based on the work performed during that period.

CHAPTER 3

Spray-on Coatings and Linings for Renewal of Potable Water Pipe Distributions

3.1 Introduction

Spray-on coatings and linings* have been used to protect and renew pipelines and other infrastructure (tanks, reservoirs, clarifiers, primary and secondary retention and treatment basins, pump stations, diversion boxes, manholes, and other structures) for decades. Shotcrete, an air-assisted spray-on lining method for cementitious products, was developed at the beginning of the twentieth century in Allentown, Pennsylvania, and became accepted as a construction method in 1910. Today, high-tech polymer coatings and composite lining methods are used to restore, protect, repair, and renew a wide range of pipelines and concrete, masonry, and steel structures.

The principal objective of a coating or lining for potable water pipe application is to apply a monolithic layer that inhibits further deterioration. Type of deterioration is dependent upon pipeline infrastructure under consideration. In water pipes, it is characterized by tuberculation, scale buildup, and corrosion that can significantly reduce flow capability and water quality (see Fig. 3.1). In sanitary sewers, coatings and linings are effective at eliminating infiltration while providing containment. In corrosive sanitary sewer environment, pipe crowns, pump stations, and manholes can lose an inch or more of concrete in less than a year. Coatings and linings can mitigate

---

*The distinction between coatings and linings is not clear so the terms are used interchangeably in practice.

Figure 3.1   Excessive tuberculation in water pipes. (*Source: 3M Water Infrastructure.*)

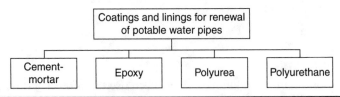

Figure 3.2   Basic coatings and linings materials for renewal of potable water pipes.[†]

further degradation and, if needed, can structurally enhance and renew severely damaged pipelines.

The primary materials used for coatings and linings fall into four broad categories of cementitious materials and polymers which include cement-mortar, epoxy, polyurea and polyurethane (see Fig. 3.2). These methods are sometimes used in conjunction with one another. For chemical resistance and monolithic coverage, adhesion is generally regarded as a required attribute of coatings and linings. For structural enhancement, adhesion may or may not be a desired property.[*] Other attributes of coatings and linings vary greatly between polymers and cementitious. Some coatings and lining materials may be excellent for bridging cracks and holes, but may have low chemical resistance owing to inherently higher porosity; others may exhibit excellent long-term strength, but poor adhesion in damp environments. As for

---

[*]Research is ongoing in this area. For more information, please refer to applicable ASTM, AWWA, and ASCE standards and practice guidelines (currently under preparation) as well manufacturers of these products.
[†]Carbon fiber is another renewal method which is mainly used for large diameter transmission pressure pipe applications.

any renewal method, true project needs should be evaluated and matched with proven product attributes.

Moisture can weaken a lining's curing process as well as its ability to bond to the existing structure. Although moisture is relatively easy to mitigate in above-surface structures, it cannot be completely avoided below grade, especially in pipeline structures. Therefore, a lining with high moisture tolerance offers an adhesion advantage for pipeline projects. Epoxies and polyureas can generally be formulated to offer the best moisture tolerance, although some polyurethanes also offer moderate tolerance or require the use of an epoxy primer. Other attributes to consider include structural enhancement, permeability and chemical resistance, quick return-to-service, future maintenance and repair requirements, and ease of tapping and service connections.

3.2   Water Distribution Pipe Applications

Many water utilities are faced with the problem of aging water pipe networks and the associated increasing costs. It is estimated that most water utilities have 20 to 30 percent unaccounted water problems due to aging and leaking water pipeline systems. Major water utilities in the United States, on average, face 20 to 60 water main breaks per year causing loss of millions of gallons of treated water and, at the same time, facing high costs of emergency repairs together with additional social costs and customer inconvenience. At the same time, increasingly limited resources require efficient use of available maintenance and renewal funds. Leakage, water quality, and structural failures are few among the various problems faced by water utilities.

Majority of these problems are caused by corrosion, as well as soil movements, traffic loads, and excessive pressures. Decay of water pipes and the need for an appropriate corrosion protection lining material is a great challenge to the water industry. The main objectives of coatings and linings are to provide corrosion protection, increase hydraulic capacity and water quality, but they can also be used for structural enhancements. An ideal coating and lining method, to provide these properties, must not contain any volatile organic compounds (VOCs), must be environmentally friendly, long-term durability, resistant to live (hoop, transverse and longitudinal stresses, vacuum pressure, water hammer) and dead loads (soil and hydrostatic pressure), chemical attack, and meet all applicable governmental, regulatory, and industry standards to be safe for potable water applications (in the United States, all potable water lining materials need to be certified according to ANSI/NSF 61). Other desirable properties of the coatings and linings may include rapid cure, smooth and pinhole-free coating, meeting adhesion requirements with the existing pipe, being locally available with certified and experienced contractors, and being cost-effective.

For structural applications in worker-entry, larger-diameter pipelines, reinforced sprayed cement mortars (shotcrete and gunite) on

| System Class | Non-structural | Semi-structural | | Structural |
	Class I	Class II	Class III	Class IV
Corrosion protection	Yes	Yes	Yes	Yes
Gap spanning capability	No	Yes	Yes	Yes
Inherent ring stiffness	No (Depends on bonding)	No (Depends on bonding)	Yes (Self support)	Yes (Self support)
Survives burst failure of existing pipe	No	No	No	Yes

TABLE 3.1   Structural Classification of Lining Systems

carbon fiber can be effectively used. Non-worker-entry water pipes may require coatings and linings to be applied with a centrifugal lining machine. In this case, lining material is pumped to a high-speed, rotating application head of the centrifugal lining machine. As the machine is pulled through the deteriorated pipe, a uniform thickness liner is applied. American Water Works Association (AWWA) classifies water pipe renewal methods into four categories as summarized in Table 3.1.

For structural applications (category IV, in Table 3.1), the lining material must provide a new design life to the existing pipe. For semi-structural applications (categories II and II), the liner may provide the following properties:

- Sufficient toughness to survive the dynamic loadings
- Sufficient ductility to accommodate any joint displacements
- Sufficient shear strength to maintain longitudinal continuity in the presence of unrestrained ground movements
- Sufficient flexural strength to provide long-term corrosion voids spanning capability

3.3   Selecting a Water Pipe Renewal Method

The key elements for the selection of a specific method for renewal of potable water pipes are:

- Nature of problem the pipe is facing and the objectives for the renewal method
- Existing pipe material, dimensions, and features (bends, alignment, joints, history of previous repairs, depth, degree of corrosion, and so on)

- Types and locations of appurtenances such as valves, fittings, and fire hydrants
- Number and type of service connections
- Length of time the pipe can be out of service or bypass requirements
- Site- and project-specific factors (surface and subsurface conditions)
- Overall cost of the renewal method (material and installation)

3.4　Installation Phases of Coatings and Linings

The successful installation of coatings and linings depends on many factors such as transportation of lining material from shop to the project site, ambient temperatures at the site, and cleanliness of the internal surface of the existing pipe as well as existing pipe conditions, geometry, alignment, and defects. Among these parameters, the most important factor is the proper installation technique as shown in Fig. 3.3. Figure 3.3 illustrates the various steps and sequences for successful coating and lining installations. Further details on these steps are provided in the following sections.

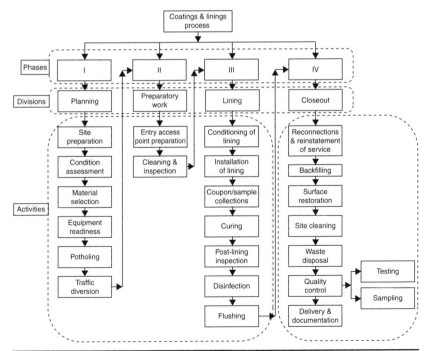

Figure 3.3　Steps and sequences for coatings and linings. (*Source: 3M Water Infrastructure.*)

3.5 Planning and Site Investigations

Every pipe renewal project is unique and requires a careful site investigation before installation can begin. In the planning phase of a project, surface and subsurface survey information provide assistance in determining the suitability of the coating and lining methods. Project owners must provide available existing pipe and site information in the bid documents, so contractors can submit realistic bids. Additionally, contractors must conduct their own surface and subsurface investigations, as bid data might be outdated at the time of bid submission and installation phase. Possibilities of unanticipated obstacles such as presence of debris, silt, and other deposits, which may make the task of lining difficult, must specifically be investigated.

Accurate information would lead to reduction of installation problems, quality issues, reworks and change orders, and associated social costs. The investigation of a lining site includes evaluation of the number and condition of service laterals, easement restrictions, potholing locations and regulations, and environmental concerns. A surface and subsurface survey may include the following steps:

- Work area requirement
- Location of all existing utilities
- Utilities and structures adjacent to proposed pit locations
- Surface features, descriptions and layouts of roadways, sidewalks, and so on
- Visible subsurface utility landmarks such as fire hydrants, valves, and so on

It is essential to determine all valve and fire hydrant locations along the full length of the existing pipe. It is recommended that contractor or installer physically check all the dimensions supplied by the project owner in the form of drawings, videos and DVD's, and reports to ensure accuracy. Once surface and subsurface surveys are completed, pit locations and thickness and section lengths of the lining installations can be determined. Proper site inspection is also important for selection of appropriate lining parameters and determination of any modifications in lining design thickness.

3.6 Pipe Inspection

The main objective of pipe inspection is to examine the condition of the existing pipe before and after liner installation. Closed-circuit television (CCTV) is usually the method of choice to inspect the interior of water pipes and also to evaluate quality of installed pipe. Valves, water hydrants or other appropriate locations may be used as inspection insertion points.

3.6.1 Objectives of Prelining Inspections

The purpose of a prelining inspection is to ensure a successful liner installation. This inspection provides a good idea of the degree of cleaning required to prepare the existing pipe before the start of a lining operation. A prelining inspection may also reveal the need for other forms of preparations required before lining, such as removal of protruding lateral service connections. The state or presence of the following issues may be revealed during the prelining inspections:

- Leaking valves and ferrules
- Leaking stop taps
- Dropped joints
- Protruding ferrules
- Structural problems (cracks, holes, and so on)
- Cleaning and possibilities for re-cleaning requirements
- Pipe bends that can affect cleaning and lining processes

3.7 Pipe Cleaning Methods

Cleaning of existing water pipes is prioritized based on the age of the pipeline and the frequency of the problems it encounters. In most of the water pipes, sediments accumulate and biofilms develop, increasing the risk of color, taste, and odor problems, along with the chance of coliform regrowth.

Cleaning must be performed on the section of the existing pipe to be lined. Before isolating and cleaning the section, water flow must be stopped or bypassed. The section of pipe will then be emptied and cleaned. The cleaning technique chosen depends on the pipe material, previous lining, and entry point locations. It must be noted that, if the structural condition of the existing pipe is poor and the pipe wall is thin, then some cleaning methods may result in damage to the pipe section being cleaned. Figure 3.4 illustrates a conventional cleaning method using a steel rod. Table 3.2 presents a summary of water pipe cleaning methods.

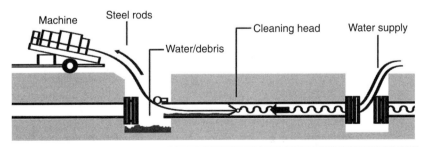

Figure 3.4 Rack-feed boring machine using steel rods. (*Source: Water UK.*)

Method	Procedures
Drag scrapers	Designed to remove hard deposits and nodules. This is made up of spring steel scrapers mounted on a central shaft. These scrapers remove deposits as they are winched through the pipe. The central shaft is fitted with a towing eye at each end to allow the scraper to be pulled back if necessary. These are available for all pipe sizes ranging from 2 in. (50 mm) to 12 in. (300 mm).
Power boring	Hydraulically powered device utilized to remove tuberculation and encrustation from the water pipes. The device is a rack-feed boring machine, a compact, diesel-powered unit using hydraulic pressure to rotate steel cutter blades at 750 rpm through the pipe.
High-pressure water jetting	High-pressure water jets are pulled through the pipe so high-velocity water blasts deposits and films from the pipe surface.
Poly pigging	This is similar to soft swab pigging except that it differs in its construction. For softer scales, the pig is simply foam covered by a plastic shell, while for harder scales abrasive materials are adhered to the pig's exterior.
Air scouring	Volumes of air and water are alternately introduced to the pipe through fire hydrants, creating turbulence that scours film, and lifts and transports sediment.
Rodding	Uses an engine and a drive unit with sectional rods. As blades rotate, they break up the deposits and loosen the debris.
Balling	A threaded rubber cleaning ball that spins and scrubs the interior of the pipe as flow increases in the pipelines.
Flushing	Introduction of alternatively high flow rate of water into the pipes, removing floatables, sands, and grids.
Soft swab pigging	Foam plug is forced through the pipe using water.
Chemical cleaning	Acid is circulated through a closed system, dissolving mineral scale, biological growth, and corrosion by-products. Finally the pipe is flushed and disinfected. Some common chemicals used are H_2S gas, bioacids, digester, enzymes, catalysts, hydroxides, and neutralizers.
Scooter	This is a hydraulic method. Round, rubber-rimmed, hinged metal shield is mounted on steel framework on small wheels, which works as a plug to build a head water. This action scours the inner walls of the pipeline.

Source: Adapted from AWWA M28, Rehabilitation of Water Mains.

TABLE 3.2   Water Pipe Cleaning Methods

3.8   Installation Considerations

Coating and lining thickness must be designed based on the existing pipe conditions, requirements and future use. Required quality assurance measures may include compliance with design specifications and expected test results in accordance with applicable contract documents and manufacturer's recommendations. The contractor and installer must be experienced and certified by the manufacturer for the specific liner application. The thickness of the lining during installation can be controlled by the flow rate of the lining material and retrieval speed of the application head. This process is usually automated and computerized for advanced applications. If desired, the actual thickness of the liner installation may be determined through coupon sampling. The contractor may need to submit lined coupons for owner's evaluations or for testing.

3.9   Disinfection Methods

Disinfecting of water pipes after renewal has been a common industry practice for many years. The first AWWA standard covering this practice was approved in September 1947 (as 7D.2-1948). In 1986, the designation of this standard was changed to AWWA C-651; the latest revision is ANSI/AWWA C-651-05. There are five main types of disinfection methods identified in AWWA standards as summarized below. It should be noted that according to project location and water utility regulatory requirements, additional steps (such as several flushings, requirements for boiling water for a certain period, etc.) may be required after lining installations.

3.9.1   Tablet Method

AWWA C-651 recommends the use of an average chlorine content of 25 to 150 mg/L for duration of 24 to 72 hours. Preferably, disinfection should be carried out overnight; however, not on a day before the weekend or holidays.

3.9.2   Continuous Feed Method

The chlorine may be added in the form of dissolved calcium hypochlorite, sodium hypochlorite, liquid chlorine, or chlorine gas. Among these, dissolved chlorine gas offers the "best" disinfection; however, environmental concerns and new regulations have made this option less desirable. The chlorine concentrations vary from 25 to 60 mg/L for durations of 24 to 72 hours.

3.9.3   Slug Method

This method is generally used in conjunction with the tablet method. After the tablet method is completed and flushed, a heavily concentrated slug of chlorine is added to the pipe and slowly forced through the system. The concentration of the slug is monitored and if the free chlorine residual drops below 50 mg/L, additional amount of chlorine is added. Several utilities use this method at a concentration of 300 to 500 mg/L. Disposal and treatment of the heavily chlorinated water can become a problem with this method (ANSI/AWWA, 2005).

3.9.4 Ozonation

Ozone being an unstable molecule of oxygen, which readily gives up one atom of oxygen providing a powerful oxidizing agent, is toxic to most waterborne organisms. Ozonation is an effective method to inactivate harmful protozoans from forming cysts. This method also works well against almost all other pathogens. Ozone gas is prepared by passing oxygen through ultraviolet light or using a "cold" electrical discharge. To use ozone as a disinfectant, it must be created on-site and added to the water by bubble contact.

3.9.5 U.K. Method

According to "Technical Guidance Notes No. 4—Distributor System (Renovated Mains)," contained in *Principles of Water Supply Hygiene and Technical Guidance Notes,* the lined pipe must be disinfected using a maximum of 0.013 oz/gal (100 mg/L) of free chlorine. After the disinfection, the lined pipe must be flushed for a minimum period of one hour at a velocity of 1.64 ft/s (0.5 m/s) for available water pipes before return to service. For lined pipe sections with dead ends, a flow regime must be established such that the residence time does not exceed one hour in the first 24 hours of service after recommissioning (Water UK, 2007).

3.10   Pipe Sample Testing

Table 3.3 presents some of the coatings and lining testing methods based on ASTM standards.

3.11   Quality Control

The liner quality must be acceptable if proper cleaning and application procedures are followed. Project documentation, for delivery to the project owner, may include testing results from laboratory (if required); printouts from the lining rig, postinspection DVD, and so on. Table 3.4 presents a description of common defects found after the application of the lining and its recommended prevention and remedies.

Test Method and Standard	Brief Description
Dry film thickness (ASTM D1186)	An electronic thickness gauge is used to measure the dry film thickness and compare with previously obtained film thicknesses, so that changes due to swelling and shrinkage can be measured to $\pm$ 0.0254 mm ($\pm$ 1 mil) in accuracy.
Knife adhesion (ASTM D6677)	Cutting the lining through the existing pipe with a utility knife to probe adhesion. Adhesion is then rated on a scale of 1–10 depending on the difficulty of removal and size of the chips. 10–Lining difficult to remove. 01–Lining can be easily peeled.
Pull-off adhesion (ASTM D4541)	This test is also utilized to verify adhesion, cohesive strength, and pull-off strength of the lining system.
Undercreep holiday (ASTM D1654)	To determine the corrosion of the existing pipe when the lining is peeled off.
Blistering (ASTM D714)	This test is used to evaluate the degree of blistering that may develop when coated sample is subjected to conditions, which will cause blistering. Blistering is rated on the size and density of the blisters.
Lining impedance measurement (ASTM G42)	An accelerated procedure for determining comparative characteristics of lining systems applied to pipe for the purpose of preventing corrosion that may occur in service where the pipe will be exposed to high temperatures and is under cathodic protection. This test measures barrier properties of the linings and coatings, and its permeability.

TABLE **3.3** Lined Pipe Testing Methods

3.12 Safety

In order to complete the work satisfactorily and accident free, it is the responsibility of the contractor to be familiar with and follow all the federal, state, and local regulatory safety requirements. All required personal protective equipment (PPE) must be fully functional and follow U.S. Occupational Safety and Health Administration (OSHA) guidelines. Contractors need to identify all possible work hazards [traffic management, pit excavation and protection (see Fig. 3.5), safe disposal of extra lining materials (see Sec. 3.15), worker safety during the existing pipe cleaning and liner installation, and so on] of lining and submit a safety management plan to the project owner before start of the lining operations.

Deficiency	Description	Prevention Method and Possible Remedial Actions
Lining over debris	Lining over corrosion, loose debris in pipe, inadequate pipe cleaning.	If the fault is localized, it is best to rectify by excavation and removal of faulty pipe section. If faulty pipe section is extensive, it must be totally replaced.
Blistering and bubbling	Round, raised sections of hardened coatings and linings caused where there is graphitization on cast iron pipe walls. It must be noted that not all instances of graphitization cause blisters.	If blisters are isolated, they may be cosmetic in nature and do not create a long-term weakening of the lining. The integrity of the lining is not compromised and no remedial action is taken.
Uncured lining	Malfunction of the lining machine. Only section of pipe lined with hardened, fully cured lining may be returned to service.	Using CCTV or other inspection methods will reveal any sections of uncured lining which must be removed and relined.
Uneven or thin lining	Machine malfunction, excessive friction along the hose during winching back through the pipe, unsatisfactory design of application head, protruding obstacles in pipe.	Short length can be replaced with new pipe and longer lengths can be relined. Protruding obstacles can be identified during CCTV inspection.
Slump	Caused by applying an excessive thickness of coating or lining to the pipe surface or overheating the components.	The extra volume of the lining material can reduce the desired pipe diameter.
Water damage	Usually caused by undetected standing water or ground water leakage through pipe.	The extent of fault must be accurately measured by CCTV inspection. Short length can be replaced with new pipe and longer lengths can be recleaned and relined.
Pinholes	Pinhole or discontinuity in the lining.	Must be repaired by relining or as directed by the owner or the city engineer.

Source: AWWA, 2002.

TABLE 3.4   Summary of Defects and Remedial Actions

FIGURE 3.5   Pit Protection.

3.13   Reconnecting Appurtenances

Dependent on liner material and installation method, fittings, valves, and fire hydrants may need to be taken out, drained; cleaned from all dirt, dust, and debris; capped; and stored as part of liner installation preparation. After lining installation, it is necessary to support heavy valves (12 in. or 300 mm and larger), with treated timbers, crushed stone, concrete pads, or a thoroughly tamped trench bottom. Fire hydrant must be installed in a manner which will provide complete accessibility and minimum possibility of damage from vehicles or injury to pedestrians. The outside of the hydrant must be above the finished ground.

All valves must be inspected upon reconnection to ensure proper working order after liner installation. Valves must be set and joined to a pipe as set forth in the AWWA standards* for the type of connection ends furnished. All valves and appurtenances must be installed at the correct alignment and rigidly supported. Any damage to these appurtenances must be repaired or replacement with new ones may be required.

Local water utilities may provide their own specifications which must be followed. Some water utilities may prefer to conduct service lateral and appurtenance reconnections with their own crews, so contractor' work ends with the liner installation.

3.14   Surface Restoration and Site Clearing

Backfilling pits are one of the most important phases of water pipe renewal. The purpose of backfill is not only to fill the pits, but also to protect the pipes and provide support for valves and hydrants.

---

*C222-08, Standard for Polyurethane Coatings for the Interior and Exterior of Steel Water Pipe and Fittings; C620-07, Standard for Spray-Applied In-Place Epoxy Lining of Water Pipelines, 3 in. (76 mm) and Larger; C602-06, Standard for Cement-Mortar Lining of Water Pipelines in Place 4 in. (100 mm) and Larger.

Backfill materials must be in good quality and free from cinders, frozen material, ashes, refuse, boulders, rocks, or organic materials. Parkways and other nonpaved areas may not need compaction, depending on the potholing size, conditions and project requirements. If needed, pit flooding may be used to obtain the necessary compaction requirements.

After pit backfilling, additional pieces of pipes, extra fittings, tools, and incidental materials, such as debris and excess spoil materials, must be removed from the jobsite or right-of-way (ROW). All undamaged walkways and pavements must be cleaned. All grass areas must be reseeded and/or replaced with sod, shrubs, trees, and other plants as per their original conditions or better. Damaged and removed pavement must be repaired/replaced according to the municipality, local government, or Department of Transportation (DOT) specifications and standards.

3.15   Waste Disposal

The contractor must be familiar with all the regulations for handling and disposing of the lining materials and chemicals, such as the following:

- Unused lining materials and other contaminated materials must be returned to shop/warehouse or be disposed of in accordance with applicable safety and waste disposal regulations.
- Uncured materials must be disposed in a facility permitted to accept chemical waste.
- Cured (or polymerized) waste must be taken to appropriate sanitary landfills. The contractor is responsible for assuring that all permits and other types of disposal documentation are completed and distributed as required by regulatory agencies.
- Since regulations vary, based on the coating and lining material, applicable federal, state, and local regulations before disposal must be followed.

3.16   Descriptions of Coatings and Linings Methods

The following sections provide specific information on each of the four basic coating and lining methods.

3.16.1   Cement-Mortar Linings

Cement-mortar spray-in liner has been in existence since the 1900s and is one of the most common lining methods used today. The first successful trial of spray applied cement lining took place in early 1930s

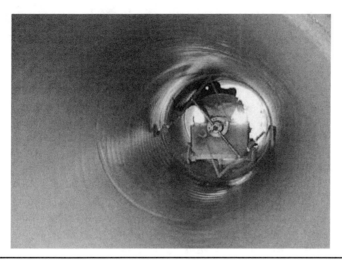

FIGURE 3.6 Cement-mortar lining troweled finish. (*Source: www.cementlining. com.*)

(AWWA-C602, 2000). Cement-mortar-lined pipes are centrifugally lined to ensure that a uniform thickness of mortar is distributed throughout the length of the existing pipe. Cement-mortar linings prevent tuberculation by creating a high pH at the pipe wall. They act as a physical barrier between water and the existing pipe with a smooth finish, which potentially may increase hydraulic capacity of the existing pipe, in spite of slight internal diameter reduction. The protective properties of cement linings are due to two properties of cement: chemically alkaline reaction of the cement and water reduction in contact with the iron pipe.

Cement-mortar lining is applied by a rotating spray head connected to a hose that is attached to the lining rig. Mortar is supplied to the machine through the high-pressure hose, and a uniform thickness of lining is applied as the machine moves through the existing pipe at a constant speed. The thickness of the applied liner is directly related to the speed at which the machine moves. After the liner has been applied, a rotating or conical drag trowel provides a smooth finish (see Fig. 3.6).

Installation Procedure of Cement-Mortar Linings*

1. Inspecting the existing pipe to determine the location of valves, hydrants, bends, level of deterioration, displacement of joints, and so on.

---

*Refer to Table 3.9 for detailed description about the steps involved during the installation of spray-on coatings and linings.

2. Bypassing water flow.

3. Thoroughly cleaning existing pipe and removing all films, loose materials, debris, and silt prior to the lining.

4. Inserting the lining head connected to the lining rig into the pipe.

5. Supplying mortar to the machine through high-pressure hoses and ensuring uniform application of lining from the sprayer at a constant speed.

6. Troweling of the applied cement-mortar with either rotating trowels or spatulas attached to the lining head.

After the completion of the application, the lining is cured in a controlled environment to prevent rapid loss of moisture. Usually CCTV or another inspection method is used to inspect quality of installation. Table 3.5 represents advantages and limitations for cement-mortar lining methods.

3.16.2 Epoxy Linings

Epoxy lining was first introduced in the United Kingdom in the late 1970s and introduced in the North America in the early 1990s.

Advantages	Limitations
1. Minimum service interruption compared to other pipe renewal methods.	1. Bends, valves, and fittings must be removed prior to lining.
2. Protects deterioration of pipe against further corrosion.	2. Bypass is required.
3. For worker-entry pipes, reinforcement can be used to provide structural support (shotcrete and gunite).	3. Curing time is dependent on curing of cement and may take a long time.
4. Usually entry and exit pit is required, no major excavation is necessary.	4. Dependent on the lining thickness, there may be a slight reduction of pipe internal diameter.
5. Usually flow capacity is improved due to smoothness of lining surface compared to the original surface of the existing pipe.	5. Does not enhance structural integrity of the pipe.
	6. Cement-mortar lining may result in high pH water.
	7. Usually plugs ¾ in. or less valves or lateral connections.
	8. Quality and durability of installation is dependent on the adhesion to the internal surface of the existing pipe, which in turn is dependent on the degree the existing pipe cleaned.

TABLE 3.5 Advantages and Limitations for Cement-Mortar Linings

American Water Works Association approved epoxy materials potable water use in 1995 (ANSI/NSF 61 Standard). The primary reason for using this type of lining technique is to overcome the water quality problems caused by corrosion of iron pipes. Epoxy linings are usually classified as a nonstructural technique. Epoxy lining is applied to the interior surface of existing pipes with a smooth surface finish that helps prevent further corrosion and tuberculation. Epoxy can effectively halt the recurrence of these problems if the existing pipe is properly cleaned and lining is adequately applied according to specifications and manufacturer's guidelines. Figure 3.7 presents removing of epoxy spray head from the pipe. Similar to cement-mortar linings, epoxy linings require use of a specialized machine and spray head (see Fig. 3.8a).

Epoxy resin products are composed of two components, base and hardener. These two components are supplied in different colors. When mixed, they form a third distinct color. Figure 3.8 (b) shows the finished epoxy lining. Table 3.6 represents advantages and limitations of epoxy linings.

3.16.3 Polyurea

Polyurea pipe lining is a rapidly growing market due to polyurea's ability to solve difficulties other lining methods face, such as the existing pipe moist or infiltration conditions, and installation speed.

FIGURE 3.7   Epoxy spray head is removed from the pipe. (*Source: CuraFlo Inc.*)

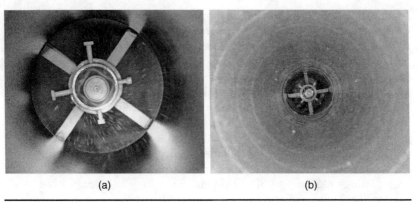

(a)   (b)

Figure 3.8   (a) Spray head for epoxy lining; (b) Finished epoxy lining. (*Source: CuraFlo Inc.*)

Advantages	Limitations
1. Has a higher "C" coefficient than cement mortar and better improves flow capacity of the existing pipe.	1. Bypass is required.
2. Provides a smooth, pinhole -free lining.	2. Pipe must be dry and free from standing water during application.
3. Resistant to corrosive water.	3. Does not enhance the structural integrity of the existing pipe.
4. Does not block service connections.	
5. Only entry and exit pit is required, no major excavation is required.	4. Tees and bends can pose installation difficulties.
6. Protects further deterioration of existing pipe against corrosion.	5. Minimum cure time can be 16 hours.
7. No major reduction in the existing pipe diameter.	

Table 3.6   Advantages and Limitations of Epoxy Linings

With a suitable existing pipe surface preparation and substrate conditions, polyurea provides fastest return to service than competing lining materials. During a field evaluation, it was shown that polyurea application does not plug service laterals and appurtenances (see Fig. 3.9).

Polyurea is formed from reaction of two components, isocyanate and amine resin. These components, unlike those created by the crystalline nature of some polyurethane hard segments, form a urea linkage, which is highly flexible. The material is moisture tolerant and has low viscosity, thus it can be easily pumped to remote spray head locations. It provides high build slump resistant linings, and dependent on manufacturer and class of material (structural or nonstructural) excellent adhesion characteristics. Finished linings are hard, glossy, and free of surface tack or greasiness. Polyurea linings

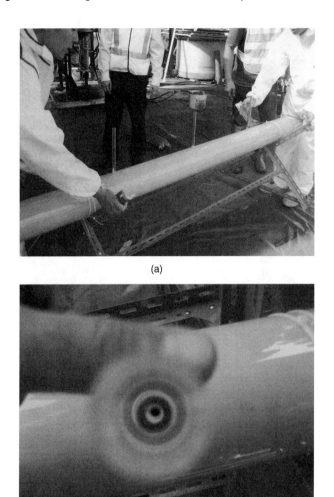

(a)

(b)

FIGURE **3.9** Polyurea application does not plug service laterals.

provide long-term corrosion protection and have excellent abrasion resistance.

Figure 3.10 illustrates the spray head for polyurea lining application. Table 3.7 represents advantages and limitations of polyurea linings.

3.16.4 Polyurethane

Polyurethanes were first used in the mid 1970s in North America to seal underground fuel tanks from corrosion. During the same time it was used in Europe to protect oil and gas pipelines. With

Figure 3.10 Polyurea lining spray head. (*Source: 3M Water Infrastructure.*)

Advantages	Limitations
1. Provides continuous barrier lining and prevents leak and further internal corrosion. 2. Smooth surface improves flow capacity. 3. Rapid setting and same day return to service. 4. Contains no volatile organic compounds (VOCs) 5. No plugging of service connections. 6. May accommodate bends exceeding 22.5°. 7. Dependent on the manufacturer and product, may provide a semi-structural lining (Class III).	1. Not fully structural lining material. 2. Pipe must be dry and free from standing water during application.

Table 3.7 Advantages and Limitations for Polyurea Linings

further development in technology in 1980s, polyurethane lining systems were used successfully in water industry (AWWA-C222, 2008). The major advantages of elastomeric polyurethanes are their excellent flexibility, elongation properties, and impact and abrasion resistance.

Polyurethane linings are based on the exothermic reaction between di- or poly-isocyanates and compounds with hydroxyl end-groups such as polyols. It is the exothermic nature of this reaction that provides fast-setting, cold-temperature curing ability, and unlimited film build-up of polyurethane linings. Polyurethane linings consist of sprayable and castable versions. The sprayable version involves lining in various formulations having a mix ratio of 1:1 with balanced viscosity between base and hardener. The castable version is basically a mix, pour, and cast method of application. Figure 3.11 shows a typical sprayer used for polyurethane application. Table 3.8 represents advantages and limitations for polyurethane linings.

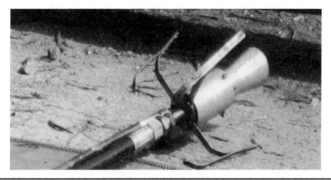

Figure 3.11 Polyurethane lining spray head. (*Source: 3M Water Infrastructure.*)

Advantages	Limitations
1. Exhibits good chemical and impact resistance.	1. Poor adhesion with the existing pipe.
2. Inner chemical structure with much less cross-linking that allows for the lining's elastic nature.	2. Polyurethane is moisture sensitive.
3. Cures in minutes to an hour.	3. Polyurethane has higher permeability rating than polyurea.
4. Offers the added benefit of better safety and environmental parameters when compared to the solvent-based materials.	4. May be more expensive compared to other linings.
5. Good abrasion resistance.	
6. Very flexible and can withstand large movements and bends before displaying cracking or disbondment.	

Table 3.8 Advantages and Limitations for Polyurethane Linings

3.17   Installation Procedures

Table 3.9 presents the steps for installation of spray-on coatings and linings. Figure 3.12 presents the schematic view of lining-installation process.

Steps	Tasks	Objectives & Considerations
1	Project Planning	• Evaluate surface and subsurface conditions: A. As-built drawings B. Crossing and nearby utilities C. Job site access and restrictions D. Traffic management E. Storage and laydown area F. Soil conditions
2	Locating Appurtenances	• Evaluate number of valves, bends and fittings • Locate the entry and exit pits
3	Excavating Pits	• Excavate and protect entry and exit pits
4	Pre-cleaning Inspection	• Check the level of cleaning required
5	Cleaning	• Remove corrosion, sediments, debris and standing water
6	Post-cleaning Inspection	• Check for leaking joints, valves, fittings, pinholes and missing pipe sections, if any
7	Lining Preparation	• Verify lining mix ratio, material temperature, and pump output as required by liner manufacturer
8	Lining Installation	• Monitor the flow rate and the retrieval speed of the spray head • The speed of the spray head must be slow-enough to produce a uniform liner as required per specifications
9	Capping Lined Pipe	• Prevent the contamination of and/or water from entering the pipe by capping the ends of the lined pipe
10	Curing	• Cure lining as required per specifications
11	Testing	• New fixture valves are installed and the system is tested to ensure no water leaks remain and that water pressure has been restored
12	Post-lining Inspection	• Verify quality of the lining work and inspect the lined pipe
13	Disinfecting Lined Pipe	• Disinfect the lined pipe with one of the methods described in Sec. 3.9
14	Restoring Water Service	• Establish connections, appurtenances, and job closeout

*Refer to Sec. 3.16.1 for detailed installation procedure of cement-mortar lining.

TABLE 3.9   Installation Steps for Spray on Coatings and Linings*

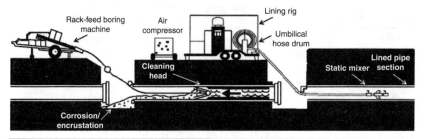

FIGURE 3.12   Schematic view of lining-installation process.

3.18   Comparison of Coating and Lining Methods

Cement-mortar, epoxy, polyurea, and polyurethane are the major materials available today for potable pipe coating and lining applications. They provide added benefits of corrosion protection and, dependent on the product and manufacturer, may provide some structural enhancements. Certain differences exist between products due to specific material properties and type and quality of applications of these products. Table 3.10 present a summary of major differences for these materials.

Parameter	Cement-Mortar	Epoxy	Polyurea	Polyurethane
Corrosion protection	Passive permeable barrier	Dielectric impermeable barrier	Effective and corrosion-resistant barrier	Effective and corrosion-resistant barrier
History	AWWA Standard since 1955 ANSI/AWWA C602-06	Introduced in U.K. water industry in the late 1970s; Standard ANSI/AWWA C-620-07	Has been in use for 10 years	Has been in use for 65 years; Standard ANSI/AWWA C222-08
Pipe preparation/ cleaning	Scraper method	Rack-feed boring	Drag scraper, power boring, jetting	Drag scraper, power boring, jetting
Lining environment	Wet or damp pipe: no standing water	Dry pipe required	Dry pipe required	Dry pipe required

TABLE 3.10   Cement-Mortar, Epoxy, Polyurea, and Polyurethane Comparison

Parameter	Cement-Mortar	Epoxy	Polyurea	Polyurethane
Typical lining thickness	4–8 mm for 4–12-in. pipe and 8–12 mm for more than 12-in. pipe	Minimum 1 mm typical 2–4 mm	Minimum 1 mm typical 2–5 mm	Minimum 1 mm typical 2–5 mm
Curing time before disinfection	Minimum 24 h	Minimum 16 h (can be less for some products)	Minimum 1 h	Minimum 2 h
VOC (lb/gal)	0.00	0.30	0.00	0.00
Application method	Centrifugal, mechanical, pneumatic, hand application	Plural component spray	Plural component spray	Plural component spray
Curing procedure	Moist curing or accelerated curing	Maintain temperature	Not required	Not required
Structural enhancement	No	No	Yes	Yes
Odor generation	No odor	Strong odor during curing	No odor	Strong odor during curing
Bonding to concrete	Good	Strong	Extremely strong	Weak

TABLE 3.10   Cement-Mortar, Epoxy, Polyurea, and Polyurethane Comparison (*Continued*)

3.19   Ongoing Research on Liner and Pipe Interactions

The decision making process for determining a renewal strategy is generally driven by a condition assessment of existing pipeline. Internal and external corrosion and pit depth measurements made on pipe samples, can provide predictions regarding the residual life of pipe. The corrosion damage is normally concentrated in localized "hotspots" rather than being universal damage. Failure of these pipelines commences with the appearance of transverse fractures or "back breaks" where soil and ground movement combined with traffic loads and thermal movements cause tensile stresses to be induced in the inherently brittle cast iron. Therefore, for structural enhancement lining to be effective, it must survive the sudden energy release imposed upon it when pipelines fail under this degree of

FIGURE 3.13   Six inch ID fractured cast iron pipe previously lined with polyurea.

FIGURE 3.14   Six inch sheared cast iron pipe previously lined with polyurea.

loading and, thereafter remain intact and bridging the crack left after fracture. This requires the structural enhancing liner to have enough elongation, as well as cohesive strength that exceeds the adhesion of the lining to the existing pipe. Class 2 and 3 should meet these requirements.

After a pipeline has failed by transverse or shear fracture, movement of the pipe can continue at that point. The lining must therefore be able to accommodate such movement which can take the form of angular deflection.

Figures 3.13 and 3.14 illustrate the results. The assembly is subjected to a three-point flexural loading in a universal testing machine. Displacement controlled loading was applied and the deflection of the joint monitored as a function of load.

3.20   Summary

This chapter focused on spray-on coating and lining methods for renewal of potable water distribution pipes. Potable water pipe renewal is relatively a new market for trenchless renewal methods. These methods have many advantages over other trenchless renewal methods, including fast return to service (and significantly reducing or potentially eliminating the need for bypass pumping), providing a barrier for corrosion protection, filling minor cracks and holes, and in certain products, enhancing the design life of the existing pipe (a semi-structural product). Polyurethane, epoxy, and polyurea lining methods do not plug the service connections, valves, and other appurtenances and can also accommodate minor bends.

CHAPTER 4

Pipe and Pipe Installation Considerations

4.1 Introduction

Pipeline construction has considerable uncertainties owing to the complex nature of the pipe-soil system. More complexity applies to trenchless methods when a machine is used to install the pipe blindly. However, trenchless technology only disturbs the ground in the area of pipe zone, but open-cut disturbs the soil from the ground surface.

Structural designs of pipelines employ models for calculation of different behavioral factors such as deflection, buckling, cracking, and so on, that are based on particular assumptions for interaction between pipe and soil. Since pipe and soil are typically designed as integral systems, it is crucial that the construction process achieves this design objective. While project plans and specifications provide the basic requirements for construction and installation, the site conditions often vary from those anticipated during design. These variations must be considered by the contractor and construction engineer. Often, alternate or additional construction considerations are necessary to address unexpected or unspecified conditions.

4.2 Pipeline Construction Using Open-Cut Method

Pipeline installation using open-cut method involves mobilization of material, labor, and machinery to obtain, assemble, install, inspect, and test the pipeline system. Underground pipeline installation may seem relatively straightforward: (1) dig a trench, (2) lay the pipe in the trench, and (3) fill the trench back in. However, in reality, pipeline installation involves many important engineering and construction considerations. An underground pipeline system incorporates both properties of pipe and soil to form a composite structure. Accordingly, structural designs of underground pipelines include specifications for the surrounding soil. The idea is for the system to be able to withstand the imposed demands through a composite action of the pipe-soil structure. If properly designed and constructed, this composite action is often advantageous and can be used to enhance the load-carrying capacity of pipelines. It follows that proper installation is extremely important to the performance of underground pipeline structures. Of particular importance are considerations for (1) proper selection and compaction of soils for foundation, bedding, embedment, and back-filling; (2) proper control over geometrical configurations of trench excavations (width, height, and others); (3) control over appropriate soil cover; (4) control of line and grade of the pipe; and (5) controlling groundwater in the trench. In addition, appropriate care must be taken during transportation, handling, and installation to avoid cracks, gouges, buckling, and other forms of structural damage.

Soils, on the other hand, require different properties for different subcomponents of an underground installation (e.g., foundation, bedding, trench walls, and so on). Some of the important properties considered in design include soil stiffness (modulus), density, type of soil, and moisture content. Achieving the design properties in the field requires careful monitoring of construction activities along with testing and evaluation. The complexity of soil behavior introduces considerable uncertainty in defining and measuring of soil proper-ties. The problem is augmented by spatial variability of natural soils combined with practical necessity to estimate the properties from very limited amount of sampling and testing. While designs use lim-ited information measured at few strategic points, construction must be carried out over the entire spectrum of variability of soils. Coeffi-cient of soil reaction (E') represents soil variability and is not directly measurable, but must be back-calculated using observed pipe deflec-tions which might be dependent on the pipe depth also.* For this reason, E' is not a property of soil alone (like Young's modulus), but is a function of the pipe-soil system. This integral treatment of pipe and soil is common in current design standards. Therefore, that the outcome of design and construction of pipelines is not the pipe alone, but the entire pipe-soil structure.

4.3 The Pipe-Soil System

As said previously, pipelines, specifically those using flexible pipe, are designed as composite structures with pipe and soil forming an integral system to resist the applied loads. The pipe-soil structure generally comprises the pipe, the in situ soil in the foundation and

---

*AWWA M45 "Fiberglass Pipe Design" Manual provides one of the most commonly used procedure for determining the E' values.

the trench walls, and placed soils (for bedding and backfilling) at locations below and around the pipe. Ideally, installation of placed soils should achieve a state as close to the in situ state of soils as possible. However, this is practically impossible to achieve and appropriate compaction methods are needed. A typical layout of a trenched pipe installation and the associated terminology are illustrated in Figs. 4.1 and 4.2 and described in Table 4.1.

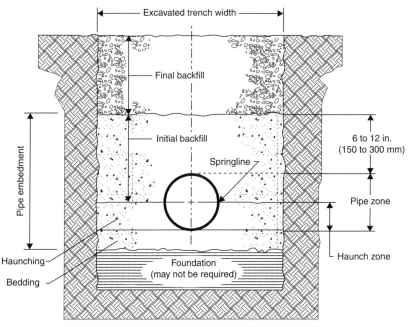

FIGURE 4.1 Typical cross-section of an underground pipeline installation for flexible pipe. (*Source: ASTM D2321, 1989.*)

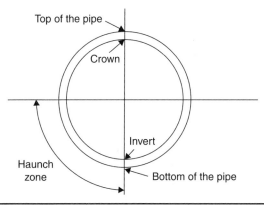

FIGURE 4.2 Pipe cross-section terminology.

Term	Description
Trench walls	The trench walls are comprised of undisturbed in situ soils through which a trench is excavated. Depending upon the type of soil they may be vertical or require sloping.
Bedding	Bedding is the soil placed on top of foundation and provides uniform support and grade for the pipe, except at pipe bells or large couplings where overexcavation is usually specified.
Foundation	The foundation may comprise of undisturbed in situ soil or have imported soils to replace any unsuitable material at the bottom of the trench.
Embedment	Embedment is comprised of material placed around the pipe providing a supporting structure. It consists of the bedding, haunching, and the initial backfill.
Haunch zone	Haunch zone is the area between the bottom of the pipe and the spring line. Backfill material in the haunch zone is critical in transferring forces in the lateral direction.
Springline	Springline is the horizontal centerline of the pipe.
Initial backfill	The initial backfill protects the pipe from final backfill placement. It typically begins at the springline of the pipe and continues about 6 to 12 in. on top of the pipe.
Final backfill	Material placed over the embedment up to the ground level.
Pipe zone	Depth of trench wall occupied by the pipe equal to the outside diameter (OD) of the pipe.
Trench width	Often specified in pipe design, this dimension must allow realistic side clearances to the outside diameter of the pipe, including belled ends and trench support systems. Excessive widths caused by careless excavation increase paving, loads on the pipe, quantities of earthwork and possibly the top width of the trench, affecting right-of-way, surface finishing, and so on.

TABLE 4.1   Explanation of Pipeline Installation Terminology (Howard, 1996)

4.3.1   Rigid Pipes and Flexible Pipes

Broadly speaking, pipe materials fall into two categories: rigid and flexible. Rigid pipes sustain applied loads by means of resistance against longitudinal and circumferential (ring) bending. Under maximum loading conditions, rigid pipes do not deform sufficiently enough to produce horizontal passive resistance from the soil

Rigid	Flexible
Concrete pipe	Steel pipe
Vitrified clay pipe	Ductile iron pipe
Prestressed concrete cylinder pipe	Polyvinyl chloride pipe
Reinforced concrete pipe	Polyethylene pipe
Bar-wrapped concrete cylinder pipe	Fiberglass reinforced plastic pipe
Asbestos-cement pipe	Acrylonitrile-butadiene styrene pipe
Fiber-cement pipe	

TABLE **4.2** Examples of Rigid and Flexible Pipes

surrounding the pipe. Typical examples of rigid pipes are clay pipes and concrete pipes. On the other hand, flexible pipes are capable of deforming (without damage to the pipe) to the extent that the passive resistance of soils on the sides is mobilized providing additional support. ASTM standards define flexible pipes as pipes that deflect more than 2 percent of their diameter without any sign of structural failure. Typical examples include ductile iron, high density polyethylene pipe (HDPE), steel, and polyvinyl chloride (PVC) pipes. Common terminology used to characterize properties of rigid and flexible pipes is strength and stiffness. While strength refers to the ability of rigid pipes in resisting loads and resulting stress in the pipe materials, stiffness refers to the ability of flexible pipes in resisting deflection. Pipes that overlap these two categories are sometimes referred to as *semirigid, semiflexible.* or *intermediate* pipes. However, such distinction is seldom made in current design standards (Moser and Folkman, 2008). Examples of different types of rigid and flexible pipes are given in Table 4.2.

Rigid and flexible pipes differ in the way they transfer the applied loads to the surrounding soil structure. Figure 4.3 gives a simplified

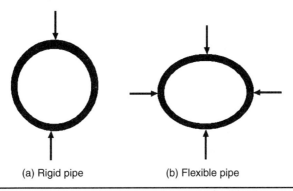

(a) Rigid pipe   (b) Flexible pipe

FIGURE **4.3** Load transfer mechanisms for rigid and flexible pipes.

illustration of the load transfer mechanism for both types of pipes due to the vertical soil pressures. As can be seen from Fig. 4.3, rigid pipes sustain vertical loads by virtue of the material strength alone and with very little deflection. On the other hand, flexible pipes tend to deflect and use the horizontal passive resistance of the soil on the sides. As described in the following sections, this difference in behavior has important consequences in analysis, design, and installation of pipelines.

4.3.2 Soils

Soils are formed over several geologic periods through physical, chemical, and temperature effects (collectively known as weathering) on rocks and minerals. For engineering purposes, soils are often classified as boulders, cobbles, gravel, sand, silt, and clay depending upon the particle sizes and Atterberg limits. Particle sizes for different soils can vary over a broad range with boulders having sizes in excess of 300 mm and clay having sizes less than 75 μm (Howard, 1996). This wide variation in particle sizes results in a wide variation of mechanical properties of soils. Furthermore, natural deposition resulting from weathering often entails a highly variable soil composition. Although, in localized areas such as along a short pipe trench, the soil composition may appear more constant, the contractor must always be alert to changes in the characteristics of soils exposed during pipeline construction.

The first step in determining properties of soil is soil identification. ASTM D2487-06 describes a system for classification of soils for construction (see Table 4.3), which is a version of the Unified Soil Classification System (USCS). According to this system, four basic soil types have been identified, gravel, sand, silt, and clay. Boulders, cobbles, and organic soils such as peat, and others, are ignored because they are generally not used in pipeline construction (Howard, 1996). The USCS classification groups soils according to particle sizes into several categories such as silty gravel (GW), silty sand (SM), elastic silt (MH), and so on (see Table 4.3). These categories range from purely gravel, sand, silt, or clay to categories that are a combination of these basic types (e.g., silty sands, [SM]). In addition to providing an estimate of soil properties, such categorization gives the experienced contractor tools to compare a site with other locations and determine the required equipment and methods of construction (ASCE, 2009).

The desired properties of soil in pipeline construction are different from other types of construction owing to the integral nature of the pipe-soil structure. In most other types of construction, soils are expected to have enough bearing capacity to sustain the imposed dead weight of the structure. In pipeline construction, however, the weight of the pipeline is typically less than the weight of the earth

Main Divisions			Group Symbol	Typical Names
Course-grained soils More than 50% retained on the 0.075 mm (No. 200) sieve	**Gravels** 50% or more of course fraction retained on the 4.75 mm (No. 4) sieve	Clean gravels	GW	Well-graded gravels and gravel-sand mixtures, little or no fines
			GP	Poorly graded gravels and gravel-sand mixtures, little or no fines
		Gravels with fines	GM	Silty gravels, gravel-sand-silt mixtures
			GC	Clayey gravels, gravel-sand-clay mixtures
	Sands 50% or more of course fraction passes the 4.75 mm (No. 4) sieve	Clean sands	SW	Well-graded sands and gravelly sands, little or no fines
			SP	Poorly graded sands and gravelly sands, little or no fines
		Sands with fines	SM	Silty sands, sand-silt mixtures
			SC	Clayey sands, sand-clay mixtures
Fine-grained soils More than 50% passes the 0.075 mm (No. 200) sieve	**Silts and clays** Liquid limit 50% or less		ML	Inorganic silts, very fine sands, rock four, silty or clayey fine sands
			CL	Inorganic clays of low to medium plasticity, gravelly/sandy/silty/lean clays
			OL	Organic silts and organic silty clays of low plasticity
	Silts and clays Liquid limit greater than 50%		MH	Inorganic silts, micaceous or diatomaceous fine sands or silts, elastic silts
			CH	Inorganic clays or high plasticity, fat clays
			OH	Organic clays of medium to high plasticity
Highly organic soils			PT	Peat, muck, and other highly organic soils

Source: ASTM D2487

TABLE 4.3 Unified Soil Classification System (USCS)

displaced by it. Therefore, if trenchless technology methods were used, where the pipe is installed with minimal disturbance to the surrounding soil, the only variable in design would be the strength of the pipe. So, for trench installations, restoring soil to its undisturbed state is desirable although it may be practically impossible (Petroff, 1995).

4.3.3   Pipe-Soil Interaction

Whether a pipe is rigid or flexible has profound effect on the way in which it interacts with the surrounding soil. The interaction between pipe and soil influences the magnitude of loads exerted on the pipe and the manner in which the pipe transfers these loads to the surrounding soils. Calculation of loads exerted on underground pipelines can be traced back to the studies conducted by Anson Marston during the early part of the twentieth century. The results were later expanded by M. G. Spangler and R. K. Watkins and are still in use today (Moser and Folkman, 2008). Figure 4.4 provides an illustration of the soil load distribution on rigid and flexible pipes. In the case of rigid pipes, the theory proposes that the soil in the side prism tends to settle relative to the central prism. This causes the pipe to assume full load of the central prism and a portion of the load from the side prisms. In contrast, a flexible pipe tends to deflect, which result in a lowering of the pressure from the central prism.

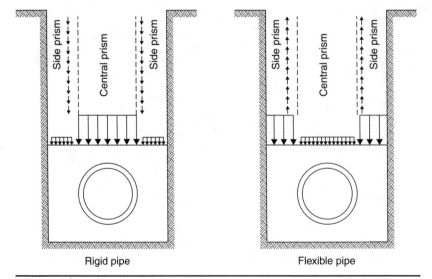

Rigid pipe   Flexible pipe

Figure 4.4 Trench load comparisons for rigid and flexible pipe. (*Najafi, 2005.*)

Soils Arching Effect

Marston's theory (Spangler, 1982) proposes that the load due to the weight of the soil prism above an underground pipeline is modified by arching action. According to this theory, part of the soil weight is transferred between side prisms and the central prism, resulting in either an increase or decrease in the effective weight of soil above the pipe. Arching effects could be classified as either positive or negative. Positive arching effects decrease the vertical pressure exerted on the pipeline whereas negative arching effects increase the vertical pressure.

For pipelines installed in a trench, insufficiently compacted backfill and embedment material are more compressible than the adjacent native soil that has become well-compacted through natural consolidation. The more compressible backfill and embedment material has a tendency to consolidate and settle more than the native soils. As a result, some of the vertical soil load is transferred through shearing stresses between the side prisms and the central prism and creates positive arching. The degree of transfer depends upon the type of backfill material and how well it is compacted. This arching action can be used advantageously for both rigid and flexible pipes by making the bedding immediately underneath the pipe more compressible than the adjacent bedding. However, care must be taken in order to avoid differential settlement of the pipe, especially for rigid pipes.

Assuming well-compacted soils in bedding and backfill, positive arching effects occur in the case of flexible pipes that deflect owing to their lower stiffness. This phenomenon is expected to naturally occur in trenchless installation of pipelines. The relative downward movement of the central prism within the trench mobilizes upward shearing stresses along the sides and creates an arching action that partially supports the soil column weight above the structure (see Fig. 4.5). In addition to this action, passive resistance of the soils adjacent to the pipe is mobilized and aids in transfer of loads. In contrast and again assuming well-compacted soils in bedding and backfill, negative arching occurs in the case of rigid pipes that do not deflect owing to their high stiffness. Owing to the relatively low stiffness of soils on the sides, shearing forces are transferred from the side prisms to the central prism increasing the effective vertical soil load on the pipe (see Fig. 4.5).

Installation conditions can have a significant effect on the soil loads acting on underground pipelines. In a long-term behavior study in Norway (Vaslestad et al., 1994), a 5.25 ft (1.6 m) diameter circular concrete culvert was constructed with imperfect trench condition (the compressible material was expanded polystyrene foam) with 46 ft (14 m) embankment height. Vertical earth pressure directly above the culvert was only 25 percent of the soil prism weight. Horizontal earth pressure at the mid elevation of the culvert, however, was 73 percent of the soil column weight above that elevation. In the same research,

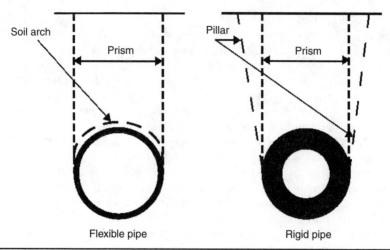

FIGURE 4.5   Arching effects for flexible and rigid pipes. (*Gabriel, 2006.*)

the vertical pressure above a box culvert with 32.8 ft (10 m) backfill height but with 1.7 ft (0.5 m) of expanded polystyrene foam placed immediately above the culvert was 50 percent of the pressure due to the weight of the soil prism above. However, the vertical pressure above an identical box culvert under the same embankment height without the foam, and with normally compacted backfill, was about 120 percent of the soil prism weight. These examples indicate that installation methods and culvert shape can strongly influence the magnitude and distribution of earth pressures on rigid culverts.

The magnitudes of the loads exerted on culverts depend on arching effects, which are the result of relative deformation of the backfill in a certain zone above the culverts. This deformation is related to both the soil and the structural stiffness. In the case of flexible culverts, culvert deformation results in arching effects, which reduce the vertical loading regardless of the installation method. Typically, the vertical earth pressure on flexible culverts is less than that due to the weight of the prism of soil above the culvert. For rigid culvert installation, arching effects can also be achieved by introducing compressible material into the backfill. In the case of positive projecting embankments, arching results in vertical earth pressures that are greater than that due to the weight of the prism of soil above the culvert. Thus, according to the Marston's theory, the vertical earth pressures on culverts are a function of the installation method, the soil and structural stiffness, the geometry of the structure, and the boundary condition with the natural ground. It is worth noting that even for flexible culverts, a load reduction is achieved as differential settlements transfer loads to the surrounding soil adjacent to the structure.

4.3.4 Behavior of Rigid Pipes

As noted earlier, rigid pipes are designed to transmit loads through their material strength. Since rigid pipes do not deflect appreciably, their design does not consider the horizontal passive resistance of the side soils. Therefore, the allowable load on rigid pipes depends only on the strength of the pipe and the strength of the soil below the pipe. This is usually represented by the following relationship (Howard, 1996):

$$\text{Load on the pipe} = \frac{\text{pipe strength} \times \text{soil strength}}{\text{safety factor}} \tag{4.1}$$

Typically, the strength of the material of a rigid pipe is determined in the laboratory by the three-edge bearing test (Fig. 4.6). The three-edge bearing strength is load per unit length required to cause either crushing or critical cracking of the pipe (Moser and Folkman, 2008).

Strength of soil refers to the amount of support on the bottom half of the pipe and depends on the properties of soil and the contact area. Figure 4.7 illustrates two extremes of contact area available for pressure

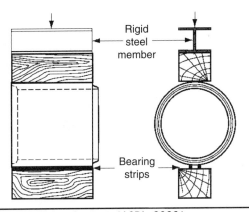

FIGURE 4.6   Three-edge bearing test. (*ACPA, 2000.*)

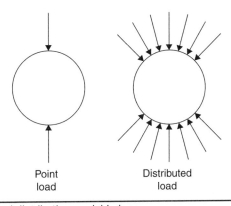

Point load   Distributed load

FIGURE 4.7   Load distribution on rigid pipes.

distribution below a rigid pipe (Howard, 1996). The point load is the worst possible support case, which gives the strength equal to the three-edge bearing test. For distributed load case, as for trenchless installation, the strength of the pipe may be enhanced due to uniform distribution, of stress as shown in Fig. 4.7. Research has shown that the load required to cause failure in installation conditions is typically greater than the three-edge bearing strength owing to distribution of forces (Moser and Folkman, 2008). To ensure appropriate distribution it is vitally important to achieve design densities in the haunch area of the pipe.

The ratio of field strength of a rigid pipe to the three-edge bearing strength is called a bedding factor and is given by

$$\text{Bedding factor} = \frac{\text{field strength}}{\text{three-edge bearing strength}} \tag{4.2}$$

Bedding factors are specified by different manufacturers for different types of pipes based on the placement methods and materials used.

4.3.5 Behavior of Flexible Pipes

Unlike rigid pipes, flexible pipes are designed to transmit part of the load to the side soils. As the load on the pipe increases, the vertical diameter of the pipe decreases and the horizontal diameter increases. The increase in horizontal diameter mobilizes the lateral resistance of the soils as shown in Fig. 4.8. Change in vertical or horizontal dimension of a pipe is usually represented as a percent change and is given by

$$\text{Percent deflection} = \frac{\text{change in diameter}}{\text{pipe diameter}} \times 100 \tag{4.3}$$

Consider the flexible steel pipe, which is a perfect circle when it is laid on top of the bedding and no soil load has been placed. Steel is a linearly elastic (not viscoelastic) material. After backfilling, though, the steel pipe deflects. When first deflection takes place, two things happen. First, soil arching reduces the soil load on the steel pipe. So the load the pipe is resisting has decreased. Second, the material in the haunch zone has been further compacted by the expansion of the horizontal diameter. In other words, using Eq. (4.4), the numerator has decreased because of soil arching and the denominator has increased because the soil's stiffness has increased due to compaction from the pipe's horizontal expansion.

$$\text{Pipe deflection} = \frac{\text{load on the pipe}}{\text{pipe stiffness} + \text{soil stiffness}} \tag{4.4}$$

Nevertheless, the load on the pipe has not been reduced sufficiently yet. The pipe further deforms to the shape given by the second deflection in Fig. 4.8. The load on the pipe reduces further

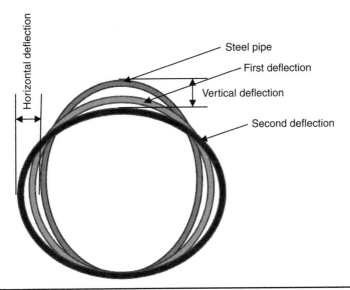

Horizontal deflection

Steel pipe

First deflection

Vertical deflection

Second deflection

FIGURE 4.8   Load distribution on flexible pipes.

due to soil arching. Soil stiffness increases because of greater density of the embedment material in the haunch.

Now, the resistance of the combination of the pipe's stiffness and the soil's stiffness is great enough to prevent further deflection. That is how the pipe-soil system stabilizes.

4.4   Common Modes of Pipeline Failures

Underground pipeline failures can occur in several different modes and owing to different types of causes. Rajani (2001) presents three principal aspects of the physical failure mechanisms in pipelines:

1. Pipe properties, material type, pipe-soil interaction, and quality of installation

2. Internal loads due to operational pressure and external loads due to soil overburden, traffic loads, frost loads, and third-party interference

3. Material deterioration due largely to external and internal chemical, biochemical and electrochemical environment

While the third aspect is usually relevant only in service conditions, the first two are very much applicable during and immediately after pipeline installation. Table 4.4 presents some of the factors that affect failure mechanisms in pipelines. For more information on pipeline deterioration, refer to Najafi (2005).

Pipe Factors	Environmental Factors
Type of pipe and pipe material	Defects during manufacturing
Location of the pipe	Damage during transportation, handling and installation
Diameter	Soil loads (which depend on the type of soil, density, level of compaction, etc.)
Length	Point loads from projecting rocks, etc.
Type of soil and embedment	Internal pressure loads
Joining method	Axial loads due to temperature, water hammer, etc.
Internal/external corrosion Protection	Frost loads in soils
Wall thickness	Freezing and expansion of water
Depth of installation	Loads due to expansive soils
Bedding conditions	Third-party damage
Foundation conditions	Traffic loads

TABLE 4.4   Factors Affecting Pipeline Failures

Failure mechanisms for rigid pipes and flexible pipes differ in several respects. In general, rigid pipes fail in tension and crack rather than deform, if the imposed loads exceed the pipe's inherent strength. Clarke (1968) reports the following major causes of failures in rigid pipelines:

- Inadequate load-carrying capacity of pipes
- Nonuniform bedding
- Inappropriate construction methods (e.g., excessive trench widths)
- Use of rigid jointing material resulting in a lack of axial flexibility and extensibility in pipeline
- Differential thermal deformation or moisture movements
- Differential settlement

Flexible pipes, in general, do not crack but fail by excessive deformation, buckling, or pipe flattening. Also, flexible pipes are more accommodating of faulty installation of embedment, bedding, or foundation because of their ability to deform. However, improperly placed embedment material could lead to loss of side support, which is vital for flexible pipes and could result in overdeflection or flattening. Farshad (2006) reports the following major causes of failure in flexible pipeline:

- Fracture
- Buckling
- Weathering, color, and dimensional changes
- Voids, blisters, and delaminations
- Fatigue and corrosion
- Clogging of the pipe system

Specific mechanisms for failure of a pipeline also depend upon the type of material of the pipe. Structurally, a pipeline is said to have failed when the performance limits of its material have been reached. Although underground pipelines and the soil embedment around them are designed as an integral system, structural failure of the embedment soil leads to eventual failure of the pipe material. Therefore, the following discussion on failure mechanisms for rigid and flexible pipes focuses only on the failure modes of the pipe itself.

4.4.1 Failure Modes in Rigid Pipes

Rigid pipe failures occur when the performance limits for the pipe material are reached. Performance limits for rigid pipes may be categorized into

- Ring flexure
- Longitudinal flexure
- Shear
- Radial tension
- Longitudinal tension
- Cracking
- Wall crushing

Depending on the specific pipe material, actual mechanisms of failure may vary. For instance, radial tension due to hoop stresses may cause rupture of reinforcement in prestressed concrete cylinder pipe (ASCE, 2009). Ring flexure in concrete and clay pipes causes quarter point cracking as shown in Fig 4.9. In cast-iron pipes, longitudinal bending forces could cause circumferential failure of the pipe as shown in Fig. 4.10 (Makar, 2001).

Other mechanisms of failure include longitudinal cracking caused by pipe wall thinning and pressure surges in cast-iron pipe as shown in Fig. 4.11 and spiral cracking of cast-iron pipe as shown in Fig. 4.12 (Makar, 2001).

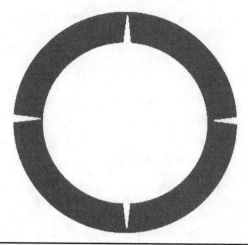

FIGURE 4.9   Flexural failure in rigid pipe resulting in quarter point cracking.

Ring flexure refers to cross-sectional bending of a pipe where the vertical diameter tends to decrease and horizontal diameter tends to increase. Owing to the high stiffness of rigid pipes, cracking at quarter points occurs before the pipe is able to deform (see Fig. 4.9). Longitudinal flexure is caused when a pipe is forced to behave as a beam resulting in axial bending stresses (see Fig. 4.13). Major causes for bending action include nonuniform bedding support, differential settlement, and ground movement due to external forces such as earthquake, frost, expansive soils, and so on (Moser and Folkman, 2008).

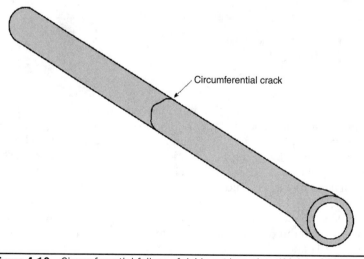

Circumferential crack

FIGURE 4.10   Circumferential failure of rigid cast-iron pipe. (*Makar, 2000.*)

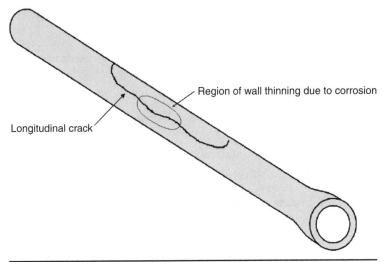

FIGURE 4.11 Longitudinal splitting of rigid cast-iron pipe. (*Makar, 2000.*)

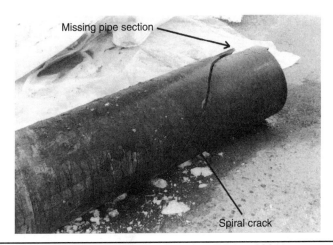

FIGURE 4.12 Spiral crack in rigid cast-iron pipe. (*Makar, 2001.*)

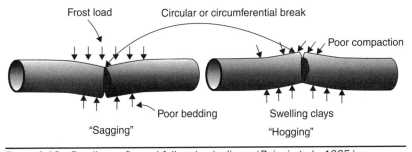

FIGURE 4.13 Bending or flexural failure in pipelines. (*Rajani et al., 1995.*)

Some of the major causes of longitudinal bending or beam effect are (Moser and Folkman, 2008):

- Differential settlement of manholes or other structures that the pipe is connected to
- Nonuniform settlement or erosion of bedding
- Expansive soils
- Frost loading
- Nonuniform foundation
- Tree root growth pressure

Shear failure occurs in large-diameter rigid pipelines owing to excessive shear stresses. Figure 4.14 illustrates an example of shearing of bell joints in large-diameter (> 18 in.) cast-iron pipe. Longitudinal tension may be caused due to temperature effects or shrinkage of the embedment soil around the pipe (Moser and Folkman, 2008).

Figure 4.14   Bell shearing in large-diameter cast-iron pipes. (*Makar, 2001.*)

4.4.2   Failure Modes in Flexible Pipe

Flexible pipe failures occur when the performance limits for the pipe material are reached. Performance limits for flexible pipes refer to (Moser and Folkman, 2008):

- Wall crushing
- Wall buckling
- Deflection
- Strain limit
- Longitudinal stress
- Wall loss (due to corrosion)

Wall Crushing

Wall crushing occurs when the in-wall compressive stresses reach the yield stress or the ultimate stress of the pipe material. This type of failure occurs when pipe is embedded in very stiff material and subjected to very high loads (Massicotte, 2000). It is characterized by localized yielding at the springline of the pipe and usually occurs in smaller-diameter pipes with thicker walls. Figure 4.15 provides an illustration of this type of failure. Normally ring compression is the influential stress causing this type of failure, although bending stresses may also cause wall crushing (Moser and Folkman, 2008).

Wall Buckling

This type of failure occurs in pipes with low stiffness embedded in low stiffness soils. It is usually caused due to the pressure around the pipe or creation of vacuum inside (Moser and Folkman, 2008). Design of large-diameter pipes may be governed by buckling, especially in conditions of high soil pressure and low stiffness soil (Reddy, 2002). Figure 4.16 gives an illustration of this phenomenon. Buckling is primarily dependant on the stiffness of the pipe, which in turn depends

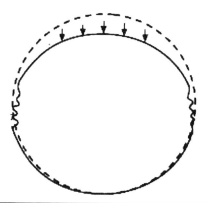

Figure 4.15   Wall crushing of flexible pipes. (*Moser and Folkman, 2008.*)

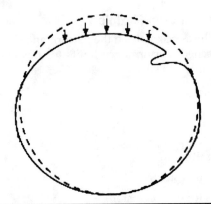

Figure 4.16   Wall buckling of flexible pipes. (*Moser and Folkman, 2008.*)

on the type of material and the pipe geometry (diameter, wall thickness, and so on). Resistance against buckling may be increased by stiffening the pipe walls by providing corrugations or by improving the soil stiffness. The soil stiffness may be improved by using a better embedment material, by providing better compaction, or both.

Deflection

Flexible pipes have relatively low bending stiffness when compared with rigid pipes. Owing to this fact, deflection is an important factor in design and installation of a flexible pipeline. Although flexible pipes are designed to deflect in order to initiate the passive soil resistance, deflection control is necessary to avoid the pipe from flattening or reversing its curvature. Deflection control plays an important role in the determination of quality of pipeline installation and can be a good indication of its long-term performance (Reddy, 2002). Deflection limits for flexible pipes usually range from 2 to 7.5 percent depending upon the pipe material (Howard, 1996). Typical deflection scenarios are illustrated in Figs. 4.17 and 4.18. The factors that affect pipe deflection include:

- Pipe stiffness
- Soil resistance (type, density, modulus of elasticity, and moisture content)
- Applied loads
- Trench configuration (geometry and embedment)
- Haunch support
- Construction stages
- Time
- Temperature
- Variability in construction procedures and in soil characteristics

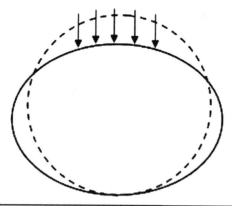

Figure 4.17 Ring deflection of a flexible pipe. (*Moser and Folkman, 2008.*)

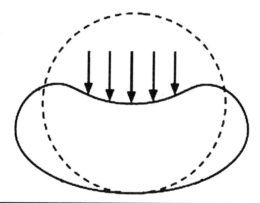

Figure 4.18 Reversal of curvature due to overdeflection in flexible pipes. (*Moser and Folkman, 2008.*)

Longitudinal Tensile Failure

In general, rigid or flexible pipe products are not designed to withstand longitudinal tension. Longitudinal failure in pipes can occur due to several factors that are given below (Moser and Folkman, 2008):

- Thermal expansion or contraction
- Longitudinal bending
- Poisson's ratio (due to internal pressure)

The tendency of a pipe to move in the axial direction is resisted by the frictional forces of soil surrounding the pipe. This creates longitudinal stresses that can cause a circular break at weaker cross-sections (cross-sections weakened by corrosion or cracking). Figure 4.19 illustrates longitudinal tensile failure in pipes. Figure 4.20 illustrates the effect on pipe due to the loads.

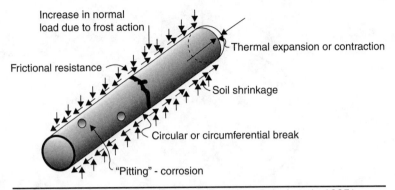

FIGURE 4.19   Longitudinal tensile failure in pipes. (*Rajani et al., 1995.*)

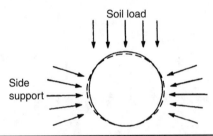

FIGURE 4.20   Side support for flexible pipes. (*NCSPA, 2000.*)

Pipe Wall Loss (Due to Corrosion)

Corrosion of metallic pipe can lead to structural failure as it affects its wall thickness. Pipe wall thickness is critical to the pipe performance specifically for pressure applications, where steel pipes are usually used.

4.5   Pipe Selection Considerations

Although economics is one of the engineering considerations during the design and selection of pipe material for a pressure or gravity sewer system, there are several other factors that are also important. These considerations include the following:

- Type of fluid to be transported (potable water, wastewater, storm, sewer, water, oil, gas, and so on.).
- Construction conditions and methods used.
- Life expectancy and related life-cycle cost analysis.
- Flow characteristics, such as corrosiveness and abrasion of wastewater.
- Ease of handling and installation of pipe.

- Pipe physical and chemical properties.
- Availability of diameter sizes, pipe section lengths, and joints (specifically for trenchless technology).
- Construction and operational stresses in the pipe. In many cases for trenchless technology projects, the construction stresses in the pipe exceeds the operational stresses.
- Location and pipe environment (inland, offshore, in-plant, corrosiveness of soil, and so on).
- Type of burial or support (underground, aboveground or elevated, underwater, and so on).
- Hydraulic properties of the pipe.

4.6   Rigid Pipes

The following sections present main types of rigid pipes.

4.6.1   Cement-Based Pipes

Pipes in this category include concrete pipe as well as asbestos-cement pipe. Both types incorporate portland cement as the base material. Concrete pipes are designed with or without reinforcements and are used for both pressure and gravity applications. Asbestos-cement pipe is composed of a mixture of portland cement, silica, and asbestos fiber and was used in North America mainly in potable water applications, though many sewer systems also have asbestos-cement pipes in service. Asbestos-cement pipes have not been manufactured or installed in the United States since the late 1980s, but large quantities of the pipe continue to be in service in water systems at various locations. It is therefore important for engineers to have a working knowledge of the properties of this material. Both concrete and asbestos-cement pipes are rigid conduits and are accordingly designed for installation.

Concrete Pipes

Trenchless construction with concrete pipe was first performed by the Northern Pacific Railroad between 1896 and 1900, when the pipe was installed by jacking. The jacking of concrete pipe is performed both for pressure and gravity applications. Concrete pressure pipes are also installed by tunneling methods. For small-diameter pipe in short tunnels, such as those under a highway or a railroad, it is a common practice to slide the pipe through a liner followed by grouting the interspace.

Manufacturing   Concrete pipes can be made with any of the five types of Portland cement. There are five manufacturing processes used in North America, four of which use mechanical means to place and compact a dry concrete mix [zero slump with low water-cement (w/c) ratios] into a form. A conventional wet mix and casting method is used in the fifth one. Each method is briefly discussed below:

1. *Centrifugal casting:* The form is rotated, acting as a centrifuge, while the concrete mix is fed into it. Vibration and compaction are also employed to consolidate the concrete mix. The centrifuge promotes extraction of water from the mix.

2. *Dry casting:* Low-frequency, high-amplitude vibrations are used to distribute and compact the dry mix in the form in this process. The form is then removed and the pipe cured.

3. *Packerhead process:* In this process, a high-speed rotating device, the packerhead, acts as the inner wall of the form and compacts the concrete against the walls of the form. As the concrete is fed from the top of the form, the packerhead gradually moves up.

4. *Tamping:* The concrete is mechanically compacted with oscillatory tampers between the walls of the outer form and an inner core. This method is declining in use.

5. *Wet casting:* This method is typically employed for the manufacture of large-diameter pipes, using a concrete mix with higher water content than the other four methods. Again, the pipe is made between the inner and outer walls of a form.

Types of Concrete Pipes   There are six different types of concrete pipes used for pressure and sewer applications:

1. *Nonreinforced concrete pipe (CP):* Used for nonpressure applications only, this pipe is manufactured in diameters of 4 through 36 in., in lengths up to 8 ft. Produced in three classes, its minimum strength requirement is the three-edge bearing per ASTM C14. Since it is nonreinforced, structurally, it can only exhibit a brittle failure mode.

2. *Reinforced concrete pipe (RCP):* Typically produced to meet ASTM C76, there are three types of reinforcements available for RCP-welded wire reinforcing, hot-rolled rod made of Grade 40 or 60 steel, and cold-drawn steel wire made from hot-rolled rod. It is used in nonpressure and low-pressure applications of up to 55 psi. There are five load-bearing capacities, ranging from 1350 to 3000 in units of lb-force/ft of diameter/ft of length known as "D-Load." Higher load capacities can be specially designed for unusual combinations of fill height or live load as per ASTM C655.

3. *Prestressed concrete cylinder pipe (PCCP):* Often used in high-pressure applications, PCCP is a composite pipe of concrete

and steel, capable of handling up to 500 psi. There are two types of PCCP-lined cylinder pipe and embedded cylinder pipe. A welded cylinder core with joint rings attached at both ends exists in both types. After curing, the pipes are wrapped with hard-drawn wire under high tensile stress, and then coated with cement slurry. Available diameters range from 24 to 144 in. AWWA C301 is the governing standard.

4. *Reinforced concrete cylinder pipe (RCCP):* This is similar to PCCP, but uses reinforcing cages in place of the hard-drawn wire. Standard AWWA C300 requires that the pipe be designed to withstand both internal and external pressure. Available diameters range from 24 to 144 in. Pressure applications are its main use.

5. *Bar-wrapped steel-cylinder concrete pipe:* Also known as pretensioned concrete cylinder pipe, the steel cylinder is internally lined with a cement mortar lining. Once cured, a steel rod, under tension, is wrapped around the cylinder. Another cement mortar lining is then placed on the wrapped cylinder. Smaller-diameter pipes are considered rigid, whereas larger diameters behave as flexible conduits. AWWA C303 is the standard governing its manufacture and testing. Like the previous two types, bar-wrapped concrete pipe is also used in pressure applications.

6. *Polymer concrete pipe (PCP):* Polymer concrete originated more than 20 years ago in Germany. This type of pipe provides a corrosion-resistant concrete as needed for piping applications that require high concrete compressive strength and resistance to corrosive chemicals. These pipes are made by mixing a high-strength, thermosetting resin with oven-dried aggregate to form a type of concrete. The resin within the mix provides for bonding the aggregate much like portland cement does in traditional concrete pipes. Figure 4.21 illustrates typical sections of polymer concrete pipes.

FIGURE 4.21 Polymer concrete pipe.

The pipe sections are cast vertically with an inner and outer form and are vibrated for compaction. After the forms are removed, the section is heated in a kiln to finish curing the resin. These pipes can typically be used to carry highly aggressive wastes, for pipe jacking as they have very high compressive strengths (up to 17,000 psi), or for microtunneling. They can also be used for gravity flow or pressure applications. Some manufacturers are also making polymer pipes in sizes appropriate for use as manholes. Also, semielliptical and circular liner pipes have been developed for sliplining sewers. There are several ASTM standards for the product.

Polymer concrete pipes have several benefits including high strength, corrosion resistance (they can be used in environments with pH ranges of 1 to 13), low wall roughness, and high abrasion resistance. The use of polymer concrete pipes is becoming more common in the following areas: direct bury, sliplining, jacking and microtunneling, tunneling, and above-the-ground applications. One of the main disadvantages in the past was the high cost associated with importing the pipe; its high cost relative to alternative materials limited its use to only niche markets where its superior qualities were needed. Another characteristic to be accounted for especially in the larger diameters is that, since it is nonreinforced, PCP can only exhibit a brittle failure mode if the external load-bearing capacity is exceeded. However, since 2002, polymer concrete pipe has been manufactured in the United States under the brand name of Meyer pipe. This is now enabling competition with products such as glass-reinforced pipe (GRP) for jacking installations.

Fiberglass sleeve joints are made separately, with elastomer sealing and spacing rings laminated into the sleeve. Factory fitting of the couplings to the end of the pipe is performed and pressure testing is done up to 35 psi. As an alternative to the couplings, stainless steel collars are also available.

Applicable Standards   Table 4.5 lists reinforced concrete pipe standards that are widely used in trenchless installations. Typical available diameter ranges are also given.

Joint Types   The concrete pipe industry has developed several different types of joints for the various types of pipes. Selection of the appropriate joint is based on the stringency of the application for water tightness.

Pressure Standards	Nonpressure Standards	Available Diameters (in.)
ASTM C361	—	12–108
AWWA C300	—	30–144
AWWA C302	—	12–144
—	ASTM C14, C76, C655	12–144

TABLE 4.5 Concrete Pipe Standards

Although bell-and-spigot type joints are available for open-trench applications, trenchless methods such as microtunneling and pipe jacking dictate that the joint be flush with the outside wall of the pipe barrel. Therefore, the tongue and groove joint or the modified tongue and groove joint is ideal. Table 4.6 shows joint types that may be used in trenchless construction. Also available are mastic or mortar joints, which are not well-suited to prevent leakage except for transient hydrostatic heads below about 5 psi. Above that pressure threshold, a gasketed joint meeting ASTM C443 is probably appropriate. ASTM C990 requires an in-plant hydrostatic test of 10 psi as "proof of design."

A proprietary concrete and PVC composite pipe, PipeForm™, with an inner and outer PVC liner that gives it the strength of concrete and the corrosion resistance of PVC, has been used in microtunneling and pipe-jacking projects for gravity applications. In addition to preventing corrosion, the PVC outer shell also reduces external skin friction during jacking (as long as it is not damaged by abrasion during the process), while the internal liner improves flow. Available diameters range from 18 through 36 in. Distribution of jacking forces at the end surface is maximized by steel or plastic collars and rubber gaskets. Details of the joint are shown in Table 4.6.

Advantages and Limitations   Table 4.7 presents a summary of advantages and limitations of concrete gravity and pressure pipes. The widespread use of large-diameter concrete pipe in water and sewer applications is an indication of the material's acceptance at the municipal level. The vast selection of available diameters and pipe lengths is a convenience to the design engineer and contractor. On the pressure side, availability of various structural and pressure strengths also makes it convenient. The ability of manufacturers to make pipes capable of handling very high pressures in large diameters has resulted in its specification on some high-profile trenchless projects in recent years. Because a specialized work crew is not needed; cost of labor for concrete pipe installation is not high. For trenchless construction, the high compressive strengths give it a definite advantage over some other pipe materials.

When used in open-cut construction, many concrete pipes installed in gravity applications 20 to 30 years ago have shown signs of poor performance. Leakage through joints and cracks in the pipe has been a constant source of inconvenience to municipalities. For contractors, the heavy weight of concrete pipes can make it difficult to install when compared to some alternative materials. Concrete pipe's sensitivity to bedding conditions in both shallow and deep installations has resulted in pipe failures by shear and beam breakage in many instances. Compared to flexible pipe types, due to their low wall stiffness, only about 10 percent of the system's capacity is attributable to the pipe wall thus requiring that a structural "exoskeleton" of carefully compacted, select material must be carefully constructed around the pipe and for an additional height above the top of the pipe.

Joint Type	Cross Section
Mastic filled tongue and groove joints (ASTM C990) for low head applications.	
Compression type rubber gasket joints with internal bell and spigot with the gasket and joint meeting ASTM C443 or C361. Intended for use with pipes meeting the loading requirements of ASTM C14, C76, and C655.	
Steel end ring joint. This is a high pressure joint, intended for use with ASTM C361. C361, Sec. 8.1, permits (3) joint material combinations, "Joints shall utilize steel joint rings, steel bells and concrete spigots, or be formed entirely of concrete." Most RCP manufacturers can make an "all concrete" rubber gasketed joint that will pass a 72 psi test for "high pressure" performance category.	
PipeForm™ joint. This proprietary product is a concrete-PVC composite, used in pipe jacking and microtunneling.	

Table 4.6 Concrete Pipe Joints

Advantages	Limitations
1. Specialized work crew not required for installation	1. In open-cut construction, pipe is sensitive to bedding conditions-shear failure and beam breakage may occur
2. Large selection of available nominal diameters	
3. Wide variety of pipe lengths available	2. Handling and installation difficulty because of heavy weight except where weight would be advantageous because of flotation concerns
4. Large selection of both structural and pressure strengths	
5. Relatively low cost of maintenance	3. Susceptible to external corrosion in acidic soil environments
6. Capability to withstand very high pressures	4. Highly vulnerable to hydrogen sulfide attacks and internal microbiological-induced corrosion at crown. A concern in sanitary applications only.
7. Ideal for pipe jacking applications owing to high compressive strengths	
8. Internal corrosion can be significantly reduced by using thermoplastic lining	5. Generally difficult to repair, particularly in cases of joint leakage or failure in pressure pipes
9. External sulfate corrosion may be reduced by an additional sacrificial wall thickness determined by the Pomeroy/Parkhurst 'AZ' design method is more commonly added to the *inside* of the pipe wall to counter the corrosion from biogenically generated H_2SO_4. Alternatively, it is possible use Type V sulfate-resistant Portland cement.	6. Tendency to leak because of high pipe wall porosity and shrinkage cracking
	7. Without internal lining, life span is significantly reduced in the case of sanitary sewer applications and only then if there is a high potential for H_2S generation
	8. Somewhat lower abrasion resistance—internal scouring can occur if solid content and flow velocities are high
	9. Reinforcements in PCCP can corrode or fail without little or no external evidence

TABLE **4.7**   Concrete Pipe Advantages and Limitations

In trenchless installation, however, concrete pipe is much less sensitive to bedding conditions in the circumferential direction. For RCP, roughly 90 percent of the load bearing capacity of the soil/structure "system" is in the pipe wall.

In terms of "beam breakage," the statement is partially correct. RCP is not designed for beam strength. But, that is the reason for the relatively short joint lengths in use—so that one length of pipe may move differentially with respect to the adjacent length. RCCP, on the other hand with their longer joint lengths are checked for longitudinal beam strength.

The susceptibility of concrete pipe to both internal and external corrosion has also come to the forefront in recent years. For sanitary sewers where the possibility of hydrogen sulfide generation is a possibility, many engineers and city authorities now require concrete pipes to be internally lined because of their experience with its deterioration. Studies suggest that unlined RCP can show signs of deterioration in as little as 5 years after installation. Failures owing to hydrogen sulfide attacks and internal microbiological-induced corrosion at the crown of RCP have also been a topic of study in recent years. In pressure applications, the loss of prestress as a result of reinforcement wire corrosion and breakage in PCCP can lead to pipe failures in high-pressure lines. Research to assess the risks of such failures has been undertaken by various engineers and entities. The preceding discussion about hydrogen sulfide corrosion is obviously not an issue in storm sewer applications.

Asbestos-Cement Pipe

With the asbestos-related litigations of the 1980s, bans were placed on all construction materials that used asbestos. Pipe was no exception. In 1986, the U.S. EPA published a proposed regulation on the commercial uses of asbestos in which a ban on the manufacture and installation of asbestos-cement (AC) pipe was proposed; this proposal was later carried out. Many AC manufacturing plants relocated to other countries, and production in the United States came to a complete stop. In the early 1990s, the ban on the pipe was lifted, but today AC pipe is no longer manufactured in the United States. Many countries of the world, including Mexico, continue to specify and install asbestos-cement pipe for both water and sewer applications.

Worldwide usage of AC pipe increased from 200,000 miles in the 1950s to 2 million miles in 1988. In the United States, it was estimated that by 1988, over 300,000 miles of AC pipe was in service in water systems. According to the Association of Asbestos Cement Product Producers, AC pipe is still used in the southwestern United States, but it must be imported from Mexico. However, this usage is certainly an anomaly. The hazards of asbestos inhalation are increased during removal of the pipe, so utilities have chosen to keep them in the ground. These pipes are regularly tapped and maintained, so a working knowledge of the material is helpful. There are no known instances of trenchless municipal piping construction with AC pipe.

Manufacturing   Asbestos-cement water and sewer pipe is a fiber-reinforced, cementitious product, composed of an intimate mixture

Figure 4.22   Joint cross-section of coupling-joint in AC pipe.

of portland cement and silica. The controlled blending of these basic raw materials is built up on a rotating steel mandrel and then compacted with steel pressure rollers into a dense homogenous structure in which a strong bond is affected between the cement and the asbestos fibers. A smooth interior surface results from this process. Final curing of the product is done in an autoclave employing high-pressure steam for dimensional and chemical stability.

Joint Types   AC pipes were joined together by means of a gasket-joint coupling (also referred to as a double-bell coupling joint). The ends of two pieces of pipe were slipped into the gasketed coupling, effectively creating two water-tight seals. Some manufacturers permitted angular movements of up to 5° at joints. Figure 4.22 shows a typical AC pipe joint.

Advantages and Limitations   Among its advantages, its long operational life, immunity to corrosion, light-weight in smaller diameters, and watertight joints may be mentioned. Known limitations of the pipe include a low flexural resistance as a whole, easily damaged by construction equipment because of its brittle nature and a low chemical resistance. Significant research was done to study the effects of dissolved asbestos fibers in drinking water on human health. AC pipe manufacturers and the American Water Works Association do not recommend the use of this piping material where water is highly aggressive as corrosive water, such as acidic water with a low pH was more likely to attack piping products.

4.6.2   Vitrified Clay Pipe

The second group of rigid pipes among piping materials is vitrified clay pipe (VCP). The use of VCP in sanitary sewer systems throughout the United States is yet to be matched by any other piping material. For two centuries, VCP was the only commercially available material capable of withstanding the chemically aggressive environments of sanitary sewers. The earliest recorded use of clay pipe in the United States was in Washington, D.C., in 1815. Clay pipe is not used in pressure applications because of its inherently low tensile strength.

Only 50 years ago, the sanitary sewer engineering community followed the philosophy that leakage of wastewater through pipe joints was an acceptable methodology for effectively transporting suspended solids and reducing excessive flows within a sewer system. Clay pipes were therefore designed with a low emphasis on the effectiveness of their joints. This philosophy soon changed as engineers realized the hazards posed by wastewater leakage to soils and groundwater sources. The EPA's role in reducing infiltration or inflow (I/I) with the passing of several congressional legislations such as the Water Pollution Control Act of 1972 and the Clean Water Act of 1977 were major factors in emphasis shifting to the requirement of watertight joints in sewer pipes. All factory-applied clay pipe jointing systems, whether on bell-and-spigot pipe or plain-end pipe, are designed to provide resilience and flexibility to accommodate minor pipe movement. All compression type jointing methods meet the requirements of ASTM C425, which requires that the joints should not leak. With the proper installation, a clay pipe sewer system can meet standard infiltration or exfiltration requirements. Furthermore, an independent study done by the University of Houston demonstrated that the joints of vitrified clay pipe exceeded the industry standards. In the arena of trenchless construction, clay pipe's ability to withstand high compressive loads and external abrasion has resulted in a significant rise in its acceptance and use in pipe-jacking and microtunneling applications.

Manufacturing

Vitrified clay pipe is made of selected clay and shale that are aged to various degrees, and blended in specified combinations. Large crushing wheels grind the clay in a heavy perforated metal pan until the finely ground clay passes through the perforations. The ground raw materials are mixed with water in a pug mill. The mixture is then forced through a vacuum, deairing chamber until a smooth, dense mixture forms. The mixture is extruded under extremely high pressures to form the pipe. After drying, the newly formed pipe is placed in kilns and heated to temperatures of approximately 2000°F. The finished pipe then undergoes a QA/QC testing.

Applicable Standards

A relatively high minimum compressive strength of 7000 psi makes clay pipe a good contender for jacking and microtunneling installation. In 1994, ASTM C1208 opened new doors for clay pipe in the trenchless construction arena (see Table 4.8).

Pressure Standards	Nonpressure Standards	Available Diameters (in.)
NA	ASTM C-1208: *Standard Specification for Vitrified Clay Pipe and Joints for Use in Microtunneling, Sliplining, Pipe Bursting, and Tunnels*	4–42

TABLE **4.8** Product Standards

Joint Types

Also known as pretensioned concrete cylinder pipe. The evolution of joints in clay pipe is a shining example of the results of research, development and innovation within the piping industry. From its earliest days of having no joints to its present-day joint types, clay pipes have seen several iterations over the course of a century. To arrive at its present-day compression joints, clay pipes went from having no joints to field-applied cement joints to field-applied bitumastic joints to factory-installed cement mortar or bitumastic joints, and finally to the compression seal joint. One manufacturer even makes a molded polyurethane joint attached to the spigot end and a PVC collar, which is shrunk and rapidly fitted and sealed to the socket end of rigid vitrified clay pipe barrel as shown in Figure 4.23.

The jointing system in ASTM C1208 clay jacking pipe is a precision ground recessed joint, ensuring dimensional accuracy and high end-bearing capacity. A stainless steel sleeve with elastomeric seals is

FIGURE **4.23** Innovative joints in clay pipe. (*Source: Can Clay Corporation.*)

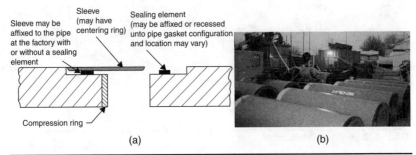

Sleeve may be affixed to the pipe at the factory with or without a sealing element

Sleeve (may have centering ring)

Sealing element (may be affixed or recessed unto pipe gasket configuration and location may vary)

Compression ring

(a)　　　　　　　　　　　　(b)

Figure 4.24 (*a*) and (*b*) Jointing system in ASTM C1208 clay jacking pipe.

used for jointing, as shown in Fig. 4.24*a* and *b*. The sealing element is compressed between bearing surfaces to promote watertight integrity. The ends of the pipe are fitted with a compression ring to distribute the jacking forces of installation.

Advantages and Limitations

Table 4.9 summarizes some of the advantages and limitations of clay pipe. The inert nature of clay pipe was the reason behind its wide acceptance for use in sanitary sewer applications in years past. Resistance to both internal and external corrosion was its main advantage over other traditional piping materials such as concrete and cast iron. Clay pipes are also abrasion resistant making it a suitable material in sewers with high solid content. Its low sensitivity to temperature differentials prevents any significant expansion and contraction when

Advantages	Limitations
1. Resistant to both internal and external corrosion	1. Available for gravity applications only because of the characteristics of the pipe
2. Proven history of long life of the pipe itself	2. Sensitive to bedding conditions – may be subject to shear and beam failure
3. Improved joints have been available since 1970	3. Poor joints in pipe installed prior to 1970 may lead to leakage and root intrusion problems
4. Ability to handle high compressive forces, making it ideal for jacking installations	4. Short lengths, resulting in more pipe joints
5. Abrasion resistant	5. Very brittle – frequently high breakage during shipping and handling

Table 4.9 Clay Pipe's Advantages and Limitations

buried in the ground. In recent years, the ability of clay pipes to withstand external abrasion and relatively high compressive forces has led to its wide usage in trenchless applications such as microtunneling and pipe jacking.

Although the inert characteristics of the material made it ideal for use in aggressive environments such as sanitary sewers, the inability of clay pipes until recently to manufacture effective joints has resulted in a drastic reduction of its use in sewer systems. Though innovation within the industry has led to better performing joints in recent years, the millions of feet of clay pipe installed in the ground decades ago are now responsible for I/I problems in sewer systems throughout the United States and Canada. Its susceptibility to shear and beam breakage owing to poor bedding conditions and ground movement has caused numerous leaks and cracks in a number of sewer systems where it was used. The short lengths in which clay pipes are manufactured (usually 8 ft or less) increases the number of joints in a sewer main line, thus raising the chances of leakage through joints. Root intrusion through clay pipe joints has led municipalities to institute annual root control measures in older systems.

4.7   Plastics Pipes

Plastic pipes were introduced in the late 1950s in North America. The three main types of plastics pipes include PVC, high-density polyethylene or cross-linked polyethylene (HDPE, PEX), and glass-reinforced pipe (also called fiberglass pipe). PVC and PE fall into the group of *thermoplastics*, while GRP and PEX is a *thermoset* pipe. The properties of thermoplastics for construction material applications are better appreciated with an understanding of *viscoelastics*, discussed in the Sec. 4.8.1. Thermoset plastics are processed by a combination of chemicals and heat, and once formed, cannot be reshaped. Both thermoplastic and thermoset pipes are considered to be flexible pipes and are designed accordingly.

In the field of municipal trenchless construction, PE has been the dominant piping material in the past decade. By butt-fusion of successive lengths of PE pipe, or by using coiled PE, a long *jointless* pipe is created, which can be installed by trenchless methods such as horizontal directional drilling (HDD). PE is also used in open-trench construction. Traditionally, PVC has been the most dominant material for open-cut installations in the North American water and sewer markets because of its bell-and-spigot gasket-joints, its light weight in smaller diameters, and ease to work with. In recent years, several manufacturers have also created proprietary PVC products with modified joints for trenchless installation. GRP pipes are used in the United States for both pressure and gravity applications, though the latter is the more prominent use.

4.7.1   Properties of Viscoelastic Pipe Materials

PVC and PE are thermoplastics, and thus, viscoelastic materials. Viscoelastic materials exhibit elastic as well as viscous-like characteristics. A material that deforms under stress, but regains its original shape and size when the load is removed is classified as elastic. Viscous materials, on the other hand, after being subjected to a deforming load, do not recover their original shape and size once the load is removed. In reality, all materials deviate from the linear relationship between stress and strain (Hooke's law) at some point in various ways.

Defining the direct relationship between stress and strain when a load is applied to a material is the most common way to evaluate the strength and stiffness of that material. Graph A in Fig. 4.25 illustrate the linear relationship between stress-strain in elastic materials. In an ideal elastic material, strain returns to zero as soon as the material is unloaded, and the linear relationship is not typically time-dependent. But it should be noted that in all materials, this behavior is valid only up to a certain stress point, called the *yield point*, after which the strain in the material will increase dramatically by creep, before finally failing.

In the set of curves, B, in Fig. 4.25, it can be seen that the stress-strain relationship is somewhat different for viscoelastic materials than it is for elastic materials. Clearly, we no longer see a directly linear relationship between stress-strain, and the gradients of the curves depend on the loading time. In other words, for a given stress level, the longer the loading time, the larger the strain reached. Creep is defined as continuing deformation (increasing strain) with time when the material is subjected to constant stress. As a

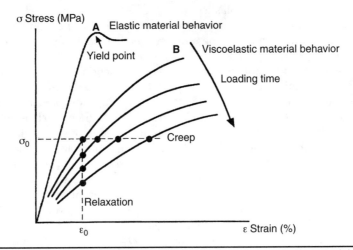

Figure 4.25   Stress-strain relationship in elastic and viscoelastic materials. (*Janson, 1999.*)

consequence of creep, failure of the material will occur after load is applied for a certain amount of time. So time dependency is a major factor to consider in viscoelastic material behavior. An important fact is that the time to failure is inversely proportional to the applied stress. In thermoplastic pressure pipe, it is therefore possible to find and apply a stress level that is low enough to ensure that the theoretical time to failure will surpass the design life of the pipeline.

In thermoplastic pipe applications, creep is prevented because the deflection of the pipe is kept constant, as is the case in buried PVC gravity (or pressure) pipe. Consequently, it can be seen from Fig. 4.25 that the initial stress decreases with time, and is referred to as the relaxation property of thermoplastic piping materials. These basic properties of viscoelastic materials, such as PVC and PE, enable engineers to design pipelines that ensure both structural integrity and the long-term design life of their municipal piping systems.

4.7.2 Polyvinyl Chloride Pipe

Polyvinyl chloride (PVC) was discovered almost accidentally in the nineteenth century when German scientists, observing a newly created organic chemical gas, vinyl chloride (C_2H_3Cl), discovered that when it is exposed to sunlight, a chemical reaction took place, resulting in the creation of an off-white accumulation of solid material. Since then, scientists had observed the first polymerization and creation of a new plastic material, PVC. In 1839, a technical paper was published detailing the observations of the process. In 1912, several decades after its accidental discovery, Fritz Klatte, another German, laid the groundwork for the technical production of PVC. The oldest known PVC pipe was manufactured and installed in the 1930s in World War II in Germany and continues to be in service today. The technology was brought to the United States following World War II and by the mid-1950s ASTM groups were organized for plastic pipe standardization.

Manufacturing

A vinyl chloride molecule comprises carbon, hydrogen, and chlorine, configured as shown in Fig. 4.26*a*. PVC is obtained by polymerization of single units of the vinyl chloride molecule, which join to create long chains, and ultimately form PVC resin, Fig. 4.26*b*.

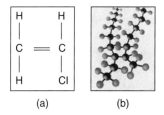

(a)   (b)

FIGURE **4.26** (*a*) and (*b*) Vinyl chloride molecule and PVC chain.

PVC pipe is manufactured by first blending PVC resin with stabi-
lizers, pigments, lubricants, processing aids, and functional additives;
and heating this mixture to a temperature in the 400°F range. This
causes the components to properly fuse and convert into a malleable
state. In this molten form, the material is mechanically extruded into
the pipe. Following the completion of the extrusion process, the pipe
is allowed to cool, after which QA/QC testing is performed, before
the final delivery to the end user.

Types of PVC Pipe

There are three distinct types of PVC pipes manufactured in the
world, each differentiated by either the way in which it is manufactured (which dictates the directional orientation of the molecules), or
by the content of modifiers in its chemical formulation (which affect
the ability of the pipe to withstand impacts by absorption and dissipation of the energy). The term PVC is a generic designation, which
includes PVC-U (unplasticized PVC), PVC-O (molecularly oriented
PVC), and PVC-M (modified PVC).

1. *PVC-U*: Unplasticized PVC is widely used piping material
   in water and sewer systems in North America. The molecular
   structure of PVC pipe is a random arrangement of long chain
   molecules, where molecular entanglement is prevalent
   throughout the length of the pipe. In general, the PVC molecules do not exhibit any definite directional orientation, and
   therefore, a generally uniform strength prevails in both the
   radial (circumferential) and longitudinal directions. Testing
   has shown that the modulus of elasticity in 15-year-old PVC
   is only slightly higher in the longitudinal direction than in the
   radial directions. For simplification, the term PVC as used in
   this chapter will denote PVC-U, the conventional type of PVC
   most used in pipe manufacture. PVC pipes are manufactured
   for both pressure and gravity applications. All PVC pressure
   pipes (ASTM and AWWA standards) must meet the cell classification 12454, indicating a tensile strength of 7000 psi, and
   a modulus of elasticity of 400,000 psi. PVC pressure pipes
   have a hydrostatic design basis (HDB) of 4000, per ASTM
   D2837, to which a design factor of 2.0 is applied. This reduces
   the long-term design stress to 2000 psi. PVC gravity pipes are
   typically manufactured to cell class 12454 or 12364, where the
   latter has a tensile strength of 6000 psi and a modulus of elasticity of at least 440,000 psi.

2. *PVC-O*: Molecularly oriented PVC is made in the United
   States by the expansion of the conventional PVC pipe; during
   the expansion process, the molecules become oriented in a
   generally radial or circumferential direction. This molecular
   reorientation increases the strength of the pipe in the hoop

direction. Also, the resulting HDB is increased from 4000 to 7100 psi. Consequently, this stronger material can have a thinner wall than a conventional PVC pipe of the same pressure capacity. The manufacture process of PVC-O in the United States uses both of the technologies available for manufacturing the product. Both the batch (offline) and continuous (inline) processes are used. PVC-O pipe is not used in gravity applications.

3. *PVC-M:* Modified PVC is produced by incorporation of additives or *impact modifiers* to enhance the toughness of the material. Resistance to fracture by absorption and dissipation of energy is an evidence of the toughness of the pipe material. PVC-M is made and used mainly in Europe and Australia, whereas only one manufacturer in the United States produces this type of pipe for nonburied applications.

Solid Wall and Profile Wall PVC Pipes

PVC pipes are available for pressure applications only as solid wall pipes. For gravity applications, both solid wall and profile wall pipes are manufactured. Solid wall pipe, as the name suggests, are made of a continuous wall of PVC of uniform thickness, as shown in Fig. 4.27*a*. Profile wall pipe, on the other hand, is braced spirally or circumferentially with structural shapes, but provides a smooth-wall interior, as shown in Fig. 4.27*b*. Profile wall pipes economize on the amount of material needed for fabrication; by altering the shape of the wall, the same stiffness as solid-wall pipe is achieved, using less material. Profile wall pipes generally fall into three categories—open profile (OP), closed profile (CP), and dual-wall corrugated profile (DWCP). OP pipes have their rib enforcements exposed on the outside of the pipe. CP pipes make use of a closed profile that provides a continuous outer wall where the wall sections are hollow and are often described as an I-beam or honeycomb. DWCP pipes have a smooth-wall waterway, braced circumferentially with an external corrugated wall.

(a)   (b)

Figure 4.27   (*a*) and (*b*) Solid wall and profile wall PVC pipe.

From a design perspective, both solid wall and profile wall PVC-gravity pipes are limited to a vertical deflection of 7.5 percent, per ASTM and Uni-Bell recommendations. For pressure pipe, AWWA M45 "Fiberglass Pipe Design" Manual recommends a deflection limit of 5 percent.

Generally, profile wall pipes are used for open-cut installations as well as trenchless renewal processes such as sliplining. Pipes used for sliplining have modified joints that facilitate their installation by segmental sliplining.

PVC Applicable Standards and Products for Trenchless Construction

Although there are several widely used bell-and-spigot gasket-joint PVC pressure and gravity piping standards in North America written by organizations such as AWWA, ASTM, and CSA, products discussed in this chapter are specifically for trenchless applications. They are proprietary in nature in that they have uniquely designed jointing systems that enable the pipe to be pulled or pushed for various trenchless construction methods (TCMs). Table 4.10 outlines details on each product along with pictures of joints. Also, there are several types of PVC piping products geared toward use in *trenchless renewal* market. Typically, these are profile wall pipes with proprietary joints. Information on these products is not included in this chapter but more details may be obtained from manufacturers or Uni-Bell PVC Pipe Association. Table 4.10 presents a summary of PVC pipe standards and joint systems.

1. *Fusible C-900™, Fusible C-905™, and Fusible PVC™:* Until recently, PE was the only available thermoplastic pipe option that used butt-fused joints in the United States, and this was the case for well over 30 years. In the late 2003, the water or wastewater industry saw the introduction of Fusible PVC, or Fusible C-900, or Fusible C-905. A first of its kind, this product combines a proprietary formulation and fusion procedure that allows lengths of PVC pipe to be joined together in a continuous string for installation through a variety of methods, including HDD, sliplining, pipe bursting, and direct bury applications. Although Fusible PVC is primarily for nondrinking water applications like recycled water, forcemains, gravity drains, and sewer applications, Fusible C-900 or Fusible C-905 are specifically for use in potable water systems. The pipes are manufactured to all requirements of AWWA C900 and C905, and are National Science Foundation (NSF) certified. Surface scratches of up to 10 percent of the pipes' outside diameter are accepted, per AWWA, during pull-in of all PVC pipes. A major advantage of the fusible pipe is that those municipalities already using conventional PVC can now use the same material in trenchless construction,

Pipe Type	Standard Compliance	Joint	Diameter
Fusible C-900™, Fusible C-905™, Fusible PVC™	AWWA C900/C905		4 through 48 in.
TerraBrute™	AWWA C900		4 through 12 in., capability to go up to 48 in.
CertaLok™	AWWA C900/C905		4 through 16 in.
Internally Restrained PVC Pipe	AWWA C900		4 through 12 in.

TABLE 4.10 Trenchless PVC Products and Standards

enabling the design of a *complete PVC system*. It also allows municipal entities to inventory a single type of fittings and restraints for all situations. Tapping is also performed in the same manner as gasket-joint PVC pipes.

2. *TerraBrute™:* This product meets AWWA C900 standard requirements, but has a modified bell-and-spigot joint as shown in Table 4.10. Currently manufactured in diameters of up to 12 in., this pipe has been used in several buried and at least one nonburied pressure application in Canada. The bell-and-spigot joint modification increases the tensile load capacity of the pipe joint by a factor dependant on the diameter and the wall thickness of the pipe. Also, controlling the number of pins, pin diameter, depth of groove in the spigot, and wall thickness of the internal ring can optimize the tensile capacity of the joint. Installation involves insertion of the spigot into the bell, alignment of the pins with the metal band built into a groove on the spigot, and then hammering down of the pins. Once again, utilities in Canada have the ability to build complete PVC systems, even when portions of a project comprise trenchless construction.

3. *Certa-Lok C900/RJ™:* This is another product conforming to AWWA C900 and C905 standards, with a proprietary joint type that makes the product ideal for trenchless installation via HDD. The pipe is also NSF certified for potable water usage. This was the only PVC alternative for trenchless construction in the past decade. The Certa-Lok C900/RJ is used in the new construction of water distribution or transmission lines or sewer force mains, or for new gravity sewer main installation. The joint has a groove machined in the pipe and in the coupling to allow the insertion of a flexible thermoplastic spline that provides a full 360° restrained joint with evenly distributed loading. Available diameters currently include 4 through 16 in., in both DR18 and DR14. The recent introduction of 16-in. pipe uses a fiberglass-based coupling joint system. According to contractors, availability of 16 in. pipe has been an issue since it was introduced.

4. *Internally restrained PVC pipe:* This product also conforms to AWWA C900. The joint is patented and is licensed to several PVC pipe and fabricated fitting manufacturers. The mechanism consists of a metal casing that sits adjacent to the Rieber gasket in the bell; and the casing is molded into the "raceway" of the bell during pipe belling. A C-shaped grip-ring with several rows of unidirectional serrations is manually inserted into the casing at the manufacturing facility. Both the casing and the grip ring are made of ductile iron that has been

coated using an electrocoating process that achieves a uniform thickness and provides superior corrosion resistance. When the pipe arrives at a jobsite, the bell already contains the casing with the grip ring inserted in it, and no additional hardware is needed to provide a restrained joint. Internally restrained PVC pipe is suitable for HDD or for direct-bury applications that require a restrained joint. The brand names for the product are Bulldog™, Diamond Lok 21™, Eagle Loc 900™, and Royal Bulldog™.

Advantages and Limitations

The greatest contributor to the rapid adoption of PVC pipe in water and sewer systems is its inherent ability to withstand both internal and external corrosion. Vast amounts of corrodible piping materials are being replaced each year by PVC. Pipes with proprietary joints for trenchless installation have high tensile strengths, allowing for long lengths to be pulled in at a time. Abrasion resistance has been an advantage over alternative cementitious piping materials. Low internal friction enables the use of smaller-diameter pipe in both pressure and gravity applications.

The sensitivity to temperatures on a long-term basis requires thermoplastics such as PVC to be derated in pressure applications. Pipe that is not formulated with a higher amount of ultraviolet inhibitor results in lowered impact strength after 2 years of continuous exposure to sunlight. PVC cannot be deflected longitudinally as much as alternative thermoplastics. The minimum bending radius for fusible PVC pipe is about 250 times the pipe diameter. Poor bedding can cause excessive deflection and failure in thinner wall pipe. PVC pipe is more susceptible than alternate thermoplastics to rapid crack propagation. Rapid crack propagation could result in long running cracks (hundreds of feet) in the event pressurized pipe is impacted or improperly tapped. Table 4.11 presents a summary of the advantages and limitations of PVC pipe.

4.7.3 Polyethylene Pipe

Polyethylene (PE) belongs to a group of thermoplastics known as polyolefins, materials made by polymerization of *olefin gases* including ethylene, propylene, and butylene.

Polyethylene (PE) was first discovered by a German scientist, Hans von Pechman in 1898, but the discovery was never commercialized. PE was rediscovered in the United Kingdom in 1933 by the Imperial Chemical Company and later commercialized in 1939 to manufacture insulation for telephone and coaxial cables. The development of low-pressure reactors in the late 1950s greatly improved the commercial manufacture of PE resins, and led to the commercial

Advantages	Limitations
1. Resistant to both internal and external corrosion 2. Gasket-joints and fusible joints have an excellent track record of leak-free performance. 3. All four restrained-joint PVC products have high tensile strengths for HDD and other trenchless processes 4. Highly abrasion resistant for sewer applications 5. Low internal frictional resistance for both pressure and nonpressure applications 6. At least 2.5 times stronger than other thermoplastic pipe (higher stiffness, higher HDB) 7. Expansion is significantly lower than in alternative thermoplastic piping material	1. Sensitive to operating temperature, must be derated in case of long-term exposure to temperatures above room temperature 2. Sensitive to ultraviolet light if exposure is greater than 2 years (unless pipe is formulated with higher ultraviolet [UV]-inhibitor level) 3. Less longitudinal flexibility than alternative thermoplastic piping material 4. Thinner-walled sewer pipe is sensitive to bedding conditions. 5. Susceptible to chemical permeation in cases of gross contamination 6. Susceptible to impact damage in cold temperatures 7. Susceptible to rapid crack propagation failure. Tapping of fused PVC pipe must be done with extreme caution.

TABLE **4.11**   Advantages and Limitations of PVC Pipe

development of PE pipe in the 1960s. Early PE pipe use was mainly in the gas industry. Today, more than 95 percent of new and replacement gas distribution pipe in the North American is PE pipe. Other applications of PE pipe gradually spread to industrial and municipal systems. Since the development of high-density PE materials in the 1970s, a significant use for PE is sliplining renewal of nonpressure gravity sanitary sewers. The development of ASTM and AWWA pressure piping standards led to the widespread use of PE pressure pipe in municipal applications in North America over the past decades. This is due to the recognition of the advantage of a corrosion resistant, fused pipeline in reducing water loss and pipeline ruptures. Many municipalities that have already been specifying another thermoplastic material, are using PE in trenchless and open-trench installations. Open-cut waterworks projects are being performed with PE pipe at an increasing rate. Also, the wider acceptance of trenchless technology construction methods for new construction and renewal of water and sewer systems has placed PE pipe at the forefront of the no-dig industry. Today, PE is the most widely used piping material for trenchless methods such as horizontal directional drilling (HDD) and pipe bursting (PB) for new installation and renewal of municipal pressure and nonpressure piping systems. PE used for pipes in North America are generally classified, on basis of density (crystallinity) and suitability for pressure service. Higher density provides greater hardness, stiffness, and tensile strength.

ASTM classifies PE piping materials having specified density ranges as low density (LDPE), MDPE, or HDPE. HDPE displays the highest stiffness whereas LDPE is the most flexible. LDPE is not used for pressure piping, but due to its flexibility is used for pneumatic instrument controls. MDPE pressure piping is predominantly used for gas distribution. HDPE nonpressure piping materials are used for electrical and communications conduit and nonpressure corrugated and profile wall pipes. HDPE pressure piping materials are used for solid-wall pressure pipes for gas, water, force mains, nuclear and industrial process piping, and for solid-wall nonpressure applications such as sanitary sewers and culverts. Gas pipes are generally made from MDPE. Municipal water and sewer pipes and industrial pipes are generally made from HDPE. At 73°F the pressure rating of HDPE is about 25 percent higher than MDPE pipes. In 2005 ASTM standards were modified to allow the introduction of an improved HDPE with significantly improved performance capabilities. HDPE materials commonly used for municipal piping have a code of either PE3608 or PE4710. The PE4710 material has a pressure rating 25 percent higher than PE3608 material (i.e., 25% higher pressure than PE3608 material). With few exceptions, both materials have a design service life of 100 years in typical municipal water or sewer applications.

Manufacturing

The manufacturing process for HDPE pipe is plastic extrusion where molten PE resin extruded under pressure through specially designed extrusion machines and dies that form the melt into pipe. The pipe is drawn into cooling tanks to solidify and cool the pipe. A second method for producing large-diameter PE pipe is to extrude a wall profile onto a rotating mandrel. Wall profiles can have hollow sections to reduce the amount of material but maintain circumferential stiffness. Once the pipe section is created, the mandrel is collapsed and removed from the pipe ID. A third method for corrugated profile wall PE pipe is to extrude a thin wall pipe into a corrugator that has vacuum blocks that draw the pipe into a corrugated wall profile and cool it. Corrugated pipes can be produced with an ID liner for a smooth waterway, and a noncorrugated OD layer for a smooth OD. Solid wall, PE pipes are typically designed for internal pressure, but are used for pressure and nonpressure services. Spiral wound and corrugated profile wall pipes are for nonpressure service. A third

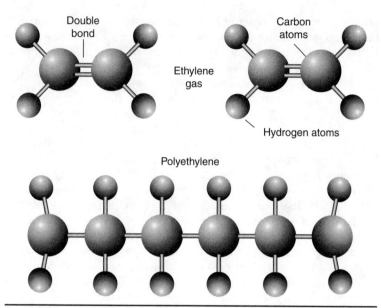

FIGURE 4.28   Ethylene molecule and polyethylene chain. (*Source: Plastics Pipe Institute, 2000.*)

technique is corrugated vacuum block extrusion that is used for corrugated profile wall HDPE pipe.

PE resin is manufactured from the base ethylene molecule (Fig. 4.28), a colorless gas composed of two double-bonded carbon atoms. Polymerization of the ethylene is performed with various catalysts, under heat and pressure, during which the double bond between the carbons is broken, resulting in the formation of a bond with another carbon atom. Polymerization produces long carbon-carbon chain molecules.

Copolymer PE pipes resins are formed by copolymerization with other monomers such as butene, hexene, and octane. Copolymerization produces a molecular structure with copolymer side chain branching. Side chain branching and molecule length significantly affect short- and long-term properties, and have a lesser effect on mechanical properties such as density, ductility, and hardness. Homoploymers, resins manufactured without comonomers, are not used for PE piping.

Engineering Properties

PE pipe materials are identified by categorizing physical property performance into numerically identified ranges in accordance with ASTM D3350. The resulting cell classification identifies cell ranges for density, melt flow rate, flexural modulus, tensile strength, slow

Material Designation Code	Typical ASTM D3350 Cell Classifications
PE3408	PE345444C[a]
PE3608	PE345464C
PE4710	PE445474C or PE445574C

[a] C identifies minimum 2 percent carbon black in the PE compound for UV resistance. For color products, E identifies color with UV resistance. C compounds are suitable for long surface or above grade service. E compounds are sufficiently stabilized for storage, but must be buried.

TABLE 4.12   Material Designation Code and Cell Classification

crack growth resistance, and long-term hydrostatic strength. Higher numerical cell classifications represent higher cell property range values. UV resistance is identified with a letter code at the end of the cell classification. ASTM D3350 also includes additional minimum performance requirements for thermal stability, brittleness temperature, and tensile elongation.

PE pipe materials are generically identified by a material designation code that identifies the acronym for the type of plastic, ASTM D3350 cell values for density and slow crack growth (SCG) resistance, and the hydrostatic design stress (HDS) rating for 73°F water in hundreds of psi with tens and units dropped. For example, PE3408 is a polyethylene material having density and SCG resistance meeting ASTM D3350 density cells 3 and 4, and a HDS rating of 800 psi. Typical material designation codes for PE pressure piping materials* are PE3408, PE3608, and PE4710. Table 4.12 provides material designation codes and typical ASTM D3350 cell classifications. Figure 4.29 shows the HDPE pipe readied for pullback in a directional bore.

Applicable Standards

Environmental stress cracking may occur where the pipe is subjected to excessive stress concentrations due to improper embedment or excessive bending strain due to improper installation. Solid wall HDPE pipes manufactured to AWWA and ASTM standards are typically used for trenchless installation applications. OD-controlled pipes are required for butt-fusion joining that produces a continuous conduit ideal for pull-in installations such as HDD. Renewal methods such as pipe bursting also provide a good match for use of HDPE pipe. Testing has shown that surface scratches sustained during installation of up to 10 percent of the pipe's minimum wall thickness

*PE materials used for pressure piping are listed in PPI TR-4.

FIGURE 4.29   HDPE pipe readied for pullback in a directional bore.

Standard	Available Diameters	PE 4710	PE 3608
AWWA C901-08	½–3 in.	Available	Available
AWWA C906-07	4–63 in.	Standard pending	Available
ASTM D3035-08	½–24 in.	Available	Available
ASTM F714-08	3–63 in.	Available	Available

TABLE 4.13   Solid Wall HDPE Pipe Standards for Trenchless Technology

do not detrimentally affect short- or long-term performance. Table 4.13 presents solid wall HDPE pipe standard for trenchless installations.

Joints

Butt-fusion of PE pipe is performed using commercially available fusion machines and performed in accordance with ASTM F2620, "Heat Fusion Joining of Polyethylene Pipe and Fittings." The steps involve preparing ends of two pieces of pipe by planning the ends using special radial blades, heating the ends with a heat plate (after which the heat plate is removed), and finally fusing the ends together by application of pressure. Fusion joining parameters are specified in ASTM F2620. Both internal and external beads can be removed prior to installation. For pressure applications, the internal bead has a negligible effect on flow. In nonpressure gravity design, some engineers believe that the turbulent flow created by the bead is helpful in keeping solids suspended in the flow; others feel it is detrimental to the flow.

Other types of joints in HDPE pipe include electrofusion and bell-and-spigot gasket joints.

Advantages and Limitations

The continuous *jointless* conduit that results from the butt-fusion of HDPE pipe makes it an ideal piping material for pull-in installations such as horizontal directional drilling and pipe bursting. HDPE pipe's ability to withstand both the internal and external corrosion is advantageous for both water and sewer systems. Its high flexibility is a favorable characteristic during trenchless installations; there is no need for very long entry pits other than for larger diameters and deep installations. A very low internal resistance to flow makes it a good material for both pressure and nonpressure gravity systems. In pressure water systems, the expansive forces of freezing water do not cause the pipe to crack. In very cold temperatures, HDPE has a high resistance to failure by impact. The pipe also resists shatter-type or rapid crack-propagation failure.

In the early years of manufacture, environmental stress cracking was a concern for HDPE pipe. In the past 25 years, higher-quality resins have successfully overcome this problem in HDPE pressure pipes. The butt-fusion of HDPE necessitates the use of a skilled labor force. The fused joint is allowed to cool for an additional 30 minutes after removal from the fusion machine prior to installation. In the case of thick wall pipes (> 2 in.) extra cooling may be required. The high sensitivity of HDPE to temperature differentials requires special design considerations for above grade pipelines. (Transitions from HDPE pressure pipe to gasket joint pressure pipe requires a thrust anchor.) However, buried forcemain lines require no special installation for thermal affects. Buried or sliplined gravity flow pipes generally require a wall anchor where they terminate in a manhole. All pipes must be designed for the proper pressure and be derated for surge and fatigue loads, temperature, corrosion, earthquake loads, C factor, installation, and others. Table 4.14 presents the advantages and limitations of HDPE.

4.7.4   Glass-Reinforced Pipe (Fiberglass Pipe)

The third type of plastic pipe is glass-reinforced pipe (GRP), also commonly referred to as fiberglass pipe. Unlike PVC and HDPE, GRP is made of a thermoset material. GRP was first manufactured in the United States in the 1950s, as an alternative to corrosion-prone concrete and steel materials. Using a patented centrifugally cast manufacturing process, Perrault Fibercast Corporation of Oklahoma

Advantages	Limitations
1. Resistance to both internal and external corrosion. Low internal friction. Smooth interior.	1. Older PE materials are subject to environmental stress cracking due to improper embedment or excessive local bending. Newer materials, like PE 4710 have enhanced resistance.
2. Butt-fused joints effectively create a continuous jointless leak free joint.	
3. Abrasion resistant. Use to convey sand and fly ash slurry.	2. Trained labor and special equipment required for butt-fusion. However, training is available for equipment manufacturers and distributors nationwide.
4. High ductility and flexibility. Lightweight in smaller diameters. Typical minimum bend radius of 25–30 times pipe diameter.	
5. Excellent resistance to fatigue and repetitive surge pressures.	3. Slightly smaller inside diameter than other pipes of the same outside diameter size. However, proper design will minimize this issue.
6. May be repaired using mechanical couplings and saddles.	
7. High resistance to failure by impact, even at very low temperatures.	4. Cannot be located unless buried with metallic wire or tape.
8. Resists shatter-type or rapid crack-propagation failure.	5. Sensitive to temperature differentials, resulting in expansion and contraction unless restrained by soil embedment after burial.
9. Does not easily crack under expansive forces of freezing water.	
10. PE with carbon black has a long UV resistance.	6. Unprotected color products usually cannot have more than 5 years of UV resistance.

TABLE 4.14   Advantages and Limitations of HDPE

manufactured the first fiberglass-reinforced polyester resin pipe. Traditionally used in industrial applications throughout the world, GRP is rapidly gaining market shares in North America for large-diameter municipal water and sewer applications. Municipal sewer installation include direct bury (open-cut), sliplining (premier product for 48 in. and larger), direct jacking, casing carrier (tunnel liner) and above ground on piers. As a plastic material, GRP is designed as a flexible conduit and is installed accordingly.

Manufacturing

Fiberglass composites are made of glass fiber reinforcements, thermosetting resins, and other additives such as small aggregates, catalysts, hardeners, accelerators, and the like. Types of resin used include epoxy, polyester, and vinyl ester. The amount and orientation of the glass fibers increase the mechanical strength of the pipe. GRP is manufactured using one of the two methods:

1. *Centrifugal casting*: This is the most widely used method of manufacture of GRP for municipal applications in North America. In this process, glass fiber reinforcements, resins and aggregates are placed in a rotating steel mold. The centrifugal action of the mold removes air from the composite, resulting in a dense laminate, free of voids. Material properties can be altered by varying aggregate content, resin type, and cure and reinforcement type, quantity or orientation. The pipe is removed from the mold after heat curing.

2. *Filament winding:* There are two distinct processes for filament winding—continuous and discontinuous. Filament winding involves impregnating several glass reinforcing strands with a matrix resin and then the application of the wetted fibers to a mandrel under controlled tension in a predetermined pattern. Repeat of this process results in the desired wall thickness. In the continuous process, an advancing mandrel causes the pipe to form. Fiber rovings, resin, and aggregate are added to make the pipe. The pipe is then cured and cut into desired lengths. In the discontinuous process, a standard length of mandrel is rotated, and resin, glass, and aggregate are added to produce pipe that is helically reinforced. The process allows for the formation of a continuous bell-end, monolithic with the pipe wall.

Applicable Standards

There are currently three widely-used pressure and nonpressure piping standards, used in trenchless installations. Standard pressure classes for AWWA C950 include 50, 100, 150, 200, 250, and over 250 psi. From the available pipe diameters, it can clearly be seen that GRP offers an alternative to other large-diameter traditional piping materials (Table 4.15).

Pressure Standards	Nonpressure Standards	Available Diameters (in.)
—	ASTM D3262	8–144
ASTM D3517	—	8–144
AWWA C950	—	1–144

TABLE **4.15**   GRP Standards for Trenchless Construction

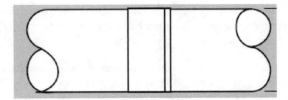

Figure 4.30   GRP flush bell-and-spigot joint for trenchless installation. (*Source: Hobas Pipe USA.*)

Joints

There are a variety of joints that have been developed over the years for GRP. The *Fiberglass Pipe Handbook* (1989) lists the following types: coupling or bell-and-spigot joints, mechanically coupled joints, restrained coupling or bell-and-spigot joints, butt and wrap joints, bell-and-spigot with laminated overlay, bell-and-spigot adhesive joint, flanged joints, and mechanical joining systems.

HOBAS, the leading manufacturer of municipal GRP, has variations of the above-described joints. For new trenchless construction, pipe-jacking in particular, they offer two separate types of joints: flush bell-spigot joint for gravity flow pipe jacking, and the flush fiber-wound collar (FWC) coupling for pressure pipe jacking. A third type of joint exists for use in the renewal of gravity pipe by sliplining.

In the flush bell-and-spigot joint type, as shown in Fig. 4.30, the sleeve is fitted to the pipe end, which has been machined-down so that the joint outside the diameter is same as the pipe itself. An elastomeric gasket, contained in a groove on the spigot end of the pipe, effectively seals the joint. The flush FWC coupling is a modified version of the FWC coupling (a filament-wound sleeve with an EPDM elastomeric membrane, the coupling is bonded to one end of the pipe during manufacture).

Advantages and Limitations

Table 4.16 outlines some advantages and limitations of GRP. Excellent internal and external corrosion resistance in natural soils and corrosive wastewater and industrial applications has given GRP an advantage over traditional piping materials in large diameters. It also displays better abrasion resistance than cement-based pipes. It is significantly lighter than a concrete pipe of the same diameter. For pressure applications, a variety of available pressure classes makes it widely desirable to engineers and contractors.

Although GRP is noncorrodible, it is susceptible to strain corrosion in the presence of certain chemicals, such as those found in the sanitary sewers, where pH is less than 4. This can be overcome during design, by ensuring that the stresses are kept within a certain limit. Certain chemical contaminants can permeate the pipe. During installation, pipe may be damaged by a severe impact force.

Advantages	Limitations
1. Good internal and external corrosion resistance in ordinary soils	1. Susceptible to impact damage
2. Better abrasion resistance than cement-based pipes	2. Resin selection is important for some chemicals
3. Light weight compared to alternative materials	
4. Various pressure classes available for pressure pipe	
5. Excellent hydraulics due to smooth interior	
6. Cast pipe has fixed OD and gasketed joints that seal on OD	
7. Dimensionally stable for fluid service and weather exposure	
8. Very high compressive strength (10,000 to 15,000 psi) so ideal for jacking	

TABLE **4.16**   GRP Advantages and Limitations

4.7.5   Metallic Pipes

Metallic pipes in present-day use include ductile iron pipe and steel pipe. The precursor to ductile iron was cast-iron pipe. Metallic pipes have traditionally been used in pressure applications, but some areas of North America use them for gravity sewer applications also. Most metallic pipes used in the past were installed by open-cut methods. Recent acceptance of trenchless technologies has led to the manufacture of metallic pipes with joints suited for pull-in or jacking installations. Steel pipe with welded joints is a product of choice for large-diameter HDD applications (usually more than 22 in. diameters), horizontal auger boring (HAB), and pipe-ramming (PR) projects. See specific chapters in this book for more information on these methods.

Ductile Iron Pipe

Ductile iron pipe was developed in the 1940s from grey cast iron by distributing the graphite into a spherical form instead of a flake form. This was achieved by the addition of inoculants such as magnesium to molten iron. It resulted in the ductile nature of the new pipe, in addition to higher strength, impact resistance, and other improved properties. The commercial production of ductile iron pipe begun in 1955, and by the 1970s, it had almost completely replaced cast-iron pipe in municipal applications. Since 1980, gray cast-iron pipe for pressure pipe applications (not soil pipe) has not been produced in the United States.

The main use of ductile iron pipe is in water systems. However, the availability of thermoplastic and thermoset pipes and newer substitute materials represents some significant competition to ductile iron since the late 1960s. The newer materials are lighter (where this might be judged an advantage), in some cases less expensive, and perform many of the required functions within a distribution system. Above all, plastic pipes are inherently suited to withstand most forms of both internal and external corrosion that can be a concern to unprotected ductile iron pipe in some service environments.

However, corrosion problems for ductile iron pipe can overcome by the use of appropriate corrosion protection. The protection methods include cement mortar or polymer linings for internal corrosion protection, maintenance of flow properties, and polyethylene encasement per ANSI/AWWA C105/A21.5, or in uniquely aggressive environments cathodic protection, for external corrosion. A newer approach to iron pipe corrosion control, called the *design decision model* (DDM), has been developed by the Ductile Iron Pipe Research Association (DIPRA). This method tailors the chosen corrosion protection method of the pipeline based on several factors including substantial history, the corrosion potential of the specific environment involved, and the critical nature/priority of the specific pipeline involved.

In recent years, various manufacturers have also developed restrained joints for ductile iron pipe; therefore, the industry is able to offer its products to the trenchless construction industry. Today ductile iron pipe can be used for HDD, pipe bursting, casing and carrier pipe installation, and sliplining. The use of ductile iron pipe in sewer systems is less in comparison to its use in potable water distribution and transmission. Nevertheless, some municipal agencies use ductile iron pipe for their gravity flow applications. Flexible, restrained joint ductile iron pipes are particularly attractive for some HDD installations. In HDD applications, joints can rapidly be assembled in field conditions with minimum equipment or highly skilled technicians. The pipe also can be installed with limited access, and need not be bent (nor result in high bending stress/strain) when installed in a normal curved drill path.

Ductile iron pipes are flexible conduits and are designed and installed accordingly. However, when internal cement-mortar lining is specified, the vertical deflection of the pipe is limited to 3 percent instead of 5 percent. Ductile iron pipes with flexible linings, such as proprietary ceramic, epoxy, and fusion-bonded epoxy primer/heat-fused polyethylene topcoat linings are limited by ASTM A716 and A746 standards to 5 percent deflection.

Cast-Iron Pipe   It is important to have an understanding of the predecessor of ductile iron pipe, the cast-iron pipe. Cast-iron pipe has been used throughout the world for many years in portable water systems. In the United States, cast-iron pipe was introduced in the early 1800s.

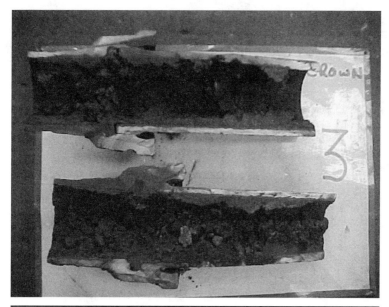

FIGURE 4.31 A sample of internal tuberculation in unlined cast iron potable water pipe. (*Source: 3M Corrosion Protection Products.*)

Also referred to as gray cast iron, it accounts for a very large portion of buried water-piping material throughout North America even today. Cast iron is a very strong, but brittle material. Early unlined installations of cast-iron pipe has lasted more than a century in some cases on account of the sheer thickness of the walls of the pipes. The thick walls played a sacrificial role as they slowly corroded over the years. Internal tuberculation in unlined cast-iron pipes has caused severe hindrance to flow in most cases. Figure 4.31 presents a sample of internal tuberculation in unlined cast-iron potable water pipe.

Manufacturing The principal raw material used in producing ductile iron pipe is recycled ferrous material, including scrap steel; scrap iron; and other ferrous materials obtained from shredded automobiles, appliances, and others. While ductile iron is very similar in basic chemical makeup to gray cast iron, ductile iron is instead produced by treating molten low-sulfur base iron with magnesium under closely controlled conditions. The startling change in the metal is characterized by the free graphite in ductile iron being deposited in nodular form, instead of flake-form as in gray iron. With the free graphite in nodular form, the continuity of the metal matrix is at a maximum, accounting for the formation of a far stronger, tougher ductile material. Ductile iron is roughly twice the strength of gray iron (ductile iron has greater yield strength than ASTM A36 carbon steel) and further surpassing gray iron in ductility and impact characteristics.

Pressure Standards	Nonpressure Standards	Available Diameters (in.)
—	ASTM A716	
—	ASTM A746	4–64
AWWA C150/C151	AWWA C150/C151	

TABLE 4.17   Applicable Ductile Iron Standards for Trenchless Technology

The pipe is cast using a centrifugal method, after which it is annealed in furnaces. An asphaltic coating is applied to the outside of the pipe, while the interior is coated with a cement-mortar lining. Unlike gray iron pipe, ductile iron pipe will bend significantly without breaking when subjected to even quite great loads, impacts, or deflections.

Applicable Standards   Table 4.17 lists a number of standards available for trenchless applications. It should be noted that these pipes can also be installed via open-cut methods. For trenchless pull-in installations, the joints must be restrained, as discussed in the following section. In recent years, ductile iron pipes with restrained joints have been used in trenchless construction methods (TCM) as well as for trenchless renewal and replacement methods (TRMs).

Joints   Restrained joints in ductile iron pipe are available primarily to accommodate the thrust forces acting on a pipeline. However, pipes with these restrained joints have been used in recent years for various types of trenchless projects, both new constructions as well as renewals. HDD and pipe bursting have been the most common applications of the restrained joint ductile iron pipe. The joints are capable of withstanding tensile forces encountered during pull-in process.

Proprietary restrained joints have been designed by various manufacturers which incorporate a push-on gasket and special bell design in conjunction with their restraint mechanisms. Because of their proprietary nature, the push-on gaskets used in these joints may not be compatible with standard push-on gaskets. In the 350 psi allowable working pressure range, the joints are suitable for pipe diameters of 4 through 24 in. In the 250 psi range, the joints are available for diameters of 30 through 64 in. Figure 4.32 shows five such proprietary joints.

Advantages and Limitations   Table 4.18 summarizes the advantages and limitations of ductile iron pipe. The high load-bearing capacity, high impact strength, and high beam strength makes it a sturdy piping material for pressure applications. There are many different types of joints, including restrained joints, which are used in trenchless applications. The long lengths of ductile iron pipe (20 ft) minimize the number of joints within a water or sewer system. The wide

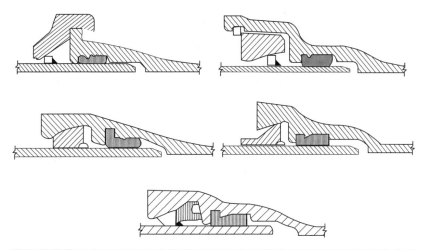

FIGURE 4.32   Proprietary ductile iron restrained joints used in trenchless construction.

Advantages	Limitations
1. Wide variety of internal and external corrosion protection systems available	1. Highly susceptible to corrosion, both internally and externally, unless protected
2. Internal cement mortar lining prevents tuberculation and enhances hydraulic capability	2. Not all available corrosion protection methods are effective
3. Strong material, with high load bearing strength, impact strength, and beam strength	3. Internal cement mortar lining is easily damaged if struck with a backhoe.
4. Wide variety of joints enable various applications, including trenchless	4. Cathodic protection is cost prohibitive and is rarely used in municipal systems
5. Available for both pressure and gravity applications	5. Polyethylene encasement is easily damaged and subject to improper installation
6. Wide range of diameters and pressure classes available	6. Heavy weight, resulting in high cost of labor
7. Long laying lengths reduce joints in the system	7. Lack of flexibility is an obstacle in trenchless installations
8. Pipe is highly resistant to chemical permeation in contaminated areas	8. Gaskets in the joints are highly vulnerable to chemical attack in contaminated soils

TABLE 4.18   Advantages and Limitations of Ductile Iron Pipe

availability of various pressure classes and diameters makes it easily desirable to engineers and specifiers. The pipe itself is highly resistant to chemical permeation. Also, there are various internal and external corrosion protection systems available.

Ductile iron is susceptible to internal and external corrosion. Although there are a variety of corrosion protection systems, they do not all perform equally. Cathodic protection is too cost prohibitive for use in the municipal systems. The internal cement-mortar lining prevents the deflection of the pipe beyond 3 percent, even though ductile iron is a flexible conduit. External polyethylene encasement is easily damaged, for example during tapping, and is subject to careful installation. The heavy weight of ductile iron pipe raises the cost of labor. Although the pipe itself resists chemical permeation, the gasket at the joint does not.

Steel Pipe

Steel pipes, made from a versatile refinement of iron, have seen a wide range of usage for more than a century and a half. The development of high-strength steel pipes has made it possible to transport fluids such as natural gas, crude oil, and petroleum products over long distances. Initially, all steel pipes had to be threaded together, which was difficult for large pipes, and they often leaked under high pressure. The development of electric arc welding machines in 1920s made it possible to construct leak-proof, high-pressure, large-diameter pipelines. One of the earliest steel water pipe installations in the United States, still in service today, was in San Francisco in 1863. Developments in technology have given way to riveted steel pipes evolving to the automatically welded steel pipes of today. Various other developments have resulted in the creation of different types of joints as well as effective mechanisms for prevention of corrosion, making steel more versatile for trenchless and open-trench applications.

In municipalities, steel pressure pipes are used today in large-diameter potable water transmission applications. In municipal trenchless construction, steel pipes are used as casing pipe in processes such as microtunneling, jacking, boring, and pipe-ramming because of their high stiffness and compressive strengths. There have even been several large-diameter spiral-welded steel pressure pipe installations in Texas, Washington, and Hawaii via horizontal directional drilling.

Corrugated Steel Pipe   Corrugated steel pipes have been used for more than a century in gravity applications such as drainage and storm sewers. Though corrugated steel pipes have been used in some sanitary sewers, this is not the case today. Due to their relatively low-compressive and tensile strengths, corrugated steel pipes are not used in trenchless or pressure applications. Therefore, a detailed discussion of corrugated steel piping products has been omitted in this chapter. The American Iron and Steel Institute's *Modern Sewer Design* is an excellent source for information on corrugated steel piping systems.

Manufacturing   Steel pipes used in municipal applications are manufactured by an automatic welding process. There are generally three types of steel pipe, each identified by the way in which it is manufactured:

1. *Rolled and welded pipe:* This is one of the oldest methods of steel pipe production, where plates of steel are rolled into cylindrical pipes, usually 6 to 12 ft in length, then welded in the circumferential and longitudinal directions. The pipes used in casing applications for trenchless technology are of this type. They are also used in other types of applications.

2. *Electric resistance welded (ERW) pipe:* ERW pipes are generally manufactured and used in smaller diameters up to 24 in. ERW is a single straight seam welding process where continuous coils of treated, low-carbon steel, called skelp, are shaped into cylindrical pipes by edge-forming, and then welded at the seam. These pipes can be manufactured in lengths of up to 100 ft. They are used in water systems, as well as other industrial applications.

3. *Spiral welded pipe:* Starting with continuous rolls of steel similar to the type used for ERW pipe, the steel is fed into a machine and spirally wrapped against buttress rolls to form the pipe. The edges of the spiral pipe are then welded in and out by a double-submerged arc process. Spiral welded steel pipes are used in municipal water transmission applications in diameters of up to 156 in. Trenchless processes such as HDD have been used to install this type of pipe in the potable water systems.

Corrosion Protection—General   There have been many approaches to corrosion-protection of all kinds of pipes through the centuries. There are decades of practical experience with many applications and corrosion-protection systems of iron and steel pipes, in particular. There are field and laboratory studies of gray and ductile iron pipes in widely divergent soil types, and also in some notably very corrosive actual soil burial test sites, by the Ductile Iron Pipe Research Association (DIPRA) working in many cases in close conjunction with the utilities involved. Both unprotected iron and steel pipes will rapidly corrode in some soil environments, and in these environments suitable corrosion protection must be provided. It is also being discovered (in more recent applications of pipes that have not been around as long) that other piping materials, such as variously reinforced concrete, plastics, and composites, as well can also undergo forms of corrosion or environment-/stress-related deterioration that are perhaps not now quite as obvious or known to many pipeline practitioners.

Corrosion Protection—Ductile Iron Pipe   American National Standard Institute (ANSI) and American Water Works Association (AWWA) have developed multiple standards for corrosion protection of iron pipe and fittings. These standards include ANSI/AWWA C104/A21.4 for internal cement-mortar lining, ANSI/AWWA C105/A21.5 for polyethylene encasement (this standard also contains a soil-evaluation procedure in the appendix that is helpful for practitioners to determine when standard pipes with thin asphaltic shop coatings can be direct buried in specific soils and when the supplementary polyethylene wrap should be applied as opposed to installing standard pipe without wrap), and AWWA C116 for fittings that in their normal production processes are coated inside and outside with fusion-bonded epoxy (FBE) instead of cement mortar.

More recently and building on these standards and other extensive experience, CORRPRO Company working with DIPRA as well as ductile iron pipe manufacturers has conducted a 2-year study of corrosion and corrosion-protection characteristics. This study has included field and laboratory evaluations related to short- and long-term polarization rates under varying conditions, corrosion-rate reduction, and corresponding cathodic current criterion. This information was then analyzed in conjunction with an extensive database from 1379 physical inspections of buried iron water lines. The result of the study is a risk-based corrosion protection design strategy for buried ductile iron pipelines referred to as the *design decision model* (Kroon et al., 2004 and Bonds et al., 2004).

Corrosion—Steel Pipe   The steel pipe industry has also been proactive in readily recommending and providing corrosion protection mechanisms to end users, despite the higher costs of some processes. According to the steel pipe industry, "Corrosion protection systems that include coatings, monitoring systems, and cathodic protection (installed incrementally as needed) are very cost effective." There are a number of standards and processes for both internal and external corrosion protection of steel pipe, including cement-mortar lining, paints and polyurethane linings, tape coatings, coal tar enamel coatings, cement-mortar coatings, and epoxy and polyurethane coatings. Furthermore, the steel pipe industry acknowledges the use of cathodic protection as an effective and/or necessary method of protection to complete the corrosion protection process.

Corrosion of Concrete, Composite, and Plastic Pipes   All common piping materials can be deteriorated or corroded and designers should consult with manufacturers to learn pipe product limitations. It should be noted that sulfates and chlorides as well as low resistivity soils in general can corrode various concrete pipes, and gasoline contamination has been known to result in swelling and bursting of some buried plastic pipes, whereas such contamination will not burst metallic pipes.

The coverage of the topic of corrosion in this chapter is intended to make the reader aware of one of the significant problems being faced by pipeline industry today. These problems include corroding, deteriorating, and/or failing water and sewer lines. There are currently huge amounts of pipeline system underground, some of which are without effective corrosion protection or other provisions for adverse environmental conditions. Pipe leakage and failure in these unprotected pipelines can significantly influence rate of unaccounted-for-water and inflow/infiltration in future. It is essential for engineers to be aware of the problems and available engineering solutions as they design and renew the pipeline systems of the future. Proper corrosion protection will ensure that the full potential of all piping materials for use in specific applications is realized in the long run.

Applicable Standards The nonpressure standard ASTM A139 Grade B is the type of casing pipe used in the gravity trenchless construction. A pipe manufactured to the ASTM A139 standard requires hydrostatic testing because it is often also used for medium internal pressure applications. As the use of the pipe for casing depends only on its structural capabilities, the hydrostatic test is not required. To exclude the hydrostatic test from the original standard, the casing pipe standard is referred to as ASTM A139, Grade B (no hydro). The pressure pipe standard AWWA C200 is used in large-diameter HDD projects for water transmission; in past years, some challenging projects in Texas, Washington, and Hawaii have used this steel pipe standard. Permalok™ steel pipe is manufactured as a proprietary joint type product for casing pipe and pressure trenchless applications. The ASTM standards to which the pipe must be compliant are listed in Table 4.19.

Pressure Standards	Nonpressure Standards	Available Diameters (in.)
AWWA C200	ASTM A139 Grade B	
ASTM A36, ASTM A515, grade 60 or ASTM A572, grade 42–with T7 type joint	ASTM A36, ASTM A515, grade 60 or ASTM A572, grade 42–with PERMALOK™ T5 type joint	3–144 and higher

Table 4.19 Applicable Standards for Steel Pipe for Trenchless Construction

Joints   There are several types of welded joints available for steel pipe, each suited for a specific application:

1. Bell-and-spigot lap-welded joints
2. Butt-welded joints (single-V butt-welded, and double-V butt-welded)
3. Butt strap welded joints
4. Mitered lap-welded joints

There are also a number of nonwelded joints:

1. Bell-and-spigot rubber gasket joints
2. Harness joints
3. Carnegie shape rubber gasket joints
4. Mechanical couplings
5. Split-sleeve mechanical coupling

For steel casing pipe used in boring and pipe-jacking applications, it is important that there are no irregularities in alignment at the joints. If the casing is not straight, the ability of the contractor to keep the pipe in-line and on grade is affected. Lap-welded joint products are not recommended for jacking and boring installations. Though spiral-welded pipes can be used for casing, it is a common practice to use a straight seam or seamless pipe. For jointing, both ends of the pipe can be beveled for welding, as in the single-V butt-welded joint, Fig. 4.33. Another recommendation for achieving a good circumferential weld at the joint is to bevel one end of the casing pipe to a standard 37° bevel, and square cutting the other end.

For casing and pressure pipe trenchless applications, a patented product line, Permalok, offers two mechanical push-on joints. It is particularly attractive to the trenchless excavation industry because the joints are designed to be flush with the interior and exterior surfaces of the pipe (see Fig. 4.34). This joint type eliminates the time required for welding traditional-type steel pipes. Although using the Permalok pipe is approximately 2.5 times faster than welding, it costs more. Permalok joints are used for HDD, pipe ramming, and horizontal auger boring. Joint quality is further enhanced because the Permalok connector is consistently round and perpendicular to the

FIGURE 4.33   Single-V butt-welded joint.

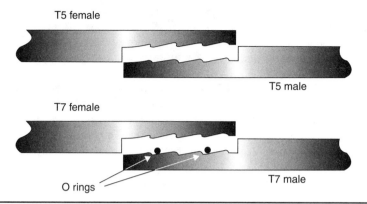

Figure 4.34   Interlocking jointing system. (*Source: Permalok.*)

pipe axis. Its unique machined groove makes stabbing and aligning easy and quick. Also, the Permalok design, combined with the use of a sealant such as RTV silicone, prevents leakage under considerable pressure. Standard specifications, listed in Table 4.19, ensure that all pipe manufactured by Permalok meet guidelines. The T5 profile was patented for steel pipe casing applications in 1993 and has been used in numerous trenchless excavation projects nationwide. The T7 profile is capable of withstanding pressures of up to 300 psi, and is used in trenchless pressure pipe applications. Figure 4.34 illustrates the Permalok interlocking jointing system.

Advantages and Limitations   Table 4.20 lists some of the advantages and limitations of steel pipe. With its high tensile strength, steel pipes are capable of handling high pressures in water transmission applications. The high compressive strength of steel pipe makes it a good material for jacking and boring. Assembly of nonweld joints is relatively easy owing to push-on bell-and-spigot gasket joints. Steel pipe has been known to adapt well in locations where ground movements occur. Good hydraulic properties of internally lined steel pipe make it ideal for use in municipal water systems. The availability of various methods and standards for both internal and external corrosion significantly prolongs the useful design life of the pipe.

Though various corrosion protection mechanisms are available, the pipe is highly prone to corrosion attacks unless the protections are used. This can be costly. In large diameters, the pipe's ability to handle external pressure is low. Air vacuum valves must be used in large-diameter lines to eliminate the possibility of pipe collapse, but this is an added cost to the system. Welding of pipe joints requires skilled labor and is time consuming.

Advantages	Limitations
1. Various standards and methods are available for internal and external corrosion protection 2. High tensile strength 3. High compressive strength 4. Easy to assemble, nonweld joints available 5. Adopts well to locations where soil movements occur 6. Good hydraulic properties when internally lined	1. Prone to internal tuberculation and external corrosion, subject to electrolysis 2. Use of internal and external corrosion protection raises price of the product 3. Low resistance to external pressures in large-diameter sizes 4. Air vacuum valves are necessary in large-diameter lines 5. Welding of joints require skilled labor and is time consuming 6. Special care required to ensure proper alignment at joint in welded pipe 7. Fully dependent on proper installation to limit deflection and collapse.

TABLE 4.20   Advantages and Limitations of Steel Pipe

4.8   Summary

This chapter presented an overview of pipe-soil interaction system describing behavior of rigid and flexible pipes under different loading and environmental conditions. To provide an aid in pipe selection, this chapter provided detail description of different types of pipes such as cement based, plastic (PVC, PE, GRP), and metallic pipes. Manufacturing process, applicable standards, joint types, and the advantages and limitations of each type of pipe were also provided.

CHAPTER 5

Project Considerations for Horizontal Directional Drilling

5.1 Introduction

Horizontal directional drilling (HDD), one of the most common trenchless installation methods, allows installation of pipelines and conduits below ground using a surface-mounted drill rig. The rig places a drill string at a shallow angle with the horizontal and has tracking and steering capabilities. Due to the importance and popularity of HDD in trenchless technology industry, a dedicated chapter is allocated in this book.

For all categories of HDD, of Maxi, Midi, and Mini (see Chap. 1), a successful installation requires the use of similar practices, including proper advance planning, equipment selection, and setup. Figure 5.1 illustrates an HDD installation.

Mini-HDD is used for boring holes several hundred feet in length, for placing pipes of up to 12 in. in diameter and at depths of up to 30 ft. Mini-HDD is appropriate for placing local distribution lines (including service lines or laterals) beneath streets, private property, and along right-of-ways. In comparison, Maxi-HDD is a relatively sophisticated class of HDD. It is employed for boring holes several thousand feet in length and placing pipes up to 60 in. diameter, at depths up to 200 ft. Maxi-HDD is appropriate for placing pipes under large rivers or other broad crossings and obstacles. Applications that are

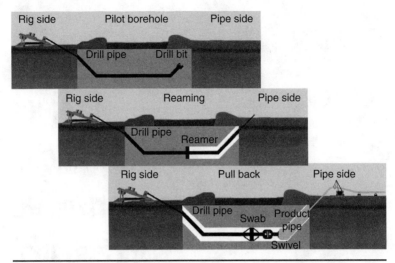

Figure 5.1 Generic HDD installation. (*Source: Trenchless Engineering Corporation.*)

intermediate to the Mini- or Maxi-HDD categories may utilize appropriate "Midi" equipment of intermediate size and capabilities.

Due to the high importance and large expense associated in accomplishing Maxi-, and possibly Midi-, HDD operations, such as major river-crossings, experienced engineers are typically engaged to help plan and design the overall installation, including use of appropriate hardware and software programs and navigational equipment. Mini-HDD is relatively a low-cost procedure, which typically does not employ the services of professional engineers. Nonetheless, since improper Mini-HDD procedures, including incorrect bore planning, pipe loading, and equipment setup, can lead to problematic installations, including inadvertent return of drilling fluids and possible damage to nearby buildings and facilities, the second portion of this chapter (Secs. 5.4 and 5.5) presents relatively an easy-to-understand bore planning guidelines and pipe loading calculations for such operations. Same concepts can be utilized for Midi- and Maxi-HDD operations.

5.2   Method Description

Directional drilling methods utilize steerable soil drilling systems to install both small- and large-diameter pipes. In most cases, HDD is a two-phase process. Phase 1 involves drilling a pilot borehole of approximately 2- to 6-in. diameter along the proposed design centerline. In phase 2, the pilot borehole is enlarged by use of a backreamer to the desired diameter to accommodate the pipeline. At the same time, the product pipe is connected to the end of the drilling rod by a swivel and pulled back through the enlarged pilot borehole. However, for large-diameter pipes, a separate (intermediate) phase comprising several passes of backreaming (or "prereaming") may be necessary to enlarge the borehole to the desired size. In this case, the pullback operation is performed simultaneously with the final back-reaming operation. The pilot borehole is drilled with a surface-launched rig with an inclined carriage, typically adjusted at an angle of 8° to 20° with the ground surface. Figure 5.1 illustrates a generic HDD installation.

The boring of the pilot borehole is accomplished by rotation of the cutterhead, assisted by the thrust force and drilling fluids trans-ferred from the drill string. The mechanical cutter may vary from a slim cutting head with a slanted face for small- and short-bore appli-cations (Mini) to a diamond-mounted roller cutter used with mud motors for large and long crossings (Maxi), as illustrated in Fig. 5.2. For Mini-HDD systems, directional steering control is accomplished mainly by the slanted cutter head face. For Maxi-HDD, a bent hous-ing (a slightly bent section between 0.5° and 2.75° of the drill rod) is used to deflect the cutterhead axis from the local direction. In both Mini- and Maxi-systems, a curved path can be followed by pushing the drill head without rotating, and a straight path can be drilled by applying simultaneous thrust and torque (rotation) to the drill head.

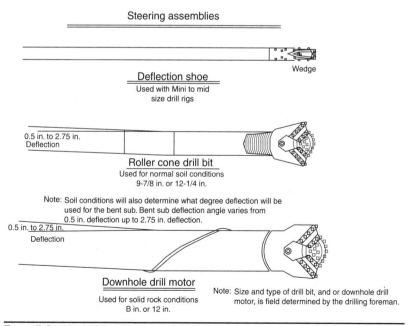

Steering assemblies

Wedge

Deflection shoe
Used with Mini to mid
size drill rigs

0.5 in. to 2.75 in.
Deflection

Roller cone drill bit
Used for normal soil conditions
9-7/8 in. or 12-1/4 in.

Note: Soil conditions will also determine what degree deflection will be
used for the bent sub. Bent sub deflection angle varies from
0.5 in. deflection up to 2.75 in. deflection.

0.5 in. to 2.75 in.
Deflection

Downhole drill motor
Used for solid rock conditions
B in. or 12 in.

Note: Size and type of drill bit, and or downhole drill
motor, is field determined by the drilling foreman.

FIGURE 5.2 Maxi-HDD drilling and steering mechanism.

5.3 Maxi-HDD Considerations

Maxi-, and larger-size Midi-, HDD operations are commonly used for obstacle crossings, such as a river, lake, roadway, an environmentally sensitive area, and shore approaches, or a combination of such situations (see Fig. 5.3). Product pipe selection (material, wall thickness, grade, coating, and others) is an important part of the HDD design process.

The American Society of Civil Engineers (ASCE) Manual of Practice (MOP) No. 108, *Pipeline Design for Installation by Horizontal Directional Drilling*, and American Society for Testing and Materials (ASTM F1962), *Standard Guide for Use of Maxi Horizontal Directional Drilling for Placement of Polyethylene Pipe or Conduit Under Obstacles, Including River Crossings*, provide overall guidelines for a Maxi-HDD operation, addressing preliminary site investigation, safety and environmental considerations, regulations and damage prevention, bore path layout and design, implementation, and inspection and site cleanup (ASCE, 2005; ASTM, 2005).

5.3.1 Site Investigation Requirements

As in any construction project, the first step in an HDD installation is a thorough investigation of the site at which the work will be undertaken. As part of this investigation, a route must be selected with sufficient space for the various HDD equipment and operations. Figures 5.4*a* and *b* illustrate the space requirements for the rig and product pipe sides of the operation. An appropriate site investigation will also include both surface and subsurface surveys. Although each survey may be performed by different specialized engineering consultants, it is important that the results be integrated onto a single plan and profile drawing, which will be used by the contractor to price, plan, and execute the HDD operations. Accurate measurements for this drawing are important, as it will be used by the contractor for downhole navigation.

Figure 5.3 Maxi-HDD installation to cross beneath obstacles.

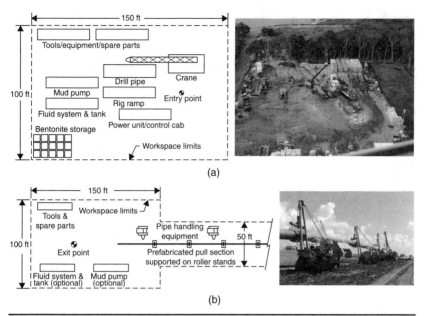

FIGURE 5.4 (a) and (b) Typical space requirements for the rig and pipe sides of the HDD operation. (*Source: ASCE, 2005.*)

Surface Survey

A topographic survey should be conducted to accurately describe the working areas where construction activities will take place. During this survey, all underground utilities and possible obstructions to the HDD operation must be identified. After proper route selection, both horizontal and vertical references must be clearly established so they can be used as a basis for specifying hydrographic and geotechnical data. A typical survey for river crossings should include overbank profiles on the crossing centerline extending from approximately 150 ft landward of the entry point (rig side) to the length of the prefabricated pull section landward of the exit point (pipe side). Survey ties should also be made to topographic features near the crossing.

For significant waterways, a hydrographic survey will be required to accurately describe the bottom contours. Typically, it should consist of readings along the crossing centerline and approximately 200 ft upstream and downstream. This scope can be expanded to include more upstream and/or downstream ranges if this data is required to analyze future river activity.

Subsurface Survey

A subsurface survey for a Maxi-HDD installation should define the geological characteristics and engineering properties of the subsurface material through which the drilled path will pass. It should include both a review of existing geological information and site-specific data obtained from exploratory soil borings. The extent of the subsurface survey should be governed by practical economic limits.

The existing geological data should be reviewed to determine the general subsurface conditions at the specified location. The site-specific geotechnical investigation should be conducted to confirm the probable subsurface conditions through which the river crossing will be installed. The number and location of borings, as well as the use of other exploratory techniques, will be based on local conditions taking into account the preliminary drilled-path design. Borings should be located approximately 50 ft off the crossing centerline and should extend to approximately 30 ft below the deepest crossing penetration depth.

The sampling interval and technique will be based on site-specific conditions and should be designed to accurately describe the subsurface materials. If rock is encountered, the borings should at a minimum penetrate the rock to a depth sufficient to confirm that it is bedrock, and preferably should extend beyond the HDD crossing's penetration depth to provide detailed information about the bedrock properties. The following data is required from the borings:

- Standard classification of soils (USCS)
- Gradation curves for granular soils
- Standard penetration test (SPT) values, where applicable
- Cored samples of rock with rock quality designation (RQD) and percent recovery
- Unconfined compressive strength for rock samples
- Moh's hardness for rock samples

The results of the geotechnical investigation should be presented in the form of a geotechnical report containing a brief description of local geology, engineering analysis, boring logs, test results, and a geotechnical profile of the subsurface conditions beneath the river or lake.

5.3.2 Drilling Operations
The drilling operations for HDD are typically divided into three phases, depending on the size of the product pipe being installed and soil conditions: pilot borehole, pream and pullback, as described in the following sections.

Pilot Borehole Phase
Pilot borehole construction in Maxi-HDD is achieved by using a nonrotating drill string with an asymmetrical leading edge. The asymmetry of the leading edge creates a steering bias, while the

nonrotating aspect of the drill string allows the steering bias to be held in a specific position while drilling. If a change in direction is required, the drill string is rotated such that the direction of bias agrees with the desired change. Leading edge asymmetry is typically accomplished with a bent sub or bent motor housing located several feet behind the bit (see Fig. 5.2). A straight path is maintained by applying simultaneous thrust and torque (rotation) to the drill head.

In soft soils, drilling progress is typically achieved by hydraulic cutting with a jet nozzle. If hard areas are encountered, the drill string may be rotated to drill without directional control until the hard spot has been penetrated. Mechanical cutting action required for harder soils or rock may be provided by a positive displacement mud motor, which converts hydraulic energy from drilling fluid to mechanical energy at the drill bit. This allows for bit rotation without drill string rotation.

The actual path of the pilot borehole is monitored during drilling using a steering tool positioned near the bit as shown in Fig. 5.5. The steering tool provides continuous readings of the inclination and azimuth at the leading edge of the drill string. These readings, in conjunction with measurements of the distance drilled, are used to calculate the horizontal and vertical coordinates of the steering tool relative to the initial entry point on the surface. When the drill bit penetrates the surface at the exit point opposite the horizontal drilling rig, the pilot borehole is complete.

Preream Phase

Enlarging the initial pilot borehole is typically accomplished using a single backreaming operation, simultaneous with pullback of the product pipe, or—for larger pipes and/or harder soils—by means of several prereaming passes prior to product pipe installation. Reaming tools consist of a circular array of cutters and drilling fluid jets, and are often custom made by contractors for a particular borehole size or type of soil (see Fig. 5.6). For a prereaming pass, a reamer attached to the drill string at the exit point is rotated and drawn back toward the drilling rig, thus enlarging the pilot borehole. Drill rod is added behind the reamer as it progresses toward the drill rig to ensure that a string of pipe is always maintained in the borehole. It is also possible to ream in the opposite direction, away from the drill rig, in which case a reamer fitted onto the drill string at the rig is rotated and thrust forward, however, in certain soil and project conditions this may cause ground disturbance and surface heave.

Pullback Phase

Pipe installation is accomplished by attaching the prefabricated product pipeline section behind a reaming assembly at the bore exit point and pulling the reaming assembly and pipeline section back toward the drilling rig. This is undertaken after completion of the prereaming phase, or, for smaller diameter lines in relatively soft soils, directly

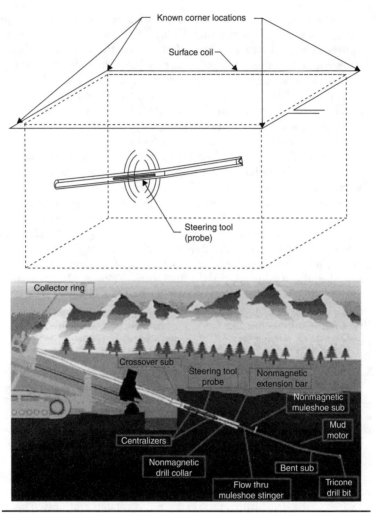

Figure 5.5 Magnetic wireline survey. (*Source: ASCE, 2005.*)

Figure 5.6 Different types of reamers. (*Source: Trenchless Engineering Corporation.*)

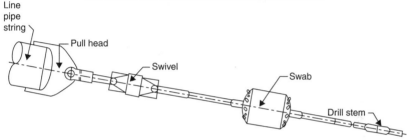

FIGURE 5.7   Pullback operation. (*Source: Trenchless Engineering Corporation.*)

after completion of the pilot borehole. A swivel is utilized to connect the pipeline section to the reaming assembly to minimize torsion transmitted to the product pipe. The aboveground pipeline section at the pipe-entry end is supported using a combination of roller stands, pipe handling equipment, or a flotation ditch to minimize tension and prevent damage to the pipe (see Fig. 5.7).

5.3.3   Drilled Path Design

To maximize the advantages offered by HDD, and help ensure a successful installation, it is important to understand the obstacle to be crossed and to carefully select the route. For example, a river is a *dynamic* entity. Not only should its width and depth be considered, the potential for bank migration and scour during the design life of the crossing should also be taken into account. It should be recognized that there is flexibility in locating a pipeline to be installed by HDD in both the horizontal plane and the vertical plane.

For the majority of drilled installations, there are six parameters which define the location and configuration of the drilled path. These are the entry and exit points, the entry and exit angles, the depth, and the radius of curvature. These parameters, or their limiting values, should be specified on the contract plan and profile drawing.

Figure 5.8   Entry and exit points. (*Source: Trenchless Engineering Corporation.*)

Bore Path Entry and Exit

Entry angles should be between 8° and 20° with the horizontal. Exit angles should be designed to facilitate breakover support during pullback (i.e., the exit angle should not be so steep that the pull section must be severely elevated in order to guide it into the drilled borehole). For relatively large-diameter pipes, the exit angle should generally be less than 10°. Figure 5.8 illustrates the entry and exit points.

Depth of Cover

Adequate cover should be provided to maintain crossing integrity throughout the design life of the pipeline being installed. Typically, Maxi-HDD crossings should be designed to provide greater than 20 ft of cover. This minimum depth aids in reducing inadvertent drilling fluid returns and provides a margin for error in existing grade elevation and pilot-borehole calculations, as well as dynamic variances in the river bottom. The depth of cover must be increased beyond these minimum criteria for installations beneath sensitive obstacles such as major waterways, highways, and railroads. Geotechnical factors should also be considered when selecting the vertical position of the pipeline.

Drill Rod Radius of Curvature

A conservative industry guideline (ASTM, 2005) indicates the minimum *radius of curvature* of the drill rods should be approximately

$$(R_{rod})_{min} = 100\, D_{rod} \tag{5.1}$$

where $(R_{rod})_{min}$ = minimum radius of curvature of drill rod, ft
D_{rod} = nominal diameter of drill rod, in.

In practice, drill rod manufacturers may allow a lower radius of curvature. The corresponding allowable degree of curvature or bending applies to bends in a vertical (profile), horizontal (plan), or inclined plane. The "Equipment and Product Restraints" section in Sec. 5.4.1 and Fig. 5.14 provide additional information regarding drill rod bending capability, and various associated terminologies. It should be noted that in Maxi-HDD operations, due to the larger size, and greater stiffness of product pipes being installed, the minimum bore path radius of curvature may depend on the product pipe diameter.

Maxi Horizontal Drilling Rig

Equipment required for a typical Maxi-horizontal drilling operation can be transported to the job site in approximately seven tractor-trailer loads. A workspace of 100 ft by 150 ft is typical (Fig. 5.4a) at the drill rig side. A similar size area at the bore path exit (pipe entry), plus a 50 ft wide segment of sufficient length to accommodate the assembled pipe and associated equipment and handling (see "Product Pipeline Fabrication" section below, and as indicated in Fig. 5.4b), will be adequate for most operations. If necessary, the workspace at the drill rig side for HDD equipment may be reduced to a minimum of 60 ft by 150 ft, but this would restrict the size and capacity of the drilling rig. Space requirements will vary depending on the make and model of the drilling rig and the position of the various components. However, the locations of the principal components (drill rig carriage, drill rod, and control trailer) are fixed by the entry point. The rig carriage must be positioned in line with the drilled segment and close to the entry point. The control trailer and drill rod must be positioned adjacent to the rig.

The workspace must be cleared and graded level. The related equipment is typically supported on the ground surface. Timber mats may be used where soft ground is encountered.

Product Pipeline Fabrication

Pipeline fabrication is accomplished using the same construction methods employed to lay a pipeline in a trench, therefore requiring similar workspace. The location of the fabrication workspace is controlled by the drilled segment exit point. Space must be available to allow the pipe to be fed into the drilled borehole. It is preferable to have workspace in line with the drilled segment and extending back from the exit point the length of the assembled pipeline plus 200 ft. This will allow the pipeline to be prefabricated in a single continuous length prior to installation. If sufficient space is not available, the pipeline may be assembled in several sections, to be welded or fused together during installation, or restraint joints can be used (see Chap. 4).

This workspace must also be cleared, but need not be graded level. Equipment is typically supported on the ground surface. Timber mats may be used where soft ground is encountered. Figure 5.4b provides additional information describing the pipe side of the operation.

5.3.4 Drilling Fluids

The primary functions of drilling fluids in Maxi-HDD systems include cutting or jetting of soft soils, transmission of rotary power to a mud motor, cleaning of cutters, cooling of downhole tool and electronics, transportation of suspended cuttings, stabilization of the drilled borehole and lubrication (reduction of friction) for drill rods tools and product pipe. The impact of HDD on the environment, including potential regulatory problems, and associated misunderstandings, typically involves the use of drilling fluids. An awareness of the function and composition of HDD drilling fluids is imperative in producing a permittable and constructable HDD crossing design. *Drilling Fluids in Pipeline Installation by Horizontal Directional Drilling* is a detailed discussion of drilling fluids relative to HDD installations (Hair, 1994).

In a recent study by the Gas Technology Institute, it was reported that inadvertent drilling fluid return occurs in the majority of HDD operations. Since this event often cannot be prevented, prior to construction, the contractor should prepare for inadvertent drilling fluid return, as discussed in the "Inadvertent Returns" section below.

Composition

The primary component of drilling fluid used in HDD pipeline installation is freshwater. In order for water to perform the necessary functions, a viscosifier is typically added to modify its properties. The viscosifier used almost exclusively on HDD installations is naturally occurring bentonite clay. The properties of bentonite used in drilling fluids are often enhanced by the addition of polymers. This enhancement typically involves increasing the yield (i.e., reducing the amount of dry bentonite required to produce a given amount of appropriate drilling fluid). For use in drilling fluids, standard bentonite yields in excess of 85 barrels of fluid per ton of material. The addition of polymers can increase the yield to more than 200 barrels per ton of material.

Mixing Water

It is standard practice on an HDD river crossing to draw water for drilling fluid directly from the waterway. Where sufficient amounts of freshwater are not available, water may be obtained from a nearby municipal source or hauled to the crossing location. Substantial amounts of water are required; therefore ample trucking and storage are critical, if water must be hauled to the site.

Disposal of Excess Drilling Fluids

The preferred method of disposal for excess drilling fluid on an HDD installation is dispersal at the drill site. As an alternative, excess fluid may be hauled to a remote disposal location. Disposal of excess drilling fluid in a waterway is not recommended and may not be allowed by regulatory agencies. The methods of disposal applied to a specific crossing will be dependent upon the size and location of the crossing as well as any applicable local, state and national regulations, and may include:

- Land farming—spread over an open area and mix with native soil

Figure 5.9 Drilling fluid disposal can be a major operation in Maxi-HDD operations. (*Source: Trenchless Engineering Corporation.*)

- Mix with backfill materials
- Transport to landfills

Proper documentation of drilling fluid disposal is usually required and includes volume removed, results of contamination testing, property owner agreement, and landfill license (see Fig. 5.9).

In addressing regulatory concerns, it is important to recognize that HDD drilling fluid is composed typically of water, bentonite, and drilled spoil. The major component of the fluid is water normally taken from a waterway or municipal source. For most HDD installations, the only foreign material introduced to the location is bentonite (a naturally occurring clay), possibly enhanced with polymers, with known characteristics and disposal recommendations. Applicable disposal regulations should be similar to those governing sedimentation and erosion control, hydrotest water disposal, or general construction spoil disposal.

Recirculation

The primary method of effectively dealing with excess drilling fluid disposal is to minimize the excess. This is accomplished by recirculating drilling fluid returns to the extent practical. Recirculation on an HDD waterway crossing is complicated by the fact that a significant portion of the drilling fluid returns often occur at the exit point, on the bank opposite the drilling rig. This requires either two drilling fluid systems to be utilized or transportation of returns from the exit point to the drill rig location. Transportation of drilling fluid returns can be accomplished by truck, barge, or a temporary recirculation line drilled beneath the bottom of the waterway. Site-specific conditions will determine which system is most advantageous.

Inadvertent Returns

An important HDD task involves management of uncontrolled subsurface discharge of drilling fluids. Under ideal circumstances,

Figure 5.10   Inadvertent fluid returns in the urban environment.

drilling fluid exhausted at the bit or reamer will flow back to the surface through the annulus between the outside of the drill rod and the drilled borehole. In practice, however, this does not always occur. Drilling fluid will tend to flow along the path of least resistance. This can lead to dispersal into the surrounding soils or possible discharge to the surface at some random location. When random flow to the surface occurs, it is referred to as an *inadvertent drilling fluid return* or *frac-out*.

Inadvertent drilling fluid return is not a critical problem in an undeveloped location. However, in an urban environment or high-profile recreational area, inadvertent returns can be a major problem (see Fig. 5.10). In addition to the obvious public nuisance, drilling fluid flow can buckle streets or wash out embankments. Drilling parameters should be adjusted to maximize circulation and minimize the risk of inadvertent returns. Nonetheless, the possibility of lost circulation and inadvertent returns cannot be eliminated. Contingency plans addressing possible remedial action should be made in advance of construction and regulatory bodies should be informed. In particular, the contractor should prepare a method of rapid detection, an inventory of containment materials, establishment of ingress/egress routes, and agreed upon cleanup methods.

5.3.5   Product Pipe Specifications

In many cases, the minimum pipe wall thickness and associated material requirements (e.g., yield strength) for safe operation during its operational (service) life will be determined by applicable codes and regulations. However, the load and stress analysis for a pipeline placed by HDD is different from that applied for the placement of conventionally buried pipelines because of the relatively high-tension loads, bending, and external fluid pressures acting on the pipeline during the installation process. These loads may be higher than the design service loads. Thus, stresses and loads imposed during the installation stage should be reviewed and analyzed in addition to the subsequent operating stresses to ensure that acceptable limits are not exceeded. ASCE 108, and ASTM F1962, are useful references for such purposes (ASCE, 2005; ASTM, 2005). Such procedures would be further refined by competent engineering expertise, including an analysis of pipe and soil characteristics and interactions, often including the use of relatively sophisticated software tools.

A brief discussion of the installation and operating loads which affect an HDD installation is presented in the following section and Sec. 5.5.1.

Installation Loads

During installation, a pipeline installed by HDD is subjected to the following loads as it is pulled into a properly prereamed borehole.

1. *Tension:* The product pipe is subjected to tension as it is pulled through the borehole by the drill rig. This tensile load is required to compensate for:

    *   Frictional drag due to contact between the outer surface of the pipe and the wall of the borehole.

    *   Fluidic drag on the pipe due to the drilling fluid in the borehole.

    *   Effective (submerged) weight of the pipe as it is pulled through elevation changes in the borehole.

    The frictional drag is due to a combination of effects leading to normal (perpendicular) pressure between the pipe and the borehole walls, including the submerged (buoyant) weight of the pipe as it rests (or is pushed) against the borehole surface in the relatively dense drilling fluid, lateral bending forces imposed on a relatively stiff product pipeline (e.g., steel) at bends along the route, and increased bearing pressure at route bends due to local tension in the pipe. Fluidic drag is due to the drilling fluid flow relative to the pipe surface. Depending upon the position of the pipe along the submerged path, the longitudinal component of the buoyant weight of the pipe locally increases or decreases the required tension.

2. *Bending:* The pipe is subjected to bending as it is forced to negotiate the curves in the borehole. Local bending stresses are proportional to the ratio of the pipe diameter to the radius of curvature. These stresses must be added to those due to the pulling tension to determine the peak tensile stress along the installed path. In general, the bending stresses will be more significant for a pipe with significant stiffness (e.g., steel pipe) than a relatively flexible product (e.g., polyethylene), depending on the pipe diameter.

3. *External pressure:* The pipe is subjected to external pressure from drilling fluid in the annulus, surrounding the pipe, due to the hydrostatic pressure head, as well as the pressure increment due to the drilling fluid introduced at the reamer. This pressure may lead to instability or collapse of the pipe during the installation stage, possibly aggravated by the simultaneous tensile stresses acting on the pipe. It is also possible that collapse may occur following installation, prior to the operational stage and corresponding internal pressurization. The drilling fluid/slurry is typically assumed to solidify over a period of weeks or months, prior to which the relatively dense slurry continues to apply external hydrostatic pressure to the pipe. Polyethylene pipe, a commonly used pipeline material for HDD installations, is characterized by reduced collapse strength over extended load duration, and this phenomenon must be considered during this preoperational stage.

In order to reduce the required tensile force, as well as the effective external hydrostatic pressure on the pipe, it is a common practice for Maxi-HDD installations to add ballast fluid (water or drilling fluid) to the interior of the pipe, particularly for plastic products. The ballast fluid will significantly reduce the buoyant weight and corresponding frictional drag forces, as well as provide internal pressure to eliminate or reduce likelihood of collapse during the preoperational stage.

Operating (Service) Loads

With one exception, the operating loads and stresses in a pipeline installed by HDD are not materially different from those experienced by pipelines installed by open-cut techniques; therefore, past procedures for calculating and limiting stresses can be applied. One exception relates to elastic bending. A pipeline installed by HDD will contain elastic bends corresponding to the route curvature. Bending stresses imposed by the HDD installation method should be checked in combination with other longitudinal and hoop stresses to ensure that acceptable limits are not exceeded. The operating loads imposed on a pipeline installed by HDD include:

- *Internal pressure:* With the exception of cable conduit, or gravity sewers, the pipeline is deliberately pressurized to transport the fluid through the interior. The pressure causes circumferential tensile stresses, as well as possible longitudinally induced tensile stresses due to Poisson's ratio effects.

- *Elastic bending:* The pipeline is subjected to elastic bending resulting from the pipe conforming to the shape of the drilled borehole, similar to that of the installation stage described above.

- *Thermal:* The difference between the temperature at initial construction and the later operating temperature of the pipeline can result in tensile or compressive longitudinal stresses.

- *External pressure:* The pipeline is subjected to long-term external pressure resulting from vertical loads that may be imposed on the pipe from the overhead soil as well as possible surface live or dead loads. These effects will typically be of greater significance for a plastic (e.g., polyethylene) pipe than a steel pipeline.

External Coating

External coatings used in HDD installations should be smooth and resistant to abrasion. Historically, pipelines installed by HDD in alluvial soils have usually been coated with corrosion coating only. Weight coating is generally not required. The deep, undisturbed cover provided by HDD installation has proven adequate to restrain buoyant pipelines.

The corrosion coating most often used on HDD crossings is thin film fusion bonded epoxy (see Fig. 5.11). This coating is popular because it is highly durable, and the field joints can be coated using a compatible fusion bonded epoxy system. For crossings installed in rock, highly abrasive soils, or soil conditions which might involve point loads, a protective coating should be used in addition to the corrosion coating. The protective coating need not have corrosion prevention properties, but must protect the underlying corrosion coating.

5.3.6 Specifications and Drawings

The HDD technical specifications should clearly define all details relative to HDD performance. Job specific details such as pilot borehole tolerances, water sources, and drilling fluid disposal requirements should be included in the specifications or an accompanying job specific section or a list of submittals. In addition to the technical specifications, the contract documents should contain a plan and profile drawing. The drawings

FIGURE 5.11   Coating inspection.

should complement the technical specifications by providing a clear presentation of the crossing design as well as the results of topographic, hydrographic, and geotechnical surveys.

5.3.7   Contractual Considerations

Once design of the crossing is complete, a set of contract documents should be produced for solicitation of bids from contractors and to govern construction of the crossing. Contract documents should be structured to clearly present technical, commercial, and legal requirements. Contract forms applied to HDD projects can be separated into four basic categories: *lump-sum, turnkey (design-build), daywork (cost-plus)* or *unit price (footage)*. For a lump-sum contract, the contractor is paid a fixed amount for delivering a drilled segment in accordance with plans and specifications. Payment is based on performance and does not vary with the time or effort expended. A turnkey contract is similar to lump-sum, but the contractor assumes responsibility for the design phase as well construction phase of the operation. For a day work contract, the contractor is paid a fixed amount per day, or possible other unit of time, for providing equipment, required personnel, and service, in accordance with the contract documents. For the unit price contract, the contractor is paid based on the footage of pipe actually installed, as measured horizontally on the surface.

Lump Sum Contracts

In most cases, Maxi-HDD installations are advantageously bid using standard lump-sum contracts. (For convenience, bid prices may be broken down for analysis.) Applicable technical specifications and drawings should be included in the contract documents. A lump sum contract allows the contractor the option of providing the most efficient, cost-effective installation, without motivation to do an unnecessarily long or more complicated installation.

Turnkey (Design-Build) Contracts

A key aspect of this type of contracting is the *partnering* concept. The owner and the contractor work closely together in planning, design, cost control, scheduling, site investigations, and possibly land acquisition or easement and project financing. A turnkey operation has the potential of significantly reducing project cost and delivery time while avoiding design-construction conflicts by having only one contractor accountable for the entire project.

Daywork (Cost-Plus) Contracts

Because of the evolving nature of HDD technology, the industry has employed many contract variations. Typically, these variations involve negotiating a completion incentive into a cost-plus (daywork) contract. Significant technical advances have been made because of owner willingness to assume the risk of cost overruns or completion delays for prospective HDD installations which were not contractually feasible. However, a day-work contract requires much greater oversight by the owner than a typical lump sum or turnkey contract.

Although a cost-plus contract may not specify contractor performance in terms of installation details, standards of contractor performance are required and should be clearly defined. The required performance primarily involves the provision of equipment of a specified capacity. The components of equipment should be listed as well as conditions with respect to downtime, maintenance, crews, fueling, etc. In additional, items in the scope of work which can be contracted on a lump-sum basis, such as mobilization and site preparation, should be broken out, priced on a lump-sum basis, and governed by appropriate performance specifications.

Unit Price (Footage) Contracts

For a unit price contract, the owner estimates the number of units included for each element (activity) of work, such as linear feet of installed product pipe. In addition, the contract may include ancillary work such as tie-ins, pit excavation, mobilization and demobilization, or traffic control—typically provided on a lump-sum basis, as well as other items the owner or engineer may require, to be listed separately.

The contractor determines the unit price bid for the various items in the contract, as listed by the owner. The contractor should include overhead and profit within the unit prices, and reflect all the costs for each element of work, such as utility locating and potholing. It should be noted that actual unit price portion may change from estimated values based on the quantity of work realigned in the field.

The unit price contracting method is most commonly used for installation of pipelines using Mini- and Midi-HDD systems.

5.3.8   Inspection and Construction Monitoring

The primary objectives of an inspector involved in construction monitoring on an HDD installation are to assist in the interpretation of the contract documents and to verify conformance, or nonconformance, by the drilling contractor. In the conduct of this task, it is important that the inspector document his or her observations and actions. Should a question or dispute arise after the installation is complete, the inspector's notes may provide the only source of confirming data and information. Since a drilled installation is typically buried with deep cover, often under an inaccessible obstacle, its installed condition cannot readily be confirmed by visual examination.

Directional Performance

The inspector should be concerned with directional drilling performance with respect to two basic measures: position and curvature. The contractor must install the pipeline such that the drilled length and depth of cover specified by the contract are satisfied. Furthermore, the contractor must not provide a path such that the pipeline will be damaged or overstressed during installation or subsequent operation. The inspector should ensure that bends are not drilled at a radius of curvature less than the recommended minimum. If an apparently overly small radius occurs, the unacceptable portion of the borehole should be redrilled. If redrilling is not practical or proves unsuccessful, the problematic section should be reviewed with the design engineers to consider the possibility the pipe can function properly and that the codes and specifications governing design of the pipeline are not violated.

The actual position of the drilled path usually cannot be visually confirmed. Therefore, it is necessary for the inspector to have a basic understanding of the downhole survey system being used by the contractor and be able to interpret its readings. It is not necessary for the inspector to continuously observe and approve the drilling operation. However, progress should be monitored periodically and problems addressed so that remedial action can be taken as soon as possible.

Drilling Fluids

The inspector should document all drilling fluid products being used, the contractor's pumping pressures and rates, and details relative to drilling fluid circulation at the endpoints of the HDD installation. The right-of-way and surrounding areas should be examined regularly for inadvertent returns. If inadvertent returns occur, they should be contained or cleaned up in accordance with the specifications and permits, and their locations monitored for continuing problems. See the "Disposal of Excess Drilling Fluids" section in Sec. 5.3.4 for additional information.

Additional Concerns

Depending upon the contractual and technical aspects of an HDD installation, there are numerous additional details which should be reviewed and documented. These may consist of gauge readings, production rates, equipment failure, downtime, etc. During pullback, it is important to record the contractor's operations relative to handling of the pipeline. Anti-buoyancy control measures, if used, should also be recorded. Following the completion of pullback, the condition of any visible pipe and coating at the leading edge of the pull section should be documented. The final steps in accepting project delivery may include:

- Preservation of entry and exit stakes until the pilot borehole is completed.

- Staking and verification of the distance, alignment, and elevation between the final as-built and initially designed entry and exit points.

- Maintenance of the following survey information for generating certified ("as-built") documentation:
  - Tabular data
  - Preparer's assumptions

- Calculation methods
- Copies of field-generated data

It is noted that the position of the desired drill path, and as-built pipeline, are only as accurate as the preconstruction survey.

5.4   Mini-HDD Considerations

Although typically a lower-cost, less critical operation than Maxi-HDD, it is nonetheless critical that proper procedures be used in the planning, equipment setup and operation for a Mini-HDD installations. Improper procedures may result in compromised utility lines, including those being installed as well as existing facilities in the vicinity, sometimes leading to safety hazards. In other cases, the final installation parameters (location, depth, etc.) may fail to meet the criteria of the inspector or utility owner. Thus, the availability of appropriate practices for Mini-HDD operations is extremely valuable. Such procedures would presumably be performed by contractors and their personnel that have the requisite skill and formal training, as well as extensive on-the-job experience.

It is also important, however, that the owners of the facility have a good understanding of the relevant HDD procedures and practices, and also participate in the planning and preliminary stages. Furthermore, it is important that owner's inspectors have sufficient knowledge to help ensure that the product pipes are placed correctly. Issues of concern include the reliability of the installed product, customer relations and safety, as well as possible environmental issues. It is noted that Rule 410C of the National Electrical Safety Code (IEEE, 2007) requires that the contractor provide a designated (qualified) person (e.g., an operator/foreman) to be in charge of the operation and equipment, including responsibility for its safe operation.

Mini-HDD is primarily used to install pipes and conduits belowground for utility distribution lines along road or street right-of-way, and to place service lines. Although there are many similarities among the various HDD categories, this section presents a set of procedures specific for Mini-HDD, and some Midi-HDD, applications, providing details for bore planning and product pipe loadings.

Similar to Maxi-HDD operations, the Mini-HDD drill string creates a pilot borehole which is subsequently enlarged to a greater diameter during a reaming operation, typically simultaneous with final pullback of the product pipe or utility line. Figures 5.12 and 5.13 illustrate these two stages of the typical Mini-HDD process. The creation of the pilot borehole and the reaming operations in Mini HDD are accomplished by fluid-assisted mechanical cutting provided by rotating the drill string. The procedure typically uses a high pressure, but low volume drilling fluid flow, to minimize the creation of voids during the initial boring and subsequent backreaming operations.

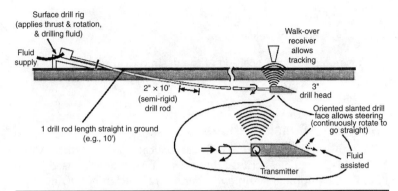

FIGURE 5.12 Mini-HDD pilot borehole installations. (*Source: Outside Plant Consulting Services.*)

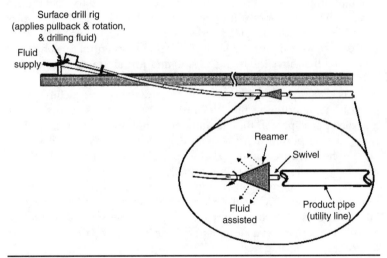

FIGURE 5.13 Mini-HDD, backream, and pullback operations. (*Source: Outside Plant Consulting Services.*)

As mentioned earlier, the drilling fluid helps stabilize the borehole, remove cuttings, provide lubricant for the drill string and product pipe, and cool the drill head. The slurry surrounds the product pipe, typically filling the annulus between the pipe and the bored cavity and providing a permanent supportive filling. In contrast, for Midi- and Maxi-HDD operations, and for hard or rocky soil conditions, "mud motors" powered by the drilling fluids may be used. The use of such mud motors would only be applicable for the larger Midi-HDD machines.

Tracking of the initial pilot bore path is accomplished by a manually operated overhead (walkover) receiver or a wireline (nonwalkover) guidance tracking system incorporated into the drill string. The walk-over system (Fig. 5.12) is more common, for which the receiver is placed above the general vicinity of the drill head to allow a determination of its location and depth, and to indicate drill head orientation for determining steering information to be implemented from the drill rig. The information may be relayed to the drill rig operator by direct communication from the tracker, or transmitted remotely. Steering is achieved by controlling the orientation of the drill head, which has a directional bias, and pushing the drill string forward with the drill head oriented in the direction desired. Continuous rotation of the drill string allows the drill head to bore a straight path.

The most common type of pipe utilized for Mini-HDD operations traditionally has been high density polyethylene (HDPE), due to its relatively high tensile strength characteristics and the absence of potentially vulnerable joints. Other products, however, have been successfully installed, including various types of polyvinyl chloride (PVC) pipes and ductile iron pipes. Field fusible and restraint joint PVC represent recently introduced products designed to be installed by HDD or other pulling type applications. Either a single pipe or a bundle of small pipes may be placed (e.g., HDPE "innerducts" for communications applications). For the latter case, the desired communication cable—fiber, coaxial, or copper—may subsequently be installed into the innerduct path, as convenient. Although it is possible to directly pull back a bare cable into the borehole, such a procedure is not recommended for most cables, unless special protection (armor, etc.) is provided (Telcordia, 2007).

5.4.1 Mini-HDD Planning

General Considerations

The owner or project engineer will provide the general requirements for the path of the pipe, including location within the right-of-way, path of laterals to residence or building, etc. The actual layout, however, will vary from case to case, depending upon the specific obstacles or specified utility line architecture. For Mini-HDD projects, the detailed bore path for each segment will be determined on-site, in advance of the operations, by the selected contractor, utility and/or regulatory engineer.

Equipment and Product Restraints

Drill-Rod Constraints   The planned path must be consistent with the steering capability of the drill string and the allowable radius of curvature of the steel drill rods based upon the corresponding bending stresses in the steel rods and joints. Although some soil conditions will inhibit sharp steering maneuvers, path limitations will often be based upon fatigue strength considerations of the rods. A given rod

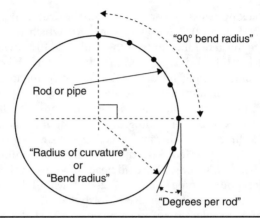

Figure 5.14 Drill rod or product pipe curvature terminology. (*Source: Outside Plant Consulting Services.*)

may be able to withstand a single bend cycle corresponding to a relatively sharp radius of curvature, but the rotation of the rod during the boring operation results in flexural cycles, which may eventually cause cumulative fatigue failure. The diameter of the drill rod is an important parameter affecting its stiffness, steering capability, and the allowable bend radii.

There are several terms that may be used to describe the bending capability of a drill rod, including the "radius of curvature," (bend radius) "90° bend radius," and "angular change per rod," as illustrated in Fig. 5.14. The significance of the radius of curvature dimension, and the 90° bend radius terms are often a source of confusion in the Mini-HDD industry. Whereas engineers are familiar with the radius of curvature (or bend radius) dimension, construction personnel often use the 90° bend radius dimension, equal to the distance along a 90° portion (quadrant) of the perimeter of the circular path, since the latter term may be more readily identified or measured during a boring operation, for a bend in a horizontal plane. Similarly, the maximum angular change per rod is a practical measure of the rod behavior. It is important that the construction personnel understand the significance of the terms in relation to drill rod specifications to avoid overstressing the rod or, conversely, underutilizing its capability.

The following relationships are applicable:

90° bend radius (ft) = 90 × rod length (ft)/angular change (deg/rod)

$$\text{(5.2a)}$$

radius of curvature (ft) = 90° bend radius (ft)/1.57   (5.2b)

radius of curvature (ft) = 57.3 × rod length (ft)/angular change (deg/rod)

$$\text{(5.2c)}$$

For a typical Mini-HDD drill rod of approximately 2-in. diameter, the corresponding radius of curvature given by Eq. (5.1) is 200 ft. In practice, manufacturers will typically allow a 100-ft radius of curvature for a 2-in. rod, confirming the conservative nature of Eq. (5.1). Such a bending capability corresponds to a *90° bend radius* of approximately 155 to 160 ft, or an *angular change* of 5.5° to 6.0° per 10 ft drill rod, consistent with Eq. (5.2). The corresponding allowable degree of curvature or bending applies to bends in all horizontal (plan) or vertical (profile)—or inclined—planes.

Product Pipe or Conduit Constraints

In general, the allowable radius of curvature for the product pipe will be provided by the pipe manufacturer. For pipes or conduits constructed from plastic or other relatively flexible material, and the relatively small diameter pipe installed by a typical Mini-HDD operations, the bending limitation of the drill rods as given in Eq. (5.1) is typically sufficiently large to be compatible with that of the product pipe. For these cases, related bending stresses need not be explicitly considered in addition to the tensile stresses, as discussed and estimated in Sec. 5.5 for polyethylene pipe. However, for steel product pipe, guideline such as that given by Eq. (5.3) would be applicable:

$$R_{min} = 100 \, D \qquad (5.3)$$

where R_{min} = minimum radius of curvature of product pipe, or bore path, ft
D = nominal diameter of product pipe, in.

For other materials, including pipe or conduit assemblies containing joints or couplings of lower strength than that of the basic pipe element, the manufacturer's recommendations should be followed regarding minimum radius of curvature. In the absence of such low strength joints or couplings, and other related product information, the following formula may be used (Hair, 1988):

$$R_{min} = E \cdot D/(SMYS \cdot F_1 \cdot 24) \qquad (5.4)$$

where E = elastic modulus, for the material and temperature of interest, psi
$SMYS$ = material specified minimum yield stress, psi
F_1 = design factor (< 1.0)

In this case, the corresponding bending stresses should be considered in addition to the estimated tensile loads, as determined for the product pipe used (ASCE, 2005). It is noted that the peak pulling load tensile stress for the pipe does not necessarily occur at the points of peak bending stresses.

5.4.2 Bore Path Layout and Design—Vertical Trajectory

Depth of Cover

The nominal depths will be specified by the owner, including minimum and maximum, and consistent with other existing utility locations, as determined by the preliminary investigations. A minimum depth of cover of 36 in. is desired to reduce the potential for drilling fluid penetration to the surface and avoid the tendency for the drill head to rise to the free surface, thereby complicating the steering operation during the initial pilot bore. In addition, typical industry guidelines recommend a minimum ratio of 10-to-1 for depth of cover to final borehole diameter to avoid surface heaving effects, for a possible compaction process during reaming and pullback in appropriate soil conditions. In such cases, spoils disposal is minimized since the displaced soil is compacted into the surrounding walls of the borehole. For shallower burial or noncompactable soil conditions, the drilling operation must use appropriate drilling fluids designed to remove at least a portion of the spoils during the initial boring or reaming operations. A greater volume of drilling fluid would generally be required to remove the soil from the borehole in a soil removal process, in comparison to a compaction process. It is the responsibility of the contractor to understand and utilize drilling fluid technology to form the borehole without surface heaving.

Figure 5.15 shows recommended minimum depth of cover for conditions compatible with the nominal 10-to-1 ratio for compactable soils, as well as for the case including a spoils removal process— assuming a 5-to-1 ratio. The increasing depth as a function of borehole (and pipe) diameter allows some degree of inefficiency of soil removal, reduced potential for any surface heave or settlement, and

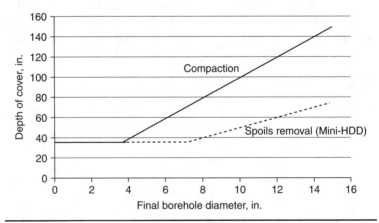

FIGURE 5.15 Recommended minimum depth of cover of borehole. (*Source: Outside Plant Consulting Services.*)

also helps offset a possible tendency for the pipe to gradually rise toward the surface due to flotation effects in saturated soil conditions. If a lower depth of cover' than that indicated in Fig. 5.15 is necessary, it is recommended that the final borehole size be gradually enlarged using several (one or more) prereaming passes, prior to the final pullback of the pipe, accompanied by careful monitoring of the drilling fluid pressures. In addition to observing the minimum depth guidelines, excessive depths may not be practical for future maintenance activities on the installed pipes or utility lines.

Bore Path Profile (Vertical Plane) Trajectory

The radius of curvature of the drill rod path and the entry angle of the rod to ground surface will determine the depths achievable at the beginning of the bore path. Figure 5.16 illustrates a Mini-HDD bore profile trajectory, including pits at the entry area and possibly along the route. These pits may be required for pipe splicing, completing lateral connections, or to expose existing utilities. The pits may also be useful for collecting drilling fluids from the boring or backreaming operations. In order to achieve a specified depth at a particular point (e.g., point 1 or 2, Fig. 5.16) at the beginning of a bore, the front of the drill rig must be setback an appropriate distance from the point of entry. This distance will also depend upon the rod entry angle, which is determined by the drill carriage angle. Typical Mini-HDD drill carriages allow an entry angle in the range of 5° to 25° (10 to 45 percent grade). Some locating systems provide the elevation angle in percent grade (vertical rise or drop per unit horizontal distance, times 100). For convenience, the angle in degrees is approximately equal to half the percent grade or pitch.

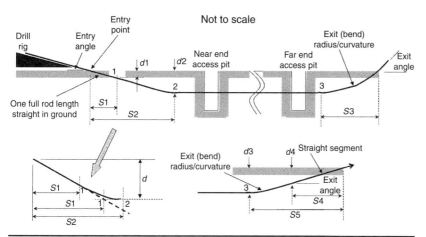

Figure 5.16 Bore path profile/trajectory (vertical plane). (*Source: Outside Plant Consulting Services.*)

Setback Distance

The shortest setback distance corresponds to a bore path segment comprising a straight line extending from the drill rig directly toward the point of entry (e.g., point 1; see Fig. 5.16):

$$S_1 = d_1 / \text{Tan } \beta$$

$$\approx d_1 / \beta \tag{5.5}$$

where S_1 = shortest setback distance from entry point to point of interest, ft

d_1 = depth of point of interest, ft

β = bore entry angle, radians or (approximately) percent grade.

This formula corresponds to a drop at a constant grade angle. It should be noted that the depth of the bore path would continue to increase beyond this point, over the distance required for the drill rods to develop a concave-up curvature and achieve a horizontal trajectory.

If it is specified that the bore path be essentially level at the depth and point of entry, a longer setback distance will be required to allow the path to reach the desired depth, and a horizontal trajectory, at a decreasing rate of descent. This corresponds to the geometry leading to point 2 in Fig. 5.16. In this case, it is assumed that the bore is initiated along a straight path, without any curvature or steering, for a distance equal to one full drill rod length (e.g., 10 ft, for typical Mini-HDD machines) in the ground. This is a recommended industry practice to avoid lateral bearing loads at the front of the drill rig. The upward desired curvature may be introduced during the placement of subsequent drill rods. The minimum setback distance corresponds to a path with the first few rods inserted such as to continue the straight (sloping) trajectory, and the subsequent rods placed with the leading rod then creating a path at the minimum allowable radius of curvature, as follows:

$$(S_2)_{min} = 1 \, C \, os \, \beta + (R_{rod})_{min} \, Sin \, \beta$$

$$+ \{d_2 - 1 \, Sin \, \beta - (R_{rod})_{min} \, (1 - Cos \, \beta)\} / Tan \, \beta$$

$$\approx 1 + (R_{rod})_{min} \, \beta + \{d_2 - 1 \, \beta - (R_{rod})_{min} \, \beta^2/2\} / \beta \tag{5.6}$$

where $(S_2)_{min}$ = minimum setback distance from entry point to point of interest, with level trajectory, ft

d_2 = depth at point of interest, ft

1 = drill rod length, ft

The setback distances of Eqs. (5.5) and (5.6) are shown in Fig. 5.17 for a typical 10 ft long drill rod and an allowable 100 ft radius of

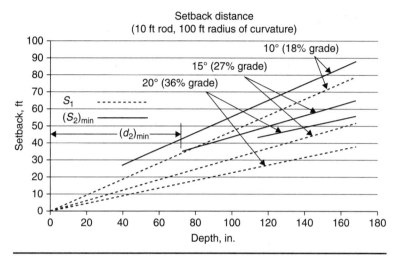

Figure 5.17   Setback distances. (*Source: Outside Plant Consulting Services.*)

curvature, for various entry angles. Figure 5.17 also indicates the minimum depths at which the trajectory may become level. Thus, for a bore entry angle of 15°, a depth of 72 in. will be achieved at a setback distance of slightly in excess of 20 ft using a straight trajectory segment. In comparison, a setback distance of at least 35 ft is required to reach the same depth with a level bore angle at that point. Figure 5.17 indicates that this particular entry angle, for the assumed radius of curvature, is not consistent with achieving a level bore at depths shallower than 72 in. In such cases, the trajectory would exceed the desired depth (i.e., beyond the point of interest, such as the entry pit). If it were necessary to remain within a specified maximum depth along the entire path, including near the entry point, a shallower entry angle or sharper (smaller) radius of curvature, or both, would be required, as quantified in the following section.

Minimum Depth at Level

Due to the recommendation that the first rod be placed in the ground such as to define a straight entry, and its subsequent path curvature consistent with the minimum allowable radius of curvature, there is a minimum depth at which the trajectory will become level, depending upon the entry angle. This depth may be calculated by:

$$(d_2)_{min} = 1 \operatorname{Sin} \beta + (R_{rod})_{min} (1 - \operatorname{Cos} \beta)$$

$$\approx 1 \beta + (R_{rod})_{min} \beta^2 / 2 \qquad (5.7)$$

Figure 5.17 shows the minimum depth value for the case of a 15° entry angle.

Depth/Setback Implications

If the determined setback distances or drill rig carriage angle consistent with the project *maximum* depth specifications are not practical, consideration should be given to receiving approval from the owner allowing increased depths in the transition area following the entry pit. If necessary, smaller diameter, more flexible, drill rods may be considered if consistent with anticipated thrust and torque loads. Smaller radii of curvature than the manufacturer's limits may be used by the contractor if it is recognized that reduced service life may result for the drill rods. If the steering conditions in the soil preclude a sufficiently sharp upward turn, mechanical assistance may be provided at the entry pit to apply an upward bending moment on the rod. It is noted that the corresponding radius of curvature is not necessarily applied to the product pipe, which may only be pulled as far as the entry pit. Otherwise, the product pipe bend limitations discussed above should be observed.

Horizontal Distance to Rise to Surface

Figure 5.16 also illustrates the horizontal distance required for the head of the drill string to reach the surface from its present bore path. The minimum distance to reach the surface from a point on a level trajectory (e.g., point 3) corresponds to that of steering upward at the minimum radius of curvature, and is given by:

$$(S_3)_{min} = \{2\,d_3\,(R_{rod})_{min}\}^{\frac{1}{2}}\,\{1 - d_3/[2(R_{rod})_{min}]\}^{\frac{1}{2}}$$

$$\approx \{2\,d_3\,(R_{rod})_{min}\}^{\frac{1}{2}} \tag{5.8}$$

where $(S_3)_{min}$ = distance to rise on arc from level trajectory, ft
d_3 = depth at point of level trajectory, ft

This distance, S_3, is shorter than the distance, S_5, corresponding to rising partially on an arc and then (e.g., point 4) continuing boring at a straight path at an upward angle to the surface:

$$(S_5)_{min} = (R_{rod})_{min}\,\mathrm{Sin}\,\alpha + \{d_3 - (R_{rod})_{min}\,(1 - \mathrm{Cos}\,\alpha)\}/\mathrm{Tan}\,\alpha$$

$$\approx (R_{rod})_{min}\,\alpha + \{d_3 - (R_{rod})_{min}\,\alpha^2/2\}/\alpha \tag{5.9}$$

where $(S_5)_{min}$ = distance to rise on arc from point on level trajectory to specified exit angle, ft
α = bore exit angle, radians or (approximately) percent grade.

The horizontal rise distances S_3 and S_5 from a level trajectory are shown in Fig. 5.18 for a 100 ft radius of curvature and various exit angles. The maximum possible exit angle is limited by the depth.

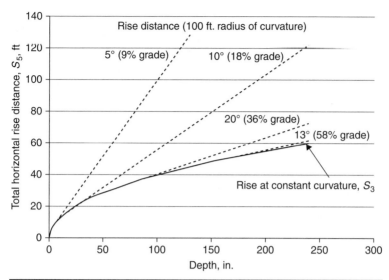

FIGURE 5.18 Horizontal rise distances. (*Source: Outside Plant Consulting Services.*)

For a drill head oriented at an upward grade, the horizontal distance to rise to the surface (e.g., from point 4) is

$$S_4 = d_4/\text{Tan } \alpha$$

$$\approx d_4/\alpha \qquad (5.10)$$

where S_4 = distance to rise to surface at constant rise angle, ft
d_4 = depth, ft

The values for the depth d_4 and elevation angle α used in the above formula may be obtained by information provided by the drill head locating system. This formula corresponds to a rise at a constant grade angle, and the results are shown in Fig. 5.19.

5.4.3 Overall Bore Path Layout and Design

Bore Path (Horizontal) Planar Trajectory
In plan view, typical bore paths are relatively straight trajectories between the entry and exit points, and pass through intermediate access points, as required. A degree of route curvature in the horizontal plane may be accommodated by the steering capability of the Mini-HDD system, consistent with the recommended bend radius of the drill rods (and product pipe) and steering capability within the soil.

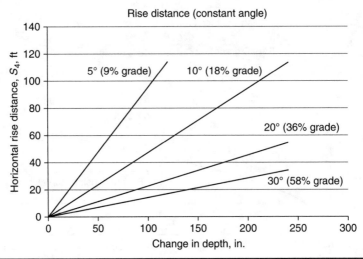

FIGURE 5.19   Horizontal rise distances at constant rise angle. (*Source: Outside Plant Consulting Services.*)

Bore Path Layout

The contractor must understand the right-of-way (ROW) constraints and the general path requirements and utility line architecture for planning the bore paths. This includes paths for pipes for distribution lines along the ROW and for services for individual residences or structures. The anticipated route must take into consideration the location of other existing utilities (Telcordia, 2007).

For a smooth level area without any underground utilities, a visual survey and simple sketch may be sufficient for defining the bore route. In more complicated situations, a transit or other type of surveying equipment may be required. In general, a proposed bore path plan view and profile layout should be prepared indicating the surface grade and important surface features, location of existing below ground utility lines, reference points, etc. The bore path layout should also show anticipated access pits for utility connections or lateral service lines, and the bore depth at critical points such as access pits or utility line, as well as at other reference points along the route. For relatively level surfaces, a taut string at a convenient height, spanning the distance between the entry and exit points, will establish the average grade of the bore, representing a basis for determining the nominal trajectory of the bore path.

The string may also provide a reference for verifying the proper depth during the actual operation in the presence of minor surface depressions or irregularities, and serves as a basis from which to interpret the guidelines of Sec. 5.4.2, which assumes a level surface grade. Thus, for the simplification purposes, the bore entry and exit angles are defined relative to the average grade. For large surface depressions or mounds (e.g., of height greater than the depth of interest and extending over a long expanse, on the order of the minimum drill rod bend radius or greater), the bore should attempt to follow a path at the nominal specified depth below the average surface profile. Available industry tools may be used to provide assistance during the planning and construction phases.

The path should avoid unnecessary bends and consecutive left and right, or upward and downward curves. Such trajectories are difficult to follow and may lead to over-steering and excessive bends, resulting in increased stresses in the drill rods and greater pulling forces during the installation of the pipe. A heavy rope or line placed on the surface may serve as a convenient means of defining a gradual curve consistent with the desired path radius of curvature. The radius of curvature of the path may be estimated by:

$$R_{avg} = \Delta S / \Delta \phi \qquad (5.11)$$

where R_{avg} = average radius of curvature along path segment, ft
ΔS = distance along path, ft
$\Delta \phi$ = angular change in direction, radians

Thus, a change of 0.1 radian (approximately 6°) per 10 ft corresponds to a radius of curvature of approximately 100 ft.

Accuracy and Tolerance

The deviation of the actual bore route from the proposed path should be within a specified tolerance and such as to meet the intent of required separations from existing facilities. Both vertical and horizontal limits should be specified by the owner, consistent with local regulations. Some regulations require that new construction activities be a minimum of 18 in. from either side of the outer edge of existing facilities (Telcordia, 2007).

Unintentional bore path deviations may result from attempting to follow a bend with sharp curvature, drill head misalignments caused by cobbles or other obstacles, or soil conditions in general. Deviations may be reduced by decreasing the interval between successive drill head location determinations. Unless otherwise specified, the drill head should exit at the surface or access pit within a 12-in. radius of the target point. It is generally less difficult to meet such a condition for a target point in a pit than at the surface in a gradually rising bore profile (see Fig. 5.16). The allowable horizontal and vertical deviations along the bore route should also be specified by the owner. Typically, horizontal and vertical deviations along the bore route should be within ± 18 in. of the proposed path.

Separation from Existing Utilities

To help maintain an 18 in. separation from an existing facility along the route, the proposed bore path, including the radius of the widest cutter/reamer, should be at least 36 in. laterally offset from, or below or above the verified depth of the closest edge of the facility (Telcordia, 2007). An exception includes the case of the bore path crossing an exposed utility, at which point the desired physical separation, or noninterference, can be positively verified (IEEE, 2007).

5.5   Pipe Load Calculations

As said earlier, the product pipe must be capable of withstanding the corresponding various loads experienced during the operational (service) phase, and during the installation and preoperational phases. It is therefore recommended that the product pipe be independently verified to be able to withstand the service loads corresponding to internal pressurization, if appropriate, as well as soil and surface loads, such as experienced in a conventional trench installation; see *The Plastics Pipe Institute Handbook of Polyethylene Pipe,* Plastics Pipe Institute (PPI, 2008). Regarding the installation phase, a method intended for Maxi-HDD projects installing polyethylene pipe is provided in ASTM F1962, *Standard Guide for Use of Maxi Horizontal Directional Drilling for Placement of Polyethylene Pipe or Conduit under Obstacles, Including River Crossings* (ASTM, 2005). In comparison, a simpler technique would be more appropriate for typical lower cost, less sophisticated Mini-HDD operations. Thus, a simplified methodology is presented in this section to estimate the required pull loads on polyethylene pipe installed by Mini-HDD systems. This simplified procedure also evaluates the potential collapse tendency of the polyethylene pipe during the installation or post-installation (preoperational) phases.

The methodology and associated formulae are based upon approximations to the more complex set of equations and procedures provided in ASTM F1962. The objective is to provide a convenient means of identifying potentially problematic Mini-HDD installations and/or to aid in the pipe selection process, in contrast to the extensive planning or analytical investigations characteristic of typical Maxi-HDD projects. The proposed mathematical model reflects the major route parameters (bore length, planned bends) and buoyant force for an empty polyethylene pipe, and also accounts for unplanned curvatures (undulations) resulting from path corrections in a typical Mini-HDD installation.

Although the methodology is primarily conveniently described with respect to Mini-HDD installations, the results are also applicable to Midi-HDD operations. The procedure is applicable to commonly used high-density polyethylene (HDPE) pipe, based on its characteristically low-bending stiffness.

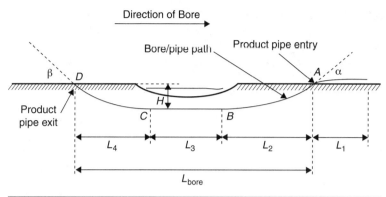

Figure 5.20 Nominal Maxi-HDD route (river crossing). (*Source: Outside Plant Consulting Services.*)

5.5.1 Pipe Load Calculations

Figure 5.20 illustrates a typical geometry for a major (Maxi-HDD) operation, corresponding to a river crossing. The indicated path corresponds to that shown in ASTM F1962 and comprises four segments, including those spanning the pipe entry to exit point (L_2, L_3, L_4) and the excess length (L_1) remaining after the span has been accomplished. Thus, the length of the actual crossing, L_{bore}, is given by

$$L_{bore} = L_2 + L_3 + L_4$$

In some cases, the intermediate horizontal segment, L_3, may be of zero length. Due to the typically low pipe entry angle, α, and exit angle, β, and gradual path curvature, the depth of the crossing, H, is small compared to the transition distances L_2 and L_4.

Pull Force

Using the above terminologies, ASTM F1962 provides a set of recursive relations to predict the required pull force —T_A, T_B, T_C, and T_D— corresponding to the leading end of the pipe reaching point A, B, C, and D (see Fig. 5.20). Thus,

$$T_A = e^{\upsilon a\,\alpha} \bullet \upsilon_a \bullet w_a \bullet (L_1 + L_2 + L_3 + L_4) \tag{5.12a}$$

$$T_B = e^{\upsilon b\,\alpha} \bullet (T_A + \upsilon_b \bullet |w_b| \bullet L_2 + w_b \bullet H - \upsilon_a \bullet w_a \bullet L_2 \bullet e^{\upsilon a\,\alpha}) \tag{5.12b}$$

$$T_C = T_B + \upsilon_b \bullet |w_b| \bullet L_3 - e^{\upsilon b\,\alpha} \bullet (\upsilon_a \bullet w_a \bullet L_3 \bullet e^{\upsilon a\,\alpha}) \tag{5.12c}$$

$$T_D = e^{\upsilon b\,\beta} \bullet (T_C + \upsilon_b \bullet |w_b| \bullet L_4 - w_b \bullet H - e^{\upsilon b\,\alpha} \bullet [\upsilon_a \bullet w_a \bullet L_4 \bullet e^{\upsilon a\,\alpha}]) \tag{5.12d}$$

where w_a and w_b are the empty (above ground) and buoyant weights of the pipe, respectively, and υ_a and υ_b. are the corresponding coefficients of friction. Equation (5.12) is sufficiently general to consider the possible implementation of anti-buoyant measures to reduce the otherwise high values of w_b for polyethylene pipe. In the absence of such anti-buoyancy measures, or with low friction pipe supports (e.g., rollers) outside the borehole, the maximum pull force will tend to occur toward the end of the installation (e.g., T_C or T_D).

Equation (5.12) is based upon conventional Coulomb friction, which assumes that drag forces on the pipe are proportional to the normal bearing forces applied at the pipe surface, with the proportionality constant designated as the "coefficient of friction." Such bearing forces may be due to the dead (empty) weight of the pipe where above ground, the buoyant weight of the submerged pipe (possibly mitigated by anti-buoyancy measures), bearing/bending forces associated with pulling a stiff pipe around a curve, or bearing forces resulting from (previously induced) axial tension tending to pull the pipe snugly against any locally curved surfaces.

For the case of polyethylene pipe, of typically low bending stiffness relative to that of the steel drill rods that created the gradually curved original borehole path, the corresponding bearing/bending forces may be ignored. However, the tension-induced bearing forces are primarily dependent upon the cumulative bend angles, which may be significant, independent of the gradual nature or variable direction of such curves or degree of pipe bending stiffness, and are included in the analysis. Such effects compounded, and in some situations may become the dominant source of drag, essentially controlling practical placement distances. This phenomenon is referred to as the "capstan effect" (i.e., the operating principle of the "capstan winch," see Fig. 5.21) and is the basis of the exponential terms in Eq. (5.12). In particular, the following

Rotating capstan/drum

Minimal
tail load
required

Pulls large load
(head tension)

Figure 5.21   Capstan winch (practical application of the capstan effect). (*Source: Outside Plant Consulting Services.*)

relationship illustrates the basic phenomenon for the idealized case of a weightless, flexible pipe:

$$F_2 = F_1 \cdot e^{\upsilon \theta} \qquad (5.13)$$

where F_1 represents axial tension at the entry point of a bend of magnitude θ (radians), υ is the local coefficient of friction between the product pipe and borehole wall surface, and F_2 is the required axial tension at the exit point of the bend. In practice, the impact of the actual weight of the pipe may be reflected in the preceding tension, F_1.

Pipe Collapse

ASTM F1962 provides the critical (buckling) pressure, P_{cr}, as given by Eq. 5.14

$$P_{cr} = 2 E \cdot f_o \cdot f_R / \{(1 - \mu^2) \cdot (DR - 1)^3\} \qquad (5.14)$$

where E = material modulus of elasticity
      μ = Poisson's ratio
      f_o = ovality compensation (reduction) factor
      f_R = tensile stress reduction factor

The dimension ratio, DR, refers to the ratio of the pipe outer diameter to its (minimum) wall thickness. In a discussion of ASTM F1962, Petroff (2006) explains the significance of these terms. The material properties, E and μ, for the viscoelastic HDPE pipe depend upon the load duration, f_o accounts for initial or subsequent out-of-roundness, and f_R recognizes a potential reduction in collapse strength in the presence of significant tensile loads during the installation phase. Petroff (2006) also provides an explanation of the possible sources and nature of the pressure loads on the pipe, including that due to hydrostatic pressure associated with drilling fluid or groundwater pressure, and asymmetric earth pressure that cause ring deformation, as well as the implications of their time dependent characteristics. In general, the detailed consideration of the interaction of the various phenomena, and the consequences for the product pipe, is relatively complex and not within the scope of this book.

5.5.2   Simplifications for Mini-HDD Applications

The detailed application of Eq. (5.12) to determine the required tensile load on the pipe during the installation phase, and Eq. (5.14) to evaluate the possibility of pipe collapse during installation or the post-installation (preoperational) phase, would be tedious for typical Mini-HDD applications. On the other hand, application to potentially problematic installations would be desirable, although not necessarily feasible, for most Mini-HDD personnel, in spite of their being otherwise well-trained in this technology. Thus, the reduction of these equations to relatively simple formulae and practical procedures, albeit at a possible loss of precision, would be beneficial, as described in the following sections.

Pull Force

In order to reduce the complexity of Eq. (5.12) for Mini-HDD installations, the procedure is limited to polyethylene pipe without the use of anti-buoyancy techniques. Such techniques are typically not employed for Mini-HDD operations. Substituting specific (conservative) values for several of the parameters, and a comparison of the typical magnitudes of the resulting calculations, allows a major simplification of the predicted pull force at the end of the installation, T_D. In particular, values of the frictional coefficients υ_a and υ_b are assumed to be equal to 0.5 and 0.3, respectively, and pipe entry and exit angles, α and β, are assumed to be 20°. Thus, it may be shown that Eq. (5.12d) can be simplified to

$$T_D \approx L_{bore} \bullet w_b \bullet (1/3) \tag{5.15}$$

It is recognized that, under appropriate conditions and actual installations, the pull force may achieve its maximum level prior to point D in Fig. 5.20. However, with the present basic theoretical model, under the assumed conditions and conservative parametric values, the predicted tension at point D would be a maximum, or reasonably close in magnitude to a previously occurring (predicted) maximum value.

For Mini-HDD installations, the above estimate T_D must be modified to account for the possibility of additional path curvature due to deliberate route bends as well as the likelihood of unplanned undulations resulting from path corrections. The presence of such characteristics in the final (as-built) path will increase the required pull force, consistent with the capstan effect described above. These effects may be conservatively estimated by the applying the exponential term in Eq. (5.13) to the tension T_D, such that

$$T_D{}^1 = T_D \bullet e^{\upsilon b \, \theta} \tag{5.16}$$

represents the net final tension, for which the angle θ is selected as equal to the total additional route curvature. The latter may be expressed as

$$\theta = n \bullet (\pi/2) \tag{5.17}$$

where n is equal to the number of additional 90° route bends due to the cumulative route curvature, as described below. Considering the assumed value of υ_b of 0.3, combining Eqs. (5.15) to (5.17) yields:

$$T_D{}^1 \approx [L_{bore} \bullet w_b \bullet (1/3)] \bullet (1.6)^n \tag{5.18}$$

It is interesting to note that Eq. (5.18) does not include the depth of the route, H, which is explicitly included in the original Eq. (5.12). The mathematical simplifications have essentially eliminated this dependency as a relatively minor effect, as appropriate for the simplification purposes.

Additional Path Curvature

The value of n in Eqs. (5.17) and (5.18) may be expressed as

$$n = n_1 + n_2 \qquad (5.19)$$

where n_1 = effective number of deliberate/planned 90° route bends
n_2 = cumulative curvature due to the unplanned undulations.

For example, if a deliberate horizontal (planar) bend of 45° to the right, in order to avoid an obstacle or follow a utility right-of-way, is followed by another 45° horizontal bend to the left, each 45° bend is equal to half of a 90° bend, corresponding to a total of ½ + ½ = 1 full 90° bend (i.e., $n_1 = 1$).

It is considerably more difficult to predict or determine the value of the cumulative unplanned curvature, n_2, since this will obviously vary among installations due to soil conditions, and expertise of the crew. However, the following rule may be used to provide a reasonable estimate for a Mini-HDD operation:

$$n_2 \approx L_{\text{bore}} \text{ (ft)}/500 \text{ ft} \qquad (5.20)$$

That is, there may be assumed to be effectively one 90° bend, due to path corrections, for each 500 ft of path length. This rule is based upon limited experiences, including analyses of sample as-built data provided in Mini-HDD equipment user manuals. The above-suggested value is consistent with the general magnitude of the corresponding curvature of the actual installed paths. It is noted that this value is not necessarily intended to be a conservative estimate, and that significant variability may be anticipated.

The magnitude of unplanned path curvature provided by Eq. (5.20) is intended to be applicable to a Mini-HDD operation, which typically uses steel drill rods of approximately 2-in. diameter. Larger diameter drill rods are stiffer and, therefore, result in more gradual path deviations and corrections, resulting in a reduced level of path undulations. Thus, when applying the above procedures to a Midi-HDD operation, a reduced value of n_2 should be used. In particular, since the rod stiffness is directly proportional to rod diameter, the following general value is implied

$$n_2 \approx [L_{\text{bore}} \text{ (ft)}/500 \text{ ft}] \cdot [2\text{-in.}/d \text{ (in.)}] \qquad (5.21)$$

where d refers to the diameter (in.) of the steel drill rod. For example, it is possible to say that a 4-in.-diameter drill rod would correspond to one 90° bend every 1000 ft.

The above linear dependence of (unplanned) curvature on rod diameter is consistent with maintaining an equivalent stress level in the steel rod, and corresponds to approximately one-third that typically allowed by bending specifications provided by drill rod manufacturers. Although, in principle, this same rule may be extrapolated to Maxi-HDD, using corresponding large diameter drill rods, it is considered excessively conservative for such well-planned, well-controlled installations.

Buoyant Weight   In order to apply Eq. (5.18), it is necessary to determine the buoyant weight, w_b, of the portion of the polyethylene pipe submerged in the drilling fluid in route segments L_2, L_3, and L_4, illustrated in Fig. 5.20. ASTM F1962 provides general formulae for calculating the effective buoyant weight of the pipe under various conditions, including empty, filled with water, and filled with drilling fluid. For Mini-HDD case of interest, for which the pipe is empty, and, as suggested in ASTM F1962, the specific gravity of the drilling fluid (mud), γ_b, is conservatively assumed to be equal to 1.5, the buoyant weight may be conveniently determined by

$$w_b \text{ (lb/ft)} = 0.5 \cdot D^2 - w_a \text{ (lb/ft)} \tag{5.22}$$

where D is the outer diameter (in.) of the product pipe. The value of w_a may be obtained from the manufacturer, specifications for each specific product pipe (diameter and DR rating).

Pipe Collapse

The critical pressure, P_{cr}, as given in Eq. (5.14), may be expressed in terms of an equivalent head (ft) of water, for idealized conditions in which the ovality reduction factor, f_o, and tension reduction factor, f_R, are assumed equal to 1.0. Since the (effective) material stiffness, E, and Poisson's ratio, μ, are dependent upon the load duration, the critical pressure is also dependent upon duration. Table 5.1 is based upon Eq. (5.14), and industry provided pipe characteristics (PPI, 2008), and is applicable to any HDPE pipe diameter.

Since the drilling fluid is of significantly greater density than water, the indicated pressure head (ft) values of Table 5.1 must be reduced by a factor of 1 divided by γ_b. The values must be further adjusted (reduced) for possible initial elevated temperature (PPI, 2008) as well as aforementioned ovality and tensile load considerations.

There are two phases to be considered with regard to possible collapse of the pipe. During the installation phase, 1- to 10-hour strength would be appropriate, in combination with anticipated values of f_o and f_R during this period. For the post-installation phase, a 1000-hour collapse strength is employed as the maximum period

	Pipe Diameter to Thickness Ratio (DR)						
Duration	7.3	9	11	13.5	15.5	17	21
Short term	2896	1414	724	371	238	177	91
1 hour	1714	837	429	219	141	105	54
10 hours	1436	702	359	184	118	88	45
100 hours	1205	588	301	154	99	74	38
1000 hours	1019	498	255	131	84	62	32
1 year	880	430	220	113	72	54	28
50 years	649	317	162	83	53	40	20

TABLE 5.1   Ideal Critical Pressure (Water Head, ft) for Unconstrained HDPE Pipe
(73° F)

during which the drilling fluid applies hydrostatic pressure on the pipe, subsequent to which it is conveniently assumed to thicken and actually provide support against possible earth-imposed loads, or by which time the soil has locally redistributed is relaxed around the pipe to provide some lateral support against buckling. For this post-installation phase, the tension reduction factor, f_R, is equal to 1.0.

Based upon the ovality reduction factor dependency trend provided in ASTM F1962, and consistent with the present simplified approach, it is reasonable to assume a maximum overall value of f_o of approximately 0.5 to account for ring deformation due to initial ovality plus the value induced by installation or post-installation loadings. Such deflections may be induced by bending, aggravated by tension-induced wall bearing pressure, or possible postinstallation soil loads.

The greater collapse strength at 1 to 10 hours relative to that at 1000 hours, would tend to be offset somewhat by the tension reduction factor, f_R, during installation as well as the degradation at possible elevated temperature as the polyethylene pipe rests on the surface prior to installation. However, these latter degrading effects would not be experienced simultaneously, or with the pipe being placed at its maximum depth and distance. Therefore, the 1000-hour (postinstallation) collapse characteristics, as adjusted, are conveniently employed to evaluate the vulnerability to collapse. Based upon the above discussion, Eq. 5.23 is derived

$$H \text{ (ft)} = 1000\text{-hour water head} \cdot f_o \cdot f_R / \gamma_b$$

$$= 1000\text{-hour water head} \cdot (0.5) \cdot (1.0)/(1.5)$$

or,

$$H \text{ (ft)} = 1000\text{-hour water head}/3.0 \qquad (5.23)$$

where H is the allowable head of drilling fluid (ft) (i.e., maximum pipe depth).

5.5.3   Applications

Pull Force

Equations (5.18) through (5.21) provide means of predicting the peak pull force, T_D^1, on the pipe during a Mini- (or Midi-) HDD installations. The predicted load should then be compared to the safe pull strength for the pipe, which is provided in Table 5.2 for HDPE for a variety of pipe sizes. The safe pull strength (lb) is based upon the safe pull tensile stress (SPS), as applied to the pipe cross-section (PPI, 2008). The SPS accounts for the effective load duration, assumed to be one hour for the Mini-HDD applications, and a significant reduction (less than half) relative to the nominal tensile test strength of HDPE (3200 lb/in.²) to limit nonrecoverable viscoelastic deformation (Petroff, 2006).

ASTM F1962 requires that the predicted peak tensile load to be no greater than the corresponding safe pull strength, without requiring additional margin or the employment of any explicit safety factor. This is considered reasonable for a typically well-planned, well-controlled Maxi-HDD operation since there is a degree of conservatism incorporated into the employed material properties, and in the other parametric values. However, for a typical Mini-HDD installation, there may be a wide variability in the as-built route characteristics, such as the degree of actual path curvature and undulations, as discussed in the "Pull Force" section in Sec. 5.5.2, and other departures from the idealizations with various approximations and assumptions incorporated into the simplifying model described herein. It is therefore required that:

$$T_D^1 \text{ [Eqs. (5.18) to (5.22)]} < \text{safe pull strength (Table 5.2)} \qquad (5.24)$$

by a reasonable margin (e.g., depending upon the application), it may be desired that the predicted tension T_D^1 be no more than approximately half the safe pull strength indicated in Table 5.2.

Nominal Size (in.)	Pipe Diameter to Thickness Ratio (DR)						
	7.3	9	11	13.5	15.5	17	21
2	2998	2505	2096	1739	1530	1404	1085
3	6511	5439	4551	3777	3324	3049	2356
4	10,762	8991	7524	6244	5494	5040	3895
6	23,327	19,488	16,307	13,533	11,909	10,924	8442
8	38,399	32,080	26,844	22,278	19,603	17,982	13,897
12	86,398	72,180	60,398	50,125	44,108	40,461	31,268

TABLE 5.2   Safe Pull Strength (lb), HDPE Pipe, 12 Hours

Pipe Collapse

ASTM F1962 specifies that the effective applied pressure to be less than the collapse load as given by Eq. (5.14), but with a safety factor, such as 2-to-1. As applied to Eq. (5.23) for Mini-HDD applications, the requirement, therefore, corresponds to:

$$H \text{ (ft)} \leq 1000 \text{ hour water head (Table 5.1)}/6.0 \qquad (5.25)$$

Such additional margin is intended to account for loads or degrading effects not previously directly considered, such as a possibly extended period for the drilling fluid/slurry to thicken and provide the anticipated lateral support.

This procedure is considered reasonably conservative for Mini-HDD installations for drilling under roads or most other obstacles. For river or creek crossings, in which the soil loads may be influenced by the overlying water, additional considerations may be warranted, such as discussed by Petroff (2006). For such cases, it may be desired to use a greater safety factor or a longer term collapse strength.

5.5.4   Design Example

The appropriate wall thickness of an HDPE pipe, for a given diameter, may be conveniently determined by the application of Eqs. (5.24) and (5.25). As an example, consider the feasibility of installing a 4-in. HDPE pipe of DR 11 rating, for a relatively long 600 ft Mini-HDD route, including one deliberate 90° planar bend, and placed at a relatively large depth of 30 ft.

The following physical properties apply to the HDPE pipe and specified route:

$$L_{bore} = 600 \text{ ft}$$
$$D = 4.50 \text{ in.}$$
$$w_a = 2.3 \text{ lb/ft}$$
$$H = 30 \text{ ft}$$

Based upon these values, the following values may be directly calculated:

$$n_1 = 1.0 \text{ (one deliberate 90° bend)}$$
$$n_2 = L_{bore} \text{ (ft)}/500 \text{ ft}$$
$$= 600 \text{ ft}/500 \text{ ft}$$
$$= 1.2 \text{ (additional equivalent 90° bend)}$$
$$n = n_1 + n_2$$
$$= 1.0 + 1.2$$
$$= 2.2$$
$$w_b = 0.5 \cdot D^2 - w_a \text{ (lb/ft)}$$
$$= 0.5 \cdot (4.50)^2 - 2.3 \text{ (lb/ft)}$$
$$= 7.8 \text{ lb/ft}$$

Thus, Eq. (5.18) predicts a peak pull load of

$$T_D{}^1 = [L_{\text{bore}} \bullet w_b \bullet (1/3)] \bullet (1.6)^n$$

$$= [600 \text{ ft} \bullet 7.8 \text{ lb/ft} \bullet (1/3)] \bullet (1.6)^{2.2}$$

$$= 4387 \text{ lb}$$

Equation (5.24) then requires that this predicted installation load, 4387 lb, be significantly less than the relevant safe pull strength (nominal 4-in. pipe, DR 11) indicated in Table 5.2 for HDPE pipe. The corresponding safe pull strength of 7524 lb, allows a safety factor of 1.72, representing a reasonable margin. Although the example considers a 4-in. pipe, for a given DR value, the predicted pull load, $T_D{}^1$, and the safe pull strength are both proportional to the square of the outer diameter. The conclusions are, therefore, independent of the pipe diameter. It is noted that the use of the DR 11 pipe in a longer, nominally straight route of 800 ft—beyond the generally accepted limit (600 ft) for Mini-HDD applications—would also be predicted to be have an equivalent margin of safety.

Regarding the potential vulnerability to collapse, either during or after installation, Eq. (5.25) requires that the peak installation depth, or 30 ft, be no greater than one-sixth the relevant head of water (1000 hours, DR 11) indicated in Table 5.1 for HDPE pipe. This corresponds to a safe depth of 255 ft divided by 6.0, or 42.5 ft, independent of pipe diameter. Thus, the relatively large 30 ft proposed installation depth is within the capability of the DR 11 wall thickness.

This relatively difficult (long, deep) installation(s) demonstrates that a DR 11 HDPE pipe represents a reliable selection for the large majority of Mini-HDD applications, and is, in fact, consistent with field experience. Thinner-walled pipe (higher DR rating) may be successful in many cases, as may be verified by specific calculations for the route of interest. It is also emphasized that the present methodology for pipe DR selection does not prove that a thinner-walled pipe, such as DR 17 commonly used, would not be successful in practice in individual installations, but as in most design procedures, it should be noted, is intended to serve as a caution that this design may be marginal (nonconservative).

5.6 Summary

This chapter has presented general background as well as basic design and project management considerations for HDD. Both Maxi-HDD and Mini-HDD operations were discussed. Although typical Mini-HDD installations are individually less critical than more complex, extensive Maxi-HDD projects, it is nonetheless important to follow proper planning and installation practices to help assure a successful projects. Improper procedures may result in damaged or compromised facilities, including those being installed as well as existing utilities in the vicinity.

Various manuals or guides, as well as software tools, are available in the industry, to help plan the bore and determine loads applied to pipelines during a Maxi-HDD installation, including during the subsequent operational stage. In particular, ASTM F1962, provides a planning and design methodology appropriate for polyethylene pipe. This chapter provided analogous information for Mini-HDD installations, including a simplified method for estimating corresponding loads and evaluating the potential for polyethylene pipe to withstand the installation and postinstallation forces. Additional details for such installations are provided in TR-46, *Guidelines for Use of Mini-Horizontal Directional Drilling for Placement of High-Density Polyethylene Pipe,* published by Plastics Pipe Institute (PPI, 2009).

Although the Mini-HDD formulae and methodology discussed in this book are only applicable to polyethylene pipe, with its characteristically low stiffness, same concept can be used for pipes with greater stiffness, such as PVC, and possibly steel or iron. It is emphasized that it is not the intention to apply the present simplified Mini-HDD methodology to a large scale, well-engineered Maxi-HDD operation. For such applications, the detailed design process, and presumably greater accuracy, associated with more conventional engineering procedures and trained construction crew, including application of ASTM F1962 is warranted. The more detailed procedures may more properly account for actual field parameters, such as bore path entry and exit angles, depth, frictional characteristics, and buoyancy effects (including possible anti-buoyancy techniques), as well as more detailed consideration of collapse potential.

CHAPTER 6

Project Considerations for Pipe Replacement Methods

6.1 Introduction

Existing sewer, water, and natural gas pipes can be replaced by three basic methods of pipe bursting—pneumatic, hydraulic, and static pull. In addition, there are proprietary trenchless pipe replacement systems that incorporate significant modifications to the basic pipe-bursting technique, including pipe reaming, the impactor method, and pipe extraction (also called pipe insertion). Pipe reaming is the most common method of *pipe removal*, where the broken pieces of the existing pipe are actually taken out of the ground with the use of slurry drilling fluids (see Sec. 6.3.1). The basic difference among these systems is in the source of energy, the method of breaking the existing pipe and some consequent differences in construction operations that are briefly described in the following sections. The selection of a specific replacement method depends on soil conditions, groundwater conditions, degree of upsizing required, type of new pipe, original construction of the existing pipeline, depth of the pipeline, availability of experienced contractors, and the like.

6.2 Pipe Bursting

Pipe bursting was first developed in the United Kingdom in the late 1970s by D. J. Ryan & Sons in conjunction with British Gas, for the replacement of small-diameter, 3- and 4-in. cast-iron gas mains (Howell, 1995). The process involved a pneumatically driven,

229

cone-shaped bursting head operated by a reciprocating impact process. This method was patented in the United Kingdom in 1981 and in the United States in 1986; but expired in April 2005. When it was first introduced, pipe bursting was used only in replacing cast-iron gas distribution pipes and later was employed to replace water and sewer pipelines. By 1985, the process was further developed to install up to 16-in.-outer-diameter (OD) medium-density polyethylene (MDPE) outside pipe.

Replacement of sewers in the United Kingdom using sectional pipes as opposed to continuously welded polyethylene pipe was described in a paper by Boot et al. (1987). Up to 2006, approximately 9000 mi of high-density polyethylene (HDPE) pipe has been installed by bursting (Najafi, 2006). Currently, pipe bursting is used to *replace* waterlines, gas lines, and sewer lines throughout the world.

Pipe bursting is too widely used to replace deteriorated pipes with new pipes of the same or larger diameter, utilizing the same existing pipe space (so more appropriate for urban conditions, where usually underground space is crowded with utilities). This method is an economic pipe replacement alternative that reduces disturbance to business and residents when compared to the open-cut technique. Pipe bursting is especially cost-effective if the existing pipe is out of capacity, deep, and/or below the groundwater table (GWT). Upsizing refers to replacement of the existing pipe with a pipe one standard size larger (e.g., replacing 8-in. pipe with a 10-in. size is one upsize). Similarly, two-size upsizing is replacement of the existing pipe with a pipe two standard sizes larger, such as, replacing 8-in. pipe with a 12-in. size.

Pipe bursting typically involves insertion of a cone-shaped bursting head into an existing pipe. The base of the cone is larger than the inside diameter of the existing pipe and slightly larger than the outside diameter of the new pipe to reduce friction and to provide space for maneuvering the pipe. The back end of the bursting head is connected to the new—for example, polyethylene (PE)—pipe and the front end is attached to a cable or pulling rod. The new pipe and bursting head are launched from the insertion shaft and the cable or pulling rod is pulled from the pulling shaft/pit or manhole (dependent on the method and upsize required to retrieve the bursting head), as shown in Fig. 6.1. The bursting head receives energy to break the existing pipe from several possible sources: a pulling cable or rod (static method), a hydraulic source (hydraulic method), or an air compressor (pneumatic method). The energy breaks the existing pipe into pieces and expands the diameter of the cavity. As the bursting head is pulled through the existing pipe debris, it creates a bigger cavity through which the new pipe is simultaneously pulled from the insertion shaft. There are many variations to this method that are presented in the following sections.

(a)   (b)

FIGURE 6.1   The pipe bursting operation. (a) Insertion Shaft, and (b) Pulling Shaft.

6.2.1   Pneumatic Bursting Systems

Currently this method, the most common pipe-bursting method is the pneumatic system (see Fig. 6.2). The bursting tool is a soil displacement hammer driven by compressed air and operated at a rate of 180 to 580 blows per minute. It is similar to a pile-driving operation, but in a horizontal orientation. The percussive action of the hammering cone-shaped head is also similar to hammering a nail into the wall; each hammer pushes the nail a short distance. With each stroke, the bursting tool cracks and breaks the existing pipe. The expander on the head, combined with the percussive action of the bursting tool, push the fragments into the surrounding soil, providing space to pull in the new pipe. The expander can be front end (attached to the front of the hammer) for pipes smaller than 12 in. or back-end (attached to the back of the hammer) for pipes larger than 12 in. The front-end expander allows withdrawing the hammer through the new pipe after removing the expander from the existing manhole or the pulling shaft. The tension applied to the cable keeps the bursting head aligned with the existing pipe and pressed against the existing pipe wall, and pulls the new pipe behind the head. An air pressure supply hose is inserted through the new pipe and connected to the bursting tool. The bursting starts once (1) the head is attached to the new pipe, (2) the winch cable is inserted through the existing pipe and attached to the head, and (3) the air compressor and the winch are set at constant pressure and tension values. The process continues with operator supervision until the head reaches the pulling shaft at which point it is separated from the new pipe and retrieved.

6.2.2   Hydraulic Bursting Systems

In the hydraulic bursting system, the pipe bursting process advances from the insertion pit to the reception (pulling) pit in sequences, which

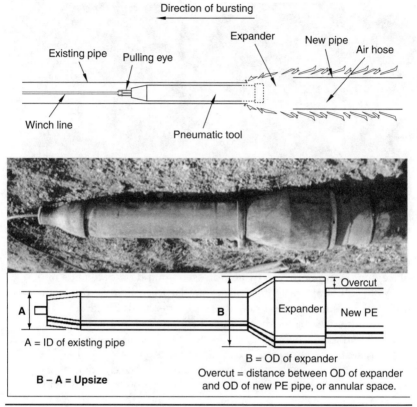

Direction of bursting

Existing pipe   Pulling eye   Expander   New pipe   Air hose

Winch line   Pneumatic tool

A = ID of existing pipe

B – A = Upsize

B = OD of expander

Overcut = distance between OD of expander and OD of new PE pipe, or annular space.

A   **B**   Expander   New PE   Overcut

FIGURE 6.2   The bursting head of the upsize concept. (*PPI, 2008.*)

are repeated until the full length of the existing pipe is replaced. In each sequence, one segment of the pipe (which matches the length of the bursting head) is burst in two steps: first the bursting head is pulled into the existing pipe for the length of the segment, and then the head is expanded laterally to break that pipe. The bursting head is pulled forward with a winch cable, which is inserted through the existing pipe from the reception pit, and attached to the front of the bursting head. The rear of the bursting head is connected to the new pipe and also hydraulic supply lines that are inserted through the new pipe. The bursting head consists of four or more interlocking segments, which are hinged at the ends and at the middle. An axially mounted hydraulic piston drives the lateral expansion and contraction of the bursting head (Najafi, 2005). This method of pipe bursting is not common.

6.2.3   Static Bursting Systems

The second common method of pipe bursting is the static pull system. In this method, a relatively large tensile force is applied to the cone-shaped expansion head through a pulling rod assembly or cable

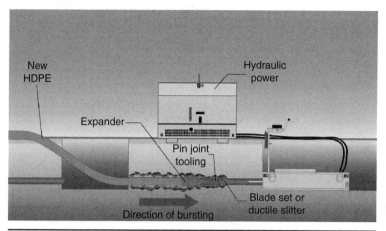

Figure 6.3   The static pull bursting. (*Source: HammerHead®.*)

inserted through the existing pipe. The cone transfers the horizontal pulling force into a radial force breaking the existing pipe and expanding the cavity, providing space for the pipe, as shown in Fig. 6.3. The steel rods, each about 4 ft in length, are inserted into the existing pipe from the pulling shaft. The rods are usually screwed together. When the rod string reaches the insertion shaft, the bursting head is connected to the rods and the new pipe is connected to the rear of the head. A hydraulic unit in the pulling shaft pulls the rods, one rod at a time, and the rod sections are unscrewed and removed. The process continues until the bursting head reaches the pulling shaft, where it is separated from the HDPE pipe. If cable is used instead of rod, the pulling process continues with minimum interruption, but the tensile force of a cable compared to a rod section is limited.

Pipe splitting is a form of static pipe-bursting method for breaking and replacing cast-iron or ductile iron pipes by longitudinal slitting. At the same time, a new pipe of the same or larger diameter may be drawn behind the splitting tool. This method has a splitting wheel or cutting knives that slit the pipe longitudinally at two more lines along the side of the pipe.

6.3   Pipe Removal Systems

Pipe removal (also known as pipe eating) is a replacement technique based on horizontal directional drilling (HDD) technology (more common), the microtunneling boring machine (MTBM), or horizontal auger boring (HAB) technology. This method excavates the existing pipe in very fine particles and *removes* them rather than displaces them into the surrounding ground, as it would be for

HDD, MTBM, or HAB methods. Pipe removal is further divided into several methods as follows:

- *Pipe reaming and impactor methods:* These techniques use a specially designed variation of the HDD process to excavate the existing pipe in fragments and remove them rather than displace them, and pull a new pipe to replace the existing pipe.

- *Pipe eating:* This technique consists of a microtunneling machine, which has been specially adapted to allow the existing pipeline to be broken up, and removed. The MTBM has a larger diameter than the existing pipeline. In the United States, due to the high cost of microtunneling for this application, this equipment rarely has not been used for this application. Horizontal auger boring equipment has successfully been experimental in several projects.

- *Pipe ejection or pipe extraction:* These techniques remove the existing pipe as whole from the ground, by pushing and/or pulling it toward a reception pit where it is broken and taken out.

6.3.1   Pipe Reaming

Pipe reaming is a pipe replacement technique that uses an HDD machine with minor modifications. After pushing the drill rods through the existing pipeline and connecting the rods to a special reamer (see Fig. 6.4), the new pipe string is attached to the reamer via a swivel and towing head. As the drill rig rotates the drill rod string and simultaneously pulls it back, the existing pipe is pulverized and replaced by the new pipe. Removal of the existing pipe is accomplished by mixing the ground material with the drilling fluid and transferring it to an exit point for collection via a vacuum truck. Directional drilling contractors can add inexpensively modified reamers of various types for different pipe materials and ground conditions. Pipe reaming is limited to nonmetallic pipeline replacement. The patented InneReam System can accomodate the surrounding environmental conditions (groundwater, sand, rock, concrete encasement, and the like) with certain modifications. Contact Nowak Pipe Reaming Inc., for more information.

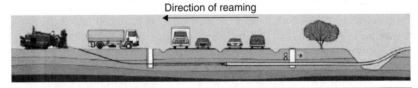

Direction of reaming

Figure 6.4   Pipe reaming method. (*Source: Nowak Pipe Reaming Inc.*)

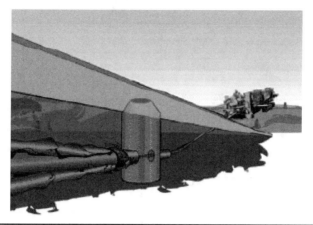

FIGURE 6.5 The impactor process combines HDD with pipe bursting. (*Source: HammerHead®.*)

6.3.2 Impactor Method

The patented impactor method is a system that combines the HDD method with pipe bursting, as shown in Fig. 6.5. The bursting head (impactor) receives air through the HDD drill rod. The drill rod is connected to an air supply and bores to an entry manhole. Then the drill rod is pushed through existing pipe to the next manhole or exit pit. Then, the impactor device, after it is attached to the drill stem and to the new pipe, is activated and pulled into the existing pipe. While pulling back, the impactor system bursts the existing pipe by combined action—of pulling using the HDD rig and hammering with the impactor device—breaking the existing pipe and replacing it with the new pipe. The impactor system overcomes blocked existing pipes. Contact Hammerhead Company for more details.

6.3.3 Pipe Ejection, Extraction, or Insertion

This method pushes or jacks a new pipe into the existing deteriorated pipe. This system utilizes the columnar strength of segmented "bell-less" jacking pipe to advance the "lead train" through the existing pipe. The lead train consists of five sections: the lead a heavy steel guide pipe which maintains the alignment within the center of the existing pipe; the cracker which fractures the existing pipe; the cone expander which radially expands the fractured line into the surrounding soil; the front jack, a hydraulic cylinder, which provides axial thrust to the penetration/compaction pieces; and the pipe adapter, which provides mating surfaces, linking the new pipe to the front jack.

The last section, the pipe adapter, is fitted with a lubricant injection port where lubricant (polymer or bentonite) can be injected into the annular space surrounding the new replacement pipe; the

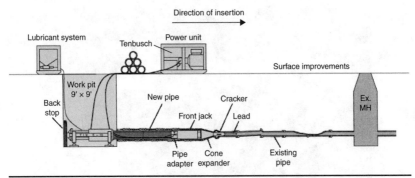

FIGURE 6.6   The Tenbusch Insertion Method™.

introduction of a lubricant allows for the efficient replacement of existing pipelines even in soft sticky clays or wet sands.

Dual flexible hose sections, which transport lubricant and hydraulic fluid to the front train, are fed through each new pipe section. Each new hose section is connected to previous sections and to the operator's control panel with quick-disconnect couplings. Using the new pipe as a support column, the front jack advances the lead train into the existing pipe independent of the advance of the new pipe column. The new pipe is jacked behind the lead train piece by piece by the jacking frame (in the jacking pit). The primary jacking frame applies the required thrust to advance the new pipe column (as the front jack is retracted). Instrumentation and controls at the operator's control panel (at the jacking frame) allow the operator to "feel" his or her way through the existing pipe as the new pipe column and front train are "inch-wormed" into the existing pipe. On completion of the pipe replacement, the lead train is disassembled inside a typical 4-ft diameter-receiving manhole and the new pipe is jacked into its final position. Figure 6.6 illustrates pipe insertion using the Tenbusch Insertion Method™.

6.3.4   Pipe Eating

The pipe removal process can potentially be executed by horizontal auger boring, a process that excavates the existing pipe and surrounding soil by using a rotating cutting head attached to an auger that continuously removes the excavated soil. Other potential methods of pipe eating are using a modified version of the MTBM or using pipe ramming equipment.

6.4   Existing Pipe Materials

In most bursting applications, the existing pipe is composed of a rigid material such as vitrified clay pipe (VCP), cast iron, plain concrete, asbestos, or some plastics. Reinforced concrete pipe (RCP) can be replaced when it is not heavily reinforced or if it is substantially deteriorated. The diameter of the existing pipe typically ranges from 2 to 30 in., although bursting of larger diameter existing pipes is becoming more common. A segment of 300 to 400 ft is a typical length for bursting, although significantly longer runs have been completed using more powerful bursting systems. However, point repairs on the existing pipe, particularly using ductile materials; can create problems. For a comprehensive discussion of existing pipe materials, refer to *Pipe Bursting Manual of Practice (MOP), No. 112*, published by the American Society of Civil Engineers (ASCE, 2007).

6.5   Replacement (New) Pipe Material

High- and medium-density polyethylene (HDPE or MDPE), collectively called "PE", have been the most widely used replacement pipes for pipe bursting applications. Polyethylene pipe may utilize high-performance PE4710 material, with somewhat greater strength capability than conventional pipe. The main advantages of PE pipes are its continuity, flexibility, and versatility. The continuity, which is obtained by butt fusing together long segments in the field, reduces the possibility of stopping the installation process. For small diameters (4 in., or less), the pipe may also be supplied in continuous lengths on a reel. The PE pipe flexibility allows bending the pipe for convenient insertion in the field. In addition, it is a versatile material that meets other requirements for gas, water, and wastewater applications. The relatively smooth interior surface reduces the friction between the fluid flow and the pipe wall, which allows increased flow capacity. The smooth exterior surface also reduces the friction between the soil and the pipe, facilitating the pulling operation (see Sec. 6.10.2). The relatively higher thermal expansion coefficients of HDPE pipes, requires proper installation and restraint.

The internal surface of PE pipe is smoother than that of the concrete or clay pipe. For gravity applications, the following Cheesy-Manning equation [Eq. (6.1)] demonstrates that the flow capacity of the PE pipe is 44 percent greater than those of the concrete or clay pipes, assuming the internal diameter for the existing clay or concrete pipe equals that of the replacement PE pipe.

$$Q = \frac{1.49}{n} A (r_H)^{2/3} \sqrt{S} \tag{6.1}$$

where Q = flow quantity (ft^3/sec)
n = Manning roughness coefficient (dimensionless)
A = internal cross-sectional area of the pipe (ft^2)
r_H = hydraulic radius (ft)
S = slope of the energy line, which is parallel to the water surface and pipe invert if the flow is uniform (dimensionless)

The *n* value for clay or concrete pipes is assumed to be between 0.012 and 0.015, in comparison to approximately 0.009 for PE (Lindeburg, 1992).

In addition to PE, other pipe materials can be polyvinyl chloride (PVC), ductile iron, VCP, or RCP. With exception of PVC and ductile iron, VCP and RCP cannot be assembled into a single pipe string, prior to the bursting operation, which allows a continuous pulling operation. However, they can be jacked into position one by one (thereby requiring less staging distance) behind the bursting head or maintained in compression by towing them via a cap connected to the cable or rod that passes through these pipes. Therefore, the static pull system is the only bursting system that can be used with RCP and VCP pipes. The joints of these pipes must be designed for jacking applications.

For a comprehensive discussion of new pipe materials, refer to *Pipe Bursting Manual of Practice (MOP), No. 112*, published by the American Society of Civil Engineers (ASCE, 2007).

6.6   When Is Pipe Bursting a Preferred Solution?

For repair and replacement, conventional techniques require open-cut excavation to expose and replace the pipe. Alternatively, the pipeline can be trenchlessly renewed by inserting a new lining, or replaced by pipe bursting. Chapter 2 described several pipe-lining methods such as cured in place pipe, close-fit pipe, and sliplining. The main advantage of the lining methods over pipe bursting is the need for little or no excavation for access to the pipeline. In contrast, pipe bursting has the advantage of increasing the pipe capacity by more than 100 percent, as described below.

Pipe bursting is most cost advantageous compared to the lining techniques when:

- There are few lateral connections to be reconnected within a replacement section.
- The existing pipe is structurally deteriorated.
- Additional capacity is needed.

For pressure applications, a 41 percent increase in the inside pipe diameter doubles the cross-sectional area of the pipe and consequently doubles the flow capacity of the pipe. For gravity applications, after some algebraic manipulation to the Chezy-Manning equation [Eq. (6.1)], it is shown that a 15 and 32 percent increase in the inside diameter of the pipe combined, with the smoother pipe surface can produce a 100 and 200 percent increase in the flow capacity, respectively.

Pipe bursting has substantial advantages over open-cut replacements; it is much faster, more efficient, and often less expensive than open-cut, especially in sewer applications due to the large depths at which gravity sewer pipes are typically installed. The large sewer depth requires extra excavation, shoring, and dewatering which substantially increases the cost of open-cut replacement. The increased depth has only a minimal effect on the cost per foot for pipe bursting, as shown in Fig. 6.7. Specific studies carried out in the United States have shown that pipe bursting cost savings are as high as 44 percent, with an average savings of 25 percent, compared to open-cut (Fraser et al., 1992). This cost saving could be much greater if the soil conditions are difficult such as, hard rock. Furthermore, open-cut excavation can cause significant damage to nearby buildings and structures (Atalah, 2004).

In addition to the direct cost advantage of pipe bursting over open-cut, pipe bursting, as a trenchless technique, has several in social cost savings, including less traffic disturbance, road or lane closing, improved productivity, less business interruption, and less

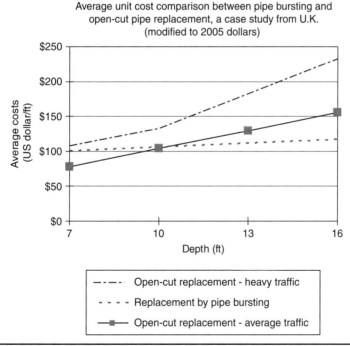

FIGURE 6.7 Cost comparison between pipe bursting and open-cut replacements. (*ASCE, 2007.*)

environmental intrusion. Pipe bursting creates minimal interference with other utilities and less safety hazards (for both operators and the public) due to reduced open excavation.

6.7   Pipe Bursting Project Classification

The *ASCE Pipe Bursting Manual of Practice* (ASCE, 2007) classifies bursting projects into three categories, in terms of difficulty: Class A—routine, Class B—moderately, difficult to challenging, and Class C—challenging to extremely challenging. The characteristics defining the appropriate classification are provided in Table 6.1, based on these criteria. Note that the degree of difficulty increases as more than one of the above criteria applies (ASCE, 2007).

6.7.1   Pipe Bursting Applicability and Limitations

Pipe bursting is used to replace waterlines, gas lines, and sewer mains, as well as sewer lateral connections. Typical replacement length is between 300 ft and 500 ft; however, in favorable conditions, longer drives have been completed successfully. The size of pipes being burst typically ranges from 2 to 30 in., although larger size

Criteria	A Routine	B Moderately	C Challenging
Depth	Less than 12 ft	12 ft–18 ft	More than 18 ft
Existing pipe	4 in.–12 in.	12 in.–20 in.	20 in.–36 in.
New pipe diameter	Same size or one diameter upsize	Two diameter upsize	Three or more diameter upsize
Burst length	Less than 350 ft	350 ft–450 ft	More than 450 ft
Trench width	Relatively wide trench compared to upsized diameter	Trench width less than 4 in. wider than upsize diameter	Incompressible soils (very dense sand, hard clay or rock) outside trench
Soil	Compressible soils outside trench (soft clay, loose sand)	Moderately compressible soils outside trench (medium dense to dense sand, medium to stiff clay)	Constricted trench geometry (width less than or equal to upsize diameter)

TABLE 6.1   Pipe Bursting Classification (ASCE, 2007)

pipes can be burst. Pipe bursting is commonly performed size-for-size and one-size upsize above the diameter of the existing pipe. Larger upsize (up to three pipe sizes) have been successful, but the larger the pipe upsizing, the more energy is required and greater ground movement will be experienced. It is important to pay close attention to the project surroundings, depth of installation, and soil conditions when replacing an existing pipe, especially in unfavorable conditions such as expansive soils, pipe repairs made with ductile material, collapsed pipe, concrete encasement, sleeves, and adjacent utility lines.

Pipe bursting has the following specific limitations:

- Insertion and pulling shafts are required, especially for larger bursts.

- Excavation for service lateral connections is required.

- Expansive soils may cause difficulties.

- A collapsed (existing) pipe segment may require excavation at that point to allow the insertion of pulling cable or rod and to repair pipe sag.

- Point repairs with ductile material can interfere with the bursting and replacement process.

- If the existing sewer line is significantly out of line and grade, the new line will also tend to be out of line and grade, although some corrections of localized sags are possible.

6.8 Design Considerations

The design phase starts with collecting information regarding the existing pipe, including current flow volume for bypass pumping, lateral connections, trench width, backfill compaction levels, and manhole locations. This phase also includes locating nearby utilities, investigating soil and trench backfill material, and developing risk assessment plans. The designer completes this phase with the development of detailed drawings and specifications and complete (final) bid documents which include a listing of required submittals. The drawings should provide all relevant information concerning the existing pipeline and environment, such as diameter and material, plan view and profile, nearby utilities and structures (crossing and parallel), repair clamps, concrete encasement, fittings, and the like. Some of this information may be collected by means of a closed circuit television (CCTV) system or equivalent.

6.8.1 Utility Survey

The presence and nature of surrounding utilities will have a significant impact on the success of the pipe-bursting operation, and the design engineer must therefore identify and locate these existing utilities as accurately as possible. However, the *exact* location of these utilities will be identified during the construction phase through visual locating, such as vacuum potholing. The identification of nearby utilities is critical for the following reasons:

- The presence of nearby utilities may eliminate pipe bursting as a viable construction method.

- Existing utilities may affect the location of insertion and pulling/jacking shafts.

- Allows protection of existing utilities from the ground movement of the bursting operation.

- Reduces the risk of injuries and fatalities to the workers and public by avoiding damage to existing utilities.

- Contractors need to know the number of utilities to be exposed, to be properly reflected in their bid.

See Chap. 9 for more information on utility locating.

6.8.2   Investigation of Existing Pipe and Site Conditions

Investigation of the existing pipe condition assists in selecting the suitable renewal technique and provides the exact location of the lateral connections. The nature and condition of the existing pipe may render pipe bursting as an unsuitable solution. The presence of sags in the pipeline may require correction prior to bursting. The existing pipe (diameter, material, and conditions) and the diameter of the new pipe guide the contractor in the selection of an appropriate type of bursting system, size, and accessories, and should, therefore, be reflected in the bid documents. The site conditions and surface features may affect the locations of the insertion and pulling shafts, staging area for fused pipe, traffic control planes, and layout for the required bursting system components.

6.8.3   Insertion and Pulling Shaft Requirements

When planning for shaft locations, the engineer identifies locations where excavation is required to replace manholes, valves, lateral connections, or fittings. If excavation at the manhole location is not necessary or feasible, shaft excavation at other locations may be considered. In selecting the location of these shafts, the engineer has to consider the following:

- Sufficient staging area for the replacement pipe to avoid blocking driveways and intersecting roads.

- Shaft length should be sufficient to allow alignment of the bursting head with existing line and for the new pipe to bend safely from the entry point to the ground surface.
- Space for the construction equipment, including backhoe, loader, crane, and the like.
- Nearby flow bypass discharge location or space to lay bypass lines without blocking driveways and intersecting roads.
- Traffic control around shafts.
- Soil borings close to proposed shafts.
- Discharge locations for dewatering, if necessary.
- Ability to use same shaft to insert or pull pipes more than once.

In general, the engineer recommends locations for the insertion and pulling shafts but leaves the final determination to the contractor (through a submittal process), subject to minimizing excavation and disturbance to the surrounding environment.

6.8.4   Geotechnical Investigation Requirements

The soil and subsurface investigations include collecting the necessary technical information to properly design the project. It assists the contractor in submitting a proper bid by selecting the appropriate bursting system (type and size), shoring of the pulling and insertion shafts, dewatering system, compacting backfill material, and others, thereby increasing the chance of success during the construction phase of the project. Thus, if the original soil borings (during the existing pipe installation) are available, they should be reviewed and provided as part of the supplemental information available to the bidders. The determination of the trench geometry and backfill material and compaction is important for the designer and contractor.

The soil investigation activities comprise performing soil borings, standard penetration tests (SPT), groundwater level determinations, trench geometry investigation, and native soil and trench backfill material classifications. If the presence of washouts or voids around the existing pipe is suspected, a ground penetrating radar (GPR) survey may assist in determining the locations and magnitudes of these voids. Special attention should be given to the presence of obstacles that may render pipe bursting not feasible, such as the presence of rock, hard cemented dense soils, very soft or loose soils, reinforced concrete encasement, very narrow trench in hard soils or rock, or ductile iron point repairs. If contaminated soil is suspected, the type and extent of contamination should be identified and indicated in the contract documents. The contractor should be instructed to take the necessary measures to handle and dispose of this contaminated soil.

The soil around the pipe (backfill and native soil) must be compressible in order to absorb the diametric expansion. Compressible soils are ideal because the outward ground displacements will be limited to an area surrounding the pipe alignment, as shown in Fig. 6.8. Soils with long "standup time" allow the overcut (expanded borehole) to remain open for most of the bursting operation, thus reducing the friction force between the soil and the pipe thereby reducing pulling forces and axial stress on the new pipe. Usually original backfill material is suitable for bursting followed by (increasing difficulty) compressible clay, loose cobble, beach and running sand, densely compacted clay, and sandstone. Somewhat less favorable ground conditions for pipe bursting also include soils below the watertable and expandable clays. Special soils such as highly expansive soils or collapsible soils may also cause problems.

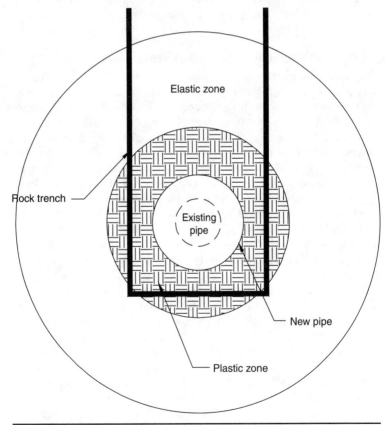

FIGURE 6.8 Cavity expansion and the plastic and elastic zones. (*Source: Handbook of Polyethylene Pipe, 2008.*)

In very soft or loose soils, significant ground movement may take place causing significant sags in the new pipeline and damage to nearby structures. In severe situations, the soil particles migrate to the interior of the existing pipe, converting the bursting operation into a piercing operation.

Although there may be some plausible advantages in performing pipe bursting below the ground watertable—for example, possible reduced friction between the pipe and soil, as well as reduced effective soil pressure on the pipe due to the buoyant force of the soil above—this situation presents significant problems. The groundwater flows toward the insertion shaft, complicating the operation. Thus, in general, dewatering should be performed.

6.8.5 Maximum Allowable Operating Pressure

For pressure applications such as water, gas, and force mains, the maximum allowable pressure should be determined based on the maximum surge pressure that pipe will be subject to and the maximum operating pressure for the pipe. The HDPE pipe should be designed to withstand the maximum allowable operating pressure (MAOP) and surge pressures according to conventional design procedures, such as provided in the Plastics Pipe Institute's *Handbook of Polyethylene Pipe*. DR 17* is typically used for bursting pressure or gravity pipe unless a higher pressure rating is required. In short, relatively shallow bursting runs, where high tensile and vertical forces are not expected, DR 21 may be used (see Sec. 6.10).

6.8.6 Risk Assessment Plan

It is important to pay close attention to the project surroundings (surface and subsurface conditions) to identify unfavorable conditions and possible risks. The risk conditions require extra attention in order to ensure the safety of the construction personnel and public, as well as surrounding facilities and infrastructure. Most underground and pipeline construction projects entail some risks due to unknown subsurface conditions. The risks associated with pipe bursting include damage to nearby facilities, failure to complete the project using the selected technology, and time and/or budget overrun. The risk of damage to nearby utilities, buried structures, and pavement increases under adverse soil conditions, improper construction techniques, design errors, inaccurate location of utilities, and the like. There are also risks associated with flow bypass, dewatering, shoring, and the like, if inappropriate procedures are employed. The additional risks

---

*The dimension ratio (DR) of the pipe is the outer diameter divided by the wall thickness.

that may halt or impede the bursting operation and/or create problems include:

- Settlement at insertion/pulling pits if the density of the backfill exceeds that of native soil.
- Bursting through curves.
- Concrete encasement or steel point repair of existing pipe.
- Excessive bursting lengths.
- Damage to the new (replacement) pipe from sharp edges or fragments of existing pipe being burst/split.
- Damage to laterals from bursting of main line.
- The presence of rock under the existing pipe may create a "bump" in the replacement pipe.
- Collapsed replacement pipe (see Sec. 6.10.4).

Projects within the Class C category (challenging to extremely challenging as indicated in Table 6.1) must be carefully examined in terms of required forces and ground displacements. In addition, due to the enlargement of the borehole and accompanying compaction of surrounding soil, the depth of the existing pipe affects the extent of ground displacement above the pipe. If the pipe is shallow, damage to the pavement may take place. Saw cutting the pavement prior to bursting might be advisable. If the existing pipe is below the groundwater table (GWT), the difficulties increase. Insertion and pulling shafts will be larger and more complex as the depth increases.

If there are unacceptable sags in the existing sewer line, these sags must be corrected before bursting. The sags can be corrected by local excavation, surface grouting, or grouting from within the pipe. Some reduction of sag magnitude may be expected (without corrective measures) from the bursting operation, but the extent to which the problem is corrected depends on the relative stiffness of the soil below the sagging section.

If there is erosion of the soil around the pipe, the bursting head and the following new pipe will tend to deviate toward the void or lower density region. Similarly, if there is a hard soil layer or rock close to the pipe, the bursting head will tend to displace toward the softer soil. In shallow conditions, the ground will deviate mostly upward toward the ground surface. If the conditions change substantially along the length of the burst, this may cause some change in the grade and/or alignment of the pipe. When the grade is critical, these possibilities should be considered.

As mentioned in Chap. 2 trenchless operations in general and pipe-bursting projects specifically can be performed successfully and safely if site and project conditions are known before bursting and appropriate measures are taken to address these conditions. There are well-known solutions to all of the above mentioned risks and problems, and experienced project engineers or construction managers identify these risks and develop risk management plans to address them. These plans include quantification of the likelihood of the identified events and their associated impact or degree of damage. It also includes measures to eliminate, mitigate or transfer these risks. One of the general measures to mitigate the project risks is building and maintaining partnerships among owners, engineers, contractors, equipment manufacturers, and pipe suppliers. The identification and development of a realistic plan to manage and share risks appropriately and setting a contingency plan, is an important part of effectively communicating responsibilities, defining roles, and building a partnering team.

6.8.7 Ground Movements

The pipe-bursting process creates a cavity in the soil around the pipe through which the new pipe is pulled. This cavity creates a compression plastic zone around the new pipe outlined by an elastic zone, as shown in Fig. 6.8. The magnitude of the compression and the dimensions of these zones correlate with the amount of upsizing, the diameter of the existing pipe, and the type of soil (Atalah, 1998). Based on investigations of the ground movements and vibrations associated with bursting small-diameter pipes in soft soils (Atalah, 1998) and with large-diameter pipes in rock conditions (Atalah, 2004),* guidelines have been developed for safe separation distances from existing nearby utilities, structures, and pavement.

There is a strong correlation between the distance from the pipe-bursting tool and the level of vibration. The vibrations due to bursting are seen to rapidly decline to levels that do not cause damage to buildings. For structurally sound residential buildings, minimum separations of 11 ft in hard soil and large-diameter bursting conditions and 8 ft in soft soils and small-diameter conditions are recommended. Corresponding minimum separations of 8 ft and 4 ft are recommended from structurally sound commercial structures. Considering pipes as "buried structures," studies by Atalah (2004) suggest that 7½ ft represents a safe separation distance for pipe-bursting operations for either small- or large-diameter *structurally sound* pipes, in various soils, with reduced spacing possible for small sizes. While below-ground pipelines where pipe bursting required may include pipes installed in the right of way, which are usually far from the residential or commercial buildings, possibility of damage to nearby buildings and adjacent pipelines must be thoroughly investigated during the planning and design phase of the project. It is also recommended to document the existing state of these structures documented (with photos and videos/DVDs) to protect the pipe-bursting project owner from unfounded claims.

---

*The investigation of the large-diameter bursting in rock conditions included upsizing 24-in.-diameter reinforced RCP pipes by 50 percent. The studies of the small-diameter bursting in soft soils included upsizing 8- and 10-in.-diameter VCP by 30 percent.

6.8.8 Drawings and Specifications

The contract documents typically include the contract agreement, general conditions, special conditions, project plans, specifications, geotechnical report, and CCTV records. The plans and specifications for pipe-bursting projects should have all the required information for a typical open-cut water or wastewater pipeline projects plus the information listed in this section. The drawings should provide information about the existing site conditions and the underground utilities. Description of site constraints (i.e., work hours, noise, etc.) and the procedures to review the CCTV data should be listed in the notes section in the drawings. The drawings may also include information to show erosion and sediment control requirements, flow bypassing plans, and service connection and reinstatement details, and should typically include:

- Limits of work; horizontal and vertical control references
- Topography and survey points of existing structures
- Boundaries, easements, and rights-of-ways (ROW)
- Existing utilities, sizes, locations, and pipe materials
- The verification requirements for existing utilities
- Plan and profile of the design alignment
- Existing point repairs, encasement, sleeves, and others
- Construction easement and the allowable work areas around the insertion and pulling pits
- Details for lateral connections and connections to the rest of the network
- Restoration plans
- Existing flow measurements for bypass pumping
- Quality control and assurance measures (Sec. 6.8.10)

The technical specifications supplement the drawings in communicating the project requirements. Information to be included in the technical specifications should include

- General considerations
  - Minimum contractor qualifications
  - Permit matrix and responsibilities
  - Safety requirements with focus on confined space entry, flow bypass, and shoring
  - Scheduling requirements and construction sequence
  - Submittals (Sec. 6.8.9)

- Pipe materials and manhole considerations
  - Standards and tolerances for pipe material, wall thickness and class, testing and certification requirements
  - Construction installation instructions for pipe joining and handling
  - Fittings, appurtenances, and connection-adaptors
  - Acceptable material performance criteria and tests
- Construction considerations
  - Flow bypassing, downtime limits, and service reinstatement requirements
  - Spill and emergency response plans
  - Traffic control requirements
  - Erosion and sediment control requirements
  - Existing conditions documentation (e.g., photographs, videos, interviews)
  - Protection plan for existing structure and utility (ground movement monitoring)
  - Accuracy requirements of the installed pipe
  - Daily construction monitoring reports
  - Field testing and follow-up requirements for pipe joining, pipe leakage, disinfection, backfill, and others
  - Site restoration and spoil material disposal requirements

6.8.9   Submittals

In addition to the submittals needed for a traditional open-cut projects, the submittals for pipe-bursting projects usually include the following: site layout plans, sequence of bursting, shoring design for all the excavations signed by a professional engineer, bypass pumping plan, manufacturers' specifications of the selected bursting system and its components, dewatering plan, new pipe material, lateral connections material and plans, contingency plans, and the like. The site layout plans would show the location of the insertion and pulling shafts, dimensions of shafts, traffic flow, safety and communication plan, storage space to store and lay the new pipe, and the like. Site restoration and clean-up plans should also be included in the submittals.

6.8.10   Quality Assurance/Quality Control Issues

The project specifications should state the quality assurance (QA) and quality control (QC) measures required to ensure that the project is executed according to the contract specifications. In addition to the quality control and quality assurance measures usually specified for a traditional open-cut projects, there are several measures specific to pipe-bursting operations. These measures can take the form of tests, certifications, inspection procedures, and the like. Extensive listing of the relevant required submittals, careful preparation of the submittals, and carefully review and approval of the submittal are significant steps in the QA/QC program. The QA/QC program should state the performance criteria for the product line and the acceptable tolerance from these criteria. For example, the invert of the new pipe should not deviate from the invert of the existing pipe by more than a specified number of inches, the depth of sags in the new pipe should not exceed one inch, and the difference in the vertical and horizontal dimensions of the new pipe diameter (ovaling) should not exceed 2 percent.

The QC program should state how these performance criteria will be measured, tested, and checked. Some of the methods for verifying performance are post-bursting CCTV inspection, pressure testing, and mandrel test. The surface and subsurface displacement-monitoring program should be outlined in the specifications along with the acceptable amount of ground movement. Certifications from the manufacturers of the bursting system, replacement pipe, and other materials may be required to verify that these products meet the contract specifications, based on tests conducted by the manufacturer or a third party. For challenging projects, the presence of the bursting system manufacturer representative at the jobsite may be required. The owner's quality assurance program should ensure that the field and management team of the contractor have the knowledge and the experience needed to complete the project successfully and are able to respond appropriately to unforeseen problems.

It is critical that the contractor ensure that the replacement pipe meets the specifications before, during, and after bursting. Adhering to the quality control and quality assurance plans during the manufacturing and shipping to the site, along with proper unloading of the pipe, reduces risk of pipe damage. Pipe fusion operations should be performed by well-trained (preferably certified) workers under appropriate supervision to reduce the risk of future pipe problems when repair would be difficult and costly. For pressure applications, the new pipe should be inspected and pressure tested prior to bursting, as well as after completion of the installation.

6.8.11 Dispute Resolution Mechanisms

As for any trenchless methods, the contract should include clauses addressing the possibility of different site conditions and unforeseen conditions that allow contract time and amount adjustment if the conditions at site *materially* differ from the conditions expected and indicated in the bid documents. See Chap. 9 for more information on geotechnical baseline reports (GBR) and other dispute resolution mechanisms.

6.8.12 Permitting Issues

As for any trenchless methods, permits from all affected parties should be secured before the start of the bursting phase. Some of these permits could be secured by the owner and its representatives, and the rest should be secured by the contractor. The responsibilities for obtaining the individual permits should be outlined in the specifications and stated on the drawings. Permits to burst under the road and to modify the regular traffic flow according to the project traffic control plan should be secured from state department of transportation (DOT) or municipality/local government having jurisdiction. If the pipe crosses beneath a runway, taxiway, drainage ditch, irrigation channel or canal, or railroad track, permits should be secured from the appropriate regulating agencies and/or owners of these facilities. Advance notices to any affected residents or businesses should take place before the bursting operation to inform them about road closures, night or weekend work, service disruptions, driveway blockings, and the like, and keep them posted with the project progress and any changes.

6.8.13 Cost Estimating

Estimating pipe-bursting projects for bidding purposes should be as detailed as possible. For each project, the contractor should estimate the labor, material, and equipment necessary for excavation and shoring of pits, pit bottom stabilization (concrete or gravel), bursting system setup, pipe fusion, lateral connection excavation, bypass pumping, bursting, service reconnection, shaft backfill, surface restoration, and potential dewatering. As shown in Fig. 6.7, the cost per foot of pipe-bursting installations are less than that of open-cut in unfavorable (average or light traffic) situations. Figure 6.7 also shows that that cost is less than that of open-cut in favorable (heavy traffic) conditions if the depth of cut is more than 10 ft.

In addition to providing the owner with the total price of the project to compare different bids, the bid form should provide the owner with a mechanism for payment based on the actual quantities during construction. The unit prices in the form can also be used to resolve disputes amicably. It is recommended that the bursting operation be measured in linear footage and segmented by classification or sections from manhole to manhole or from insertion to pulling shafts. Segmentation by run or bursting class provides the owner and the contractor with fair pricing mechanism and helps reduce disputes. Lateral connections should be measured by the number and type, required, indicated as separate bid items, and segmented by depth. Cleaning, testing, and post CCTV of the new line should also be a separate bid item. Bypass pumping can be priced as a lump sum or estimated for each section by number of days required.

6.9   Construction Considerations

Once the owner has accepted a bid, based upon the bidding process, a notice to proceed is issued. Then, typically, the contractor takes the following steps:

- Preconstruction survey
- Cleaning or pigging of line, if necessary
- CCTV inspection, if necessary
- Excavations at services for temporary bypass
- Setting up temporary bypass with connections to customers
- Excavation of insertion and pulling shafts
- Joining (fusing or assembling) individual new pipe sections, as necessary
- Setting up the winch or hydraulic pulling unit and insertion of pulling cable or pulling rods inside the existing pipe
- Installation of hoses through the new pipe to attach to bursting head (air supply hoses or hydraulic hoses for pneumatic or hydraulic systems respectively), as appropriate
- Connection of bursting head to pulling cable or rod
- Pipe bursting and replacement with new pipe
- Removal of bursting head and hoses from the pipe
- Post installation inspection
- Pipeline chlorination, flushing, and testing if it is not prechlorinated (for potable water pipes)
- Reconnection of services and reinstating manhole connections
- Site restoration and project close out

It should be noted that, if allowed by the water agency and local, city and state guidelines, prechlorination of potable pipe saves installation time and bypass pumping. In this method, prechlorination and testing is conducted above ground, and left to stand for 24 hours. Then, new pipe is flushed, and pulling head is welded to the pipe to keep pipe ends sealed hygienically. After completion of bursting operation, a super chlorinated swab is passed through the new pipe. After removal of swab, the new pipe is flushed and commissioned, and service laterals are connected. Usually one water sample is taken after commissioning and another after 24 hours of commissioning (see Fig. 6.9).

Butt fusion of HDPE replacement pipe is carried out prior to the bursting operation, so that all fused joints can be chlorinated (for waterlines), checked, and tested. The pipe should not be dragged over the ground, and rollers, pipe cutouts, or slings should be used for both insertion and transportation of the pipe. The ends of water or gas pipes should be capped to prevent the entry of contaminants.

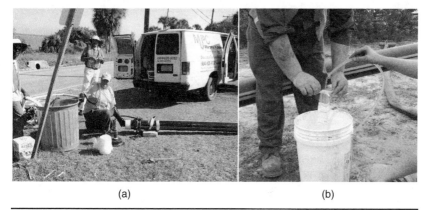

(a)   (b)

Figure 6.9   (a) Prechlorination and (b) sampling. (*Source: Murphy Pipelines and Hammer Head.*)

6.9.1   Typical Pipe-Bursting Operation Layout

The first step in planning the pipe-bursting operation is the optimization of the locations of the insertion and pulling shafts by using the insertion shafts to insert the new pipe in two (opposite) directions. This optimization reduces the amount of excavation, mobilization, and demobilization efforts. These shafts should preferably be located at manholes or lateral connections in sewer lines and at fire hydrants or gate valves in water applications. The length of the run between the insertion and pulling shafts should not generate friction forces that exceed the capabilities of the bursting system and the tensile strength of the new pipe (see Sec. 6.10). The next step is to ensure that the area around every shaft is sufficient for safe operation of the necessary equipment and material staging.

The insertion shaft contains a flat section and sloped section; the flat portion must be sufficiently long to allow alignment of the centerline of the bursting head with that of the existing pipe. The slopped section must be sufficiently long to allow the HDPE pipe to bend without damage to the pipe (i.e., accommodate the manufacturer's bending radius requirements). HDPE pipes can typically be cold bent to a radius of 25 to 30 times the OD of the pipe, depending on the DR value. For example, for 18-in diameter HDPE pipe with a DR of 17, the minimum length of the insertion shaft is a horizontal length of 12 times the diameter of the new pipe (18 ft) plus a sloped length of 2.5 times the depth of the shaft, as shown in Fig. 6.10. The width of the pit depends on the pipe diameter and required working space around the pipe. The pulling pit must be large enough to allow for operation of the winch or pull-back device, along with removal of the bursting head. Due to the flexibility of HDPE pipe, the outside lay-down area of the pipe prior to insertion does not necessarily have to be in line with the existing pipe.

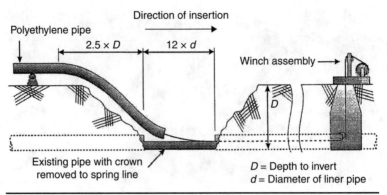

Direction of insertion

Polyethylene pipe

2.5 × D   12 × d

Winch assembly →

D

Existing pipe with crown removed to spring line

D = Depth to invert
d = Diameter of liner pipe

Figure 6.10   Insertion shaft dimensions for HDPE pipe with DR 17.

Acceptable arrangements for traffic control (based on DOT and local government regulations, and for the joined new pipe), with minimum inconvenience to nearby residents and businesses, must be carefully considered. The flow bypass pumping and pipe layout should be also planned. If it is necessary to dewater, safe and proper flow discharge plans are required. The contractor has, presumably, previously (see Sec. 6.8.9) submitted the jobsite layout plan that reflects the intended method of construction and addresses the above considerations. The contractor does not start the bursting operation before the engineer has reviewed and approved the jobsite layout plan. The site inspector enforces the adherence to this plan unless there is a valid reason for the deviation, and has been approved by the engineer or owner. If contaminated soil is excavated, the contractor must take the necessary measures for handling and disposal of the contaminated soil.

6.9.2   Shoring the Entry and Pulling Shafts

Proper shoring of the entry and pulling shafts is essential for the safety of the workers and the stability of the surrounding environment. The trench shoring or bracing must be constructed to comply with Occupational Safety and Health Administration (OSHA) standards. There are several available means of shoring these shafts: trench box, solder pile and lagging, steel sheet piles, corrugated pipes, and the like. In addition, if space is available, another option is to slope the sides of the shaft to provide stability. The judgment and the supervision of a competent person (as defined by OSHA) or a qualified geotechnical engineer is required to ensure the shoring is safe.

In static pipe bursting method, the winch will thrust against the wall of the pulling shaft on one side. This side must be capable of withstanding the corresponding pressure. Since the pulling forces in static bursting applications are relatively high, in such cases the contractor should construct a thrust block to distribute the forces over a reasonably large area. The thrust block, shoring, and soil behind the shoring must be able to withstand the stresses from the pulling system. The passive earth pressure of the soil must exceed the stresses generated by the pulling system with an acceptable factor of safety.

6.9.3 Matching System Components to Reduce Risk of Failure

One of the most critical activities prior to initiating the bursting project operation is to ensure that the system has sufficient power to burst the existing pipe segment between the insertion shafts to the pulling shaft. Additionally, the bursting system must overcome the friction between the soil and the outside surface of the new pipe with a reasonable margin of safety to account for unforeseen repair sleeves, clamps, and the like. In general, the bursting system components should be appropriately matched to the project; for example, the winch capacity should be compatible with the bursting head size and the conditions of the job.

The contractor should adhere to the sizing guidelines stated in the operations manual issued by the bursting system manufacturer to match the system with the needs of the job. The manufacturer should be consulted if there is any doubt regarding the adequacy of the system for that specific run under those particular conditions (soil, depth, type of pipe, etc.). Lubrication of the outside surface of the new pipe with polymer or bentonite (depending on the type of soil) can reduce the coefficient of friction between the pipe and the soil, and consequently, reduce the required pulling force.

6.9.4 Nearby Utilities

The contractor must perform its due diligence to identify, locate, and verify the nearby underground utilities prior to digging the shafts and initiating the bursting operation. This includes contacting the local one-call center and request representatives of the utilities mark their existing lines on the ground surface. Then the contractor should then verify the exact location and depth of these utilities via careful hand digging and/or vacuum excavation. Manual excavation may be required within a few inches from the existing utilities to avoid damage. Vacuum excavation is an excellent tool to expose utilities with minimum surface excavation and minimum risk to the existing utility.

Underground utilities that are in good conditions are unlikely to be damaged by vibrations at distances of greater than 2½ ft from the bursting head for small sizes (less than 12 in. in diameter), in soft soil, which is a typical application for pneumatic pipe-bursting operations (Atalah, 1998). The safe distance for large-diameter bursting (up to 24 in.) is about 7½ ft (see Sec. 6.8.7). These guidelines are generally consistent with that reported by Rogers (1995), indicating that ground displacements are unlikely to cause problems at distances greater than 2–3 diameters from the pipe alignment. Utilities that are closer to the bursting head than these distances should be exposed prior to bursting so the vibration from the bursting operation would be isolated or reduced prior to reaching the utility in question.

6.9.5 Bypass Pumping

One of the objectives of the bursting project team (owner, engineer, contractor, etc.) is to minimize customers' service interruptions. The key for achieving this objective is the bypass pumping system. For water applications, the system should be able to deliver the needed flow volume with the specified pressure to the customers. For gravity applications, the system should be able to adequately pump the upstream flow and discharge it to the manhole downstream of the run being burst. The plan should ensure that the bypass system has adequate pumping capacity to handle the flow with emergency backup pumps to ensure no interruption to existing services. The bypass pipes and fittings should have sufficient strength to withstand the surge water pressures. Contractual arrangements between the owner and contractor should be made regarding responsibility for maintaining service to customers.

6.9.6 Dewatering

The pulling and the insertion shafts should be dry during installation to avoid disturbing bursting operations. Installation of a dewatering sump pump at one corner of the shaft is required. If the pipe invert is significantly below the GWT in sandy or silty soils, a more elaborate dewatering system is recommended such as well-point system, deep wells, or larger sump and pump system. As the water level is drawn down, soil particles tend to travel with the water toward the dewatering system undermining utilities and structures. Therefore, as it is the case with every dewatering system, the contractor should take all necessary measures to prevent the migration of the soil particles from underneath nearby buildings and utilities. Since the discharge flow volume in this case is expected to be large, a suitable discharge permit in compliance with the EPA requirements may be necessary. If a sump pump is used, preliminary treatment of flow to reduce the sediments before discharging into water streams may be required.

6.9.7 Ground Movement Monitoring

The safety of nearby utilities, buildings, and structures due to ground movement is always important. For extremely challenging pipe-bursting operations (Class C, Table 6.1), preconstruction survey and monitoring of ground movement is advisable. A preconstruction survey should document all nearby buildings and structures that have cracks, cosmetic problems, and structural deficiencies. The elevations of carefully planned settlement points (on nearby buildings and on the ground surface) around the insertion and pulling shafts should be surveyed prior to bursting, during bursting, and after bursting. These preconstruction surveys and elevations monitoring can significantly reduce the risk of law suits to the contractor and the owner.

6.9.8   Manhole Connections

The thermal elasticity of the PE material causes changes in the pipe length, for example, 1 in. change in length per 100 ft of pipe for each 10°F temperature change. Therefore, in extreme hot or cold weather when there is significant difference between the temperature of the deep soil and the ambient air temperature, it is recommended to allow the pipe to rest for 12 to 24 hours prior to tie-ins. Also, when PE pipe has been pulled to a significant portion of its allowable tensile load, it may be prudent to let the pipe rest before connecting to other pipes, fittings, manholes, and lateral connection. This allows the pipe to recover from any stretching that may have occurred during bursting. For other pipe materials, the manufacturer should be consulted.

In most pipe bursting applications, the existing pipe is old and deteriorated, including the manholes along the line. In most cases, it is therefore economical, on a life-cycle cost basis, to replace the existing deteriorated manholes and use their locations for pulling or insertion shafts. When existing manholes are replaced with new ones, connections to HDPE pipe can be made using flexible rubber manhole connectors called "boots." A pipe clamp is used to tighten the boot around the HDPE pipe as shown in Fig. 6.11.

If the existing manhole is in reasonable conditions, and it is judged to be capable of providing adequate future service, the manhole benching is removed and the pipe opening is enlarged to allow the passage of the bursting head. Expandable urethane grout and oakum can be used to create a seal between the existing pipe opening and the HDPE pipe, as shown in Fig. 6.12. The flexibility of HDPE pipe may facilitate its movement through this opening.

Frequently, the inlets and outlets of the manhole are damaged during the bursting operation, such that the resulting inlet or outlet is no longer round. A low shrink polymer cement grout may then be used to repair the damage. To obtain a good seal to the HDPE pipe, a special PVC fitting with bell-end and sand texture outer surface (as shown in Fig. 6.13) may be used. The grout bonds to the manhole and the rough sandy surface of the PVC fitting provides a good seal. The gasket between the PVC fitting and the HDPE pipe allows the HDPE

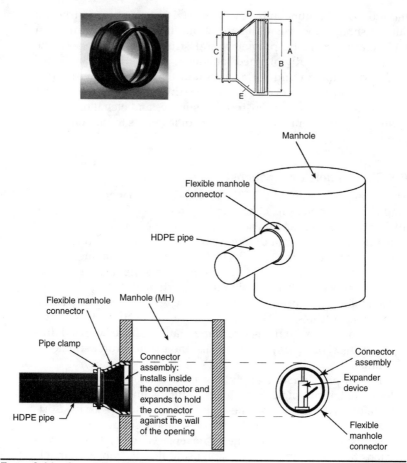

Figure 6.11   Connecting HDPE pipe to new concrete manhole. (*Source: Handbook of Polyethylene Pipe, 2008.*)

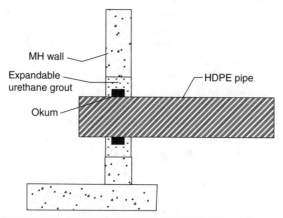

Figure 6.12   Connecting HDPE pipe to existing manhole. (*Source: Handbook of Polyethylene Pipe, 2008.*)

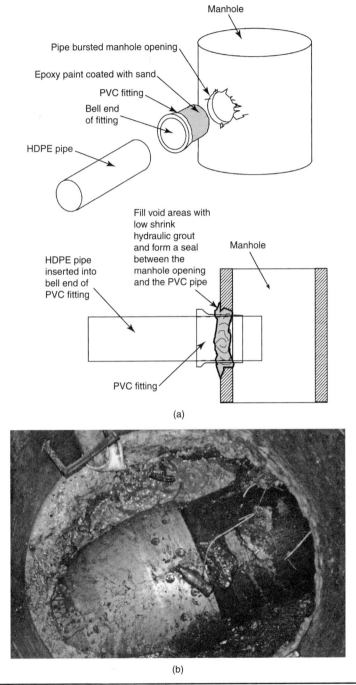

FIGURE **6.13** (*a*) and (*b*) Connecting HDPE pipe to existing manhole with damaged inlets/outlet. (*Source: Handbook of Polyethylene Pipe, 2008.*)

pipe to move if expansion or contraction occurs. The PVC fitting requires HDPE pipes with SDR of 21 or lower.

6.9.9 Pipe Connection to Other Pipes

HDPE pipes are joined to other PE fittings by heat fusion or mechanical fittings. They are joined to other material by means of compression fittings, flanges, or other qualified transition fittings (PPI, 2008).

6.9.10 Pipe Bursting Water Mains

The most common materials for existing water mains are cast iron, ductile iron, and PVC. All three can be replaced by pipe bursting, but each requires a different piping burst approach. Cast-iron pipe is a relatively brittle material, and, therefore, the basic pipe bursting system is sufficient to shatter the pipe. PVC pipes are not brittle and require multiblade cutting accessories in front of the bursting head to facilitate cutting the pipe. Ductile iron pipe is also not brittle; therefore, pipe splitting is the most suitable bursting system.

Connections between the replacement HDPE pipe and remaining segments of existing pipe (non-HDPE pipe), including valves, are accomplished by mechanical joint (MJ) adapters, which are butt fused to the HDPE pipe. A gland ring is then used to make a restrained connection between the two types of pipes. Figure 6.14 shows connections to PVC or ductile iron pipes made using a female MJ connector, which is butt fused to the HDPE pipe. This connection provides a restrained connection.

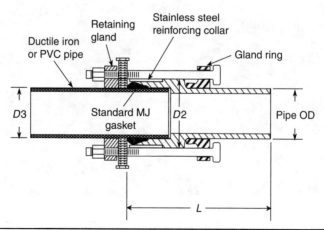

FIGURE 6.14 Connection to existing PVC or DI pipes using female MJ adapter. (*Source: Handbook of Polyethylene Pipe, 2008.*)

6.9.11   Service Connections

The service (lateral) connections stated in the submittal list should explain the proposed material and connection procedures. Inspection of the existing pipe normally provides the location of service connections. In replacing water and gas lines, metal detectors can often be used. Standard practice is to locate, expose, and disconnect services prior to pipe bursting. Service connections can be made with Inserta Tee®, specially designed fusion fittings, or strap-on saddles. Figure 6.15 shows an electrofusion saddle with a cutter attached, which is convenient to hot tap HDPE pipe lines for water or gas applications.

For sewer applications, a "window" is cut in the HDPE pipe wall, and then one of the above fittings is used to connect the new HDPE pipe to the lateral connection. The Inserta Tee connection is a three piece service fitting consisting of a PVC hub, rubber sleeve, and stainless steel band, as shown in Fig. 6.16.

The Inserta Tee is a compression fit into the cored wall of a sewer mainline and requires no special tooling. Inserta Tees are designed to connect 4- through 15-in. services to all known solid wall, profile, closed profile, and corrugated pipes. The PE lateral connection options include fusing a lateral HPDE pipe to the main line and placing an electrofusion sewer saddle. Fusing a lateral PE to the main PE line requires curved iron that allows heating the ends and exposed surfaces of both pipes. This connection requires a highly skilled fusion worker because it is usually made in a tight space.

An electrofusion saddle is mounted on the opening for fusion with the main line, as shown in Fig. 6.17. It is important to ensure that all exposed surfaces are clean and maintained in an acceptable

Figure 6.15   Electrofusion saddle with cutter attached. (*Source: Handbook of Polyethylene Pipe, 2008.*)

FIGURE 6.16 Insert a tee fittings for sewer lateral connections. (*Source: Handbook of Polyethylene Pipe, 2008.*)

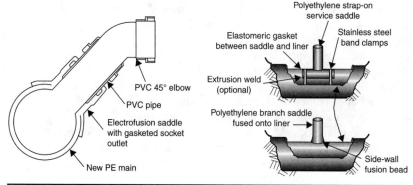

FIGURE 6.17 Pipe fusion and strap-on saddle sewer lateral connections. (*Source: Handbook of Polyethylene Pipe, 2008.*)

condition for the fusion operation. Strap-on saddles use a PE or PVC saddle that are lined with a rubber layer; the saddle is strapped around the main line using a stainless steel band, as shown in Fig. 6.17. After testing and inspecting the line and the connections, the excavation is backfilled and the line returned to service. Additional details for gravity and pressure service connections are presented in the *Handbook of Polyethylene Pipe* (2d ed., PPI, 2008).

6.9.12   Grooves on the Outside Surface of the Pipe

A common misconception regarding pipe bursting is that the pipe fragments from the existing pipe can damage the HDPE pipe as it is pulled into the cavity. British Gas conducted a detailed study on bursting cast-iron pipes and concluded that there is no significant damage to the HDPE replacement pipes. Related supporting research

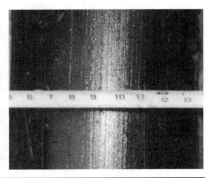

FIGURE 6.18   Grooves on the outside surface of the HDPE pipes after bursting.
(*Source: Handbook of Polyethylene Pipe, 2008.*)

conducted at Louisiana Tech University and Bowling Green State
University, Ohio, indicates any possible grooves are very shallow and
narrow when bursting clay, asbestos, and concrete pipes. The widest
groove was 0.07 in. across and the deepest groove was 0.03 in., as shown
in Fig. 6.18. Cast iron, clay, concrete, and asbestos pipes generally
shatter without producing sharp shards that may potentially produce
significant damage to HDPE pipe (Atalah, 1998, 2004). An exception
to this rule occurs when attempting to significantly upsize ductile iron
(DI) pipe. It is therefore recommended to limit PE replacement pipe to
size-on-size bursting, or a single upsize, when bursting DI pipe. If
larger upsize of DI pipe is required, HDPE pipes with a protective
outside surface, similar to that shown in Fig. 6.19, may be used.

FIGURE 6.19   Protecting the HDPE pipe from shards. (*Source: Handbook of
Polyethylene Pipe, 2008.*)

FIGURE **6.20**   Protecting the HDPE pipe from grooves using wheels on the bucket. (*Source: Handbook of Polyethylene Pipe, 2008.*)

If the PE replacement pipe will be dragged for a long distance over rough surface or pavement, the contractor can reduce the risk of scratching the pipe by placing it on cut-outs of available (scrap) PE or PVC pipe to keep the pipe off the rough surface. Sometimes it is necessary for the contractor to push down on the replacement HDPE pipe with the bucket of the excavator to ensure that the HDPE pipe is properly aligned with the existing pipe at the entry point in the insertion shaft (see Fig. 6.12), in the event the shaft is not of sufficient length. Wheels similar to those shown in Fig. 6.20 reduce the risk of grooves or scratches on the HDPE pipe. When the head of the pipe reaches the pulling shaft, the pipe should be inspected for surface damage. Surface scratches or defects in excess of 10 percent of the wall thickness should be rejected.

6.9.13   As-Built Drawings

As-built drawings are usually required for any underground utility construction including pipe-bursting projects. The bursting contractor should mark the new pipe, manholes, ancillary structure information on a copy of the drawings marked and dedicated as the as-built. The contractor should document on these drawings any changes to the original layout of the underground utilities and structures that occurred during the construction phase. For example, rerouting any utility due to the excavation of the shafts, reconfiguration of other utilities required for bypass pumping, and others, should be marked on the as-built drawings. These changes shown on the as-built drawings should be verified and used to update the corresponding, certified-record electronic files (GIS) for the utility.

6.9.14   Contingency Plan

An important submittal in trenchless technology projects is the contractor's contingency plan. Most contractors have contingency plans that include planned corrective actions if certain events take place. Pipe-bursting projects require specific additions to their standard contingency plan. Some of these specific events that are unique to pipe bursting and should be addressed include:

- Excessive ground movement or vibration
- Slow bursting progress or stuck bursting head
- Problems with bypass system and with diverting and reconnecting the services to the customers
- Damage to existing water, gas, sewer pipes and power cable, and the like
- Dewatering problem in insertion or pulling shaft or at lateral connection pit

Specific issues and potential solutions are briefly discussed in Sec. 6.9.16.

6.9.15 Safety Considerations

Standard safety procedures, as followed in typical open-cut construction, should also be followed in bursting projects. In addition, the workers should understand the components of the bursting system and how they function, with special attention to the moving parts in the system. The workers should be trained and equipped with the necessary tools for confined space entry. The winch pushes against a thrust block that (along with the soil behind it) must withstand the forces applied by the winch. The stability of the soil behind the thrust block should not be compromised. During the flow bypass, the upstream pipe will be plugged; these plugs should be braced and, preferably, remotely inflated and deflated. Prior to bursting, the contractor must ensure that there is no gas line or power line close to bursting head.

6.9.16 Potential Problems

The best method for dealing with potential pipe-bursting problems is avoidance or minimizing probability of occurrence by properly following the design and construction precautions indicated in this chapter. However, if some of these precautions are not followed or in the event of unforeseen circumstances, problems may occur. Some of the causes of such problems include excess (existing) pipe sag, relatively large soil displacement, inadequate protection of nearby utilities, inappropriate bursting system, unforeseen obstacles, and unexpected site restrictions.

If, prior to initiating the bursting operation, the existing pipeline is determined to display excessive sag, this condition may be repaired using the previously discussed techniques. However, if such a condition was discovered after initiating bursting, the contractor can repair the sag by exposing the problematic location and adjusting the soil beneath the pipe. Replacement of a section of pipe may be necessary at this excavation. If direct excavation at the location is not feasible, grouting or otherwise stabilizing the soil beneath the pipe may be possible.

If excessive ground movement is anticipated close to an existing structure, a ground movement and vibrations monitoring plan should be developed (see Sec. 6.9.7). If problematic levels (see Sec. 6.8.7) are observed, a slower the rate of bursting is mandated. If the ground movement remains excessive, the bursting operation should be halted until an analysis of the causes and corrective options is conducted. If there is a deteriorated or old gas line, waterline, or sewer line that is particularly close to the bursting head and at risk of damage, exposure of the line significantly reduces risk of damage. The corresponding excavation should be performed using means that do not damage the line, such as vacuum or manual excavation. If the existing pipe is shallow and there is a high risk of damaging the surface pavement, saw cutting the pavement prior to bursting prevents the spreading of the damage to the rest of the pavement. After the operation is completed, the pavement over the trench can be repaired. In the absence of a viable solution, one option includes abandoning the pipe-bursting method.

If the bursting is significantly slower or more difficult than anticipated, the contractor should attempt to determine the cause and consider available corrective actions. Possible reasons and solutions for low bursting progress, or excessive pull force, include:

- The bursting system does not have sufficient power relative to the application (upsize percentage, large diameter, length, etc.). If the problem becomes evident shortly after the start of the run, the bursting system should be upgraded with one that has more power. If the problem occurs when the unit is close to the pulling shaft, toward the end of the run, the operation should be continued until the bursting head reaches the pulling shaft. At that point, the system may be upgraded for performing the next run, unless it is determined that an isolated problem, such as that due to a repair ductile clamp or a fitting, and unlikely to occur again. As an alternative to employing a more powerful bursting system, the length of the runs may be reduced. If the problem arises in the middle of the run, at a location where excavation is feasible, the bursting head should be exposed by a new shaft and the system upgraded. The introduced shaft can serve as an insertion shaft for the remainder of the run.

- The soil around the pipe prone to flowing or running, possibly leading to excessive friction and associated elevated pull forces (see Sec. 6.10). Lubrication of outside surface of the pipe with suitable lubricant is an effective way to reduce required pulling force on the new pipe by reducing the friction between the pipe and the soil. The key to applying this solution is establishing a lubrication manifold and lubrication line and pumping the lubricant during the bursting operation.

- Certain components or accessories of the system (for example, the winch, air compressor, hydraulic components, cutting accessories in front of bursting head, etc.) are under sized or not compatible with the application. The addition of appropriate accessories in front of the bursting head is assumed in order to cut PVC fittings, ductile clamps or fittings, and the like, reducing the potential of stopping or slowing the bursting operation. If necessary, these components should be upsized (within the allowable range of that system).

- Breaking or shattering the existing pipe in running soil, below the GWT, may result in the pipe filled with soil, converting the bursting operation to a piercing operation. After verifying that the bursting head did not damage any nearby waterline, the site should be dewatered.

6.10 Sample Pipe Load Calculations

6.10.1 Introduction

A method for estimating pipe loads on HDPE pipe for a Mini-horizontal directional drilling (Mini-HDD) operation was described in Chap. 5. It would be advantageous to obtain similar estimates for such loads due to a pipe-bursting operation. Unfortunately, such attempts are hampered by the wide range of possible field conditions and their effect on the loadings. Thus, only a rough guide or "rule-of-thumb" may be obtained, based on a combination of theoretical predictions and review of practical field results. Nonetheless, the process of arriving at such estimates yields valuable insight into the mechanics of the pipe-bursting operation and an understanding of the significant parameters impacting such a process, aiding the user in achieving a successful installation.

6.10.2 Pulling Loads—Theoretical Considerations

In principle, the required pulling loads on the replacement pipe may be determined by estimating the normal (perpendicular) pressures applied to the pipe as it is pulled through the expanded borehole or cavity created by the bursting head. In particular, the pulling load must be equal or exceed the drag force imposed on the pipe which may be based a conventional Coulomb friction model. This theoretical model assumes that friction or drag forces on a moving body are proportional to the local normal bearing forces applied to its surface, with the proportionality constant designated as the "coefficient of friction." Such bearing forces may be due to the dead (empty) weight of the pipe, pressure on the pipe due to vertical or lateral pressure imposed by the soil, bearing/bending forces associated with pulling a stiff pipe around a curve, or bearing forces resulting from (previously induced) axial tension tending to pull the pipe snugly against any locally curved surfaces.

For the simple case of a replacement pipe pulled along the bottom of a stable cavity, with clearance between the pipe and internal cavity walls, as illustrated in Fig. 6.21, the required tension T_1 is given by

$$T_1 = w \times L \times \nu \tag{6.2}$$

where w = weight of the pipe per unit length (lb/ft)
  L = length of the pipe within the cavity (ft)
  ν = coefficient of friction between the pipe and cavity surfaces (dimensionless)

Equation (6.2) also assumes that there is no significant restraining load at the trailing end, such as due to reel resistance for a continuous length pipe. Such resistance, or tail load, would result in an equivalent increased load at the leading end.

It is a relatively simple matter to apply Eq. (6.2) to a particular pipe and application, based upon an assumed value, or range of values, for frictional characteristics. For example, 4-in. DR 17 pipe of

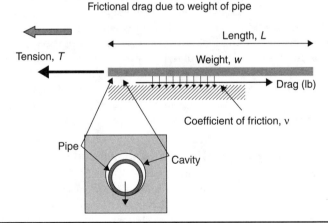

Frictional drag due to weight of pipe

FIGURE 6.21 Replacement pipe pulled through stable cavity.

Nominal	Pipe Diameter to Thickness Ratio (DR)						
Size (in.)	7.3	9	11	13.5	15.5	17	21
4	8,600	7,200	6,000	5,000	4,400	4,000	3,300
6	18,500	15,500	13,000	11,000	9,500	9,000	7,200
8	31,000	26,000	21,500	18,000	16,000	14,500	12,000
12	69,500	58,000	48,500	40,500	35,500	32,500	26,500
24	246,000	205,500	172,000	142,500	125,500	115,000	94,500

TABLE 6.2   Safe Pull Strength (lb) HDPE (PE3608) Pipe, 12 Hours

HDPE material weighing 1.53 lb/ft, an assumed coefficient of friction on the order of 0.5, and an arbitrarily long section length of 1000 ft corresponds to a drag force, or required pulling load, of 765 lb.

The predicted load should then be compared to the safe pull strength for the pipe, which is provided in Table 6.2 for HDPE for a variety of pipe sizes. The safe pull strength (lbs) is based upon the safe pull tensile stress (SPS) as applied to the pipe cross-section. The SPS is a conservative value that accounts for the load duration, assumed to be 12 hours, as well as a significant reduction (less than half) relative to the nominal tensile strength of HDPE (3200 lb/in.^{2}) to limit nonrecoverable viscoelastic deformation (Petroff, 2006). For MDPE pipe, the values in Table 6.2 must be adjusted by a factor of 0.75.

It is recognized that the new-bursting operations may impose dynamic impact loads on the pipe, not reflected in the steady-state movement assumed in the present simplified analyses. Such transient effects may be expected to increase the peak steady-state loads by approximately 25 percent for HDPE pipe (Atalah, 1998). Due to the brevity of the transient load duration, as well as the conservative nature of the SPS value, such effects are not considered significant for the present calculations.

It is readily confirmed that the pull load (765 lb) calculated before is an order of magnitude below the indicated 4000 lb safe pull strength of Table 6.2. In general, the installation distances appear to be essentially unlimited compared to that based upon actual field present experiences, as outlined in Table 6.1, with similar conclusions independent of wall thickness (DR value). It may therefore be concluded that the simple model indicated in Fig. 6.21 and Eq. (6.2) represent idealized conditions, not typically encountered in practice. In particular, the assumption that the expanded borehole remains stable, along its entire length, as illustrated in Fig. 6.21, is optimistic.

Previous studies (Atalah, 1998) indicate that the borehole may be expected to collapse along at least a portion of the length, directly

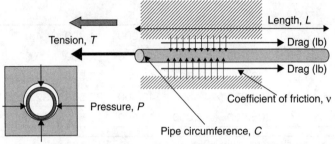

Frictional drag due to soil weight/pressure in collapsed borehole

Length, L

Tension, T

Drag (lb)

Drag (lb)

Pressure, P

Coefficient of friction, v

Pipe circumference, C

FIGURE 6.22   Replacement pipe pulled through collapsed borehole.

subjecting the pipe to pressure from the soil at the top and its sides, as illustrated in Fig. 6.22. A conservative estimate may, therefore, assume that the borehole collapses along the entire length of the bore. Furthermore, the "prism load" corresponding to the height of soil above the pipe may conservatively be used to estimate the local radial soil pressure applied around the circumference, C. Thus, the reconsidered required tension, T_2, would now be given by

$$T_2 = \gamma \times d \times C \times L \times v \qquad (6.3)$$

where d = depth of cover above the pipe (in.)
  C = pipe outer circumference (in.)
  γ = soil density (lb/ft³.)

As mentioned before, it is a relatively simple matter to apply Eq. (6.3) to the 4-in. HDPE pipe considered previously. In this case, however, a segment length of only 100 ft, as well as a relatively shallow depth of 5 ft, is considered, corresponding to that which should be a readily accomplished installation, based on Table 6.1. For the present calculations, a relatively low value of soil density (γ equal to 100 lb/ft³) and frictional coefficient (v equal to 0.3) are assumed, the latter possibly corresponding to the use of lubricant, such as bentonite. The calculation results in a drag force exceeding 17,500 lb, more than four times the 4000 safe pull strength provided in Table 6.2, for a DR 17 HDPE pipe. This broad inconsistency between actual field capabilities and the theoretical predictions confirms the degree of excessive conservatism reflected in Eq. (6.3), due to the assumed extent and magnitude of the applied soil loads for typical applications.

Consistent with the widely varying assumptions reflected in Eqs. (6.2) and (6.3), as directly related to the degree of stability of the created cavity, the corresponding required tensile forces—and placement distances—can vary over a wide range, implying that it will be difficult to confidently predict practical placement distances for a given pipe system (specified material, diameter, wall thickness, and the like).

While recognizing the inherent difficulty in providing a generally accepted, conservative design protocol for the broad range of possible pipe-bursting operations, the availability of a "rule-of-thumb" would be useful to help during the planning stages. Such a rough guide is presented in Sec. 6.10.3, as well as in Sec. 6.10.4.

6.10.3 Pulling Loads—Planning Guide

The extremes inherent in the above theoretical models are associated with the degree and extent to which the expanded borehole (cavity) may collapse, and the method by which the resulting soil loads is estimated. A reasonable approach in developing a more useful rule-of-thumb or guide would be to assume a moderate degree of cavity collapse and/or reduced soil loads that reflect a degree of soil arching, and compare the resulting predictions to the relatively recent field experiences included in Table 6.1.

Previous studies (Atalah, 1998) have indicated that the potential for collapse of the borehole encompasses a wide range of possibilities, including collapse along the entire length as well as collapse of a lesser extent, for example, approximately 25 percent of the length. The same studies included estimates of estimated soil loads somewhat less than the simple prism load, reflecting soil arching. An oversimplified, but convenient, means of attempting to account for these effects is to use a modified version of Eq. (6.3), based on an effective length equal to 25 percent that of the actual section[*,†], and an effective depth given by

$$d_{eff} = d \qquad \text{for} \qquad d/D < 5$$
$$= \tfrac{2}{3} d \qquad \text{for} \qquad 5 \le d/D < 10$$
$$= \tfrac{1}{2} d \qquad \text{for} \qquad 10 \le d/D < 20$$
$$= \tfrac{1}{3} d \qquad \text{for} \qquad 20 \le d/D \qquad (6.4)$$

where D is the nominal (trade size) pipe diameter of the HDPE replacement pipe (in.). (The approximate nature of this simplification obviates the need to carefully distinguish between the nominal and the actual outer diameters of the product pipe.)

---

*The weight of the pipe itself, along the entire length L, is insignificant compared to the soil loads, and may be ignored for these calculations.
†The 25 percent factor also reflects the possible (likely) reduced radial soil pressure due to variation around the pipe circumference.

Based on these assumptions, and the preceding assumed values for soil density and frictional coefficient, the estimated pull force Eq. (6.3), leads to:

$$T_{guide} \text{ (lb)} = d_{eff} \text{ (in.)} \times D \text{ (in.)} \times L \text{ (ft)}/6 \tag{6.5}$$

where d_{eff} is given by Eq. (6.4) and the conversion units for the various parameters are reflected in the "6" in the denominator of Eq. (6.5).

For example, for the 100-ft-long, 5-ft-deep installation of the 4-in. HDPE pipe considered in Sec. 6.10.2, for which Eq. (6.3) yielded a pull force in excess of 17,500 lb, Eq. (6.5) predicts a required pull force on the order of only 2000 lb, based on d_{eff} of half the actual depth (for d/D = 60 in./4 in. = 15). Thus, this operation would appear to be well within the capability of the DR 17 HDPE pipe under consideration.

As another example, consider a 250-ft section of 12-in. HDPE DR 17 pipe, placed with 10-ft cover. Based on Eq. (6.5), a required pull force of approximately 30,000 lb is predicted, within the tabulated safe pull strength of 32,500 lb. This result is consistent with field experiences indicating a "routine" (see Table 6.1) operation for an installation of this approximation geometry.

6.10.4 Pipe Collapse Conditions

In addition to potential failure by excessive pulling tension, it is possible that the HDPE product pipe can significantly deform or collapse during the installation phase or during the postinstallation (e.g., operational) stage. A relationship for critical (buckling) pressure, P_{cr}, is given in ASTM F1962-05 (ASTM, 2005) for unconstrained collapse under uniform external (hydrostatic) pressure:

$$P_{cr} = 2 E f_o \bullet f_R / \{(1 - \mu^2) \bullet (DR - 1)^3\} \tag{6.6}$$

where E = material modulus of elasticity (psi)
   μ = Poisson's ratio (dimensionless)
   f_o = ovality compensation (reduction) factor (dimensionless)
   f_R = tensile stress reduction factor (dimensionless)

In a discussion of ASTM F1962-05, Petroff (2006) explains the significance of these terms. The material properties, E and μ, for the viscoelastic HDPE pipe depend on the load duration, f_o accounts for initial or subsequent out-of-roundness, and f_R recognizes a potential reduction in collapse strength in the presence of significant tensile loads during the installation phase.

In the case of pipe bursting, and consistent with above discussion, it must be anticipated that at least a portion of the expanded bore path will be unstable and, therefore, tends to collapse with the soil descending, applying vertical (and possibly a degree of lateral) pressure on the pipe. In general, the detailed consideration of the interaction of the various phenomena, and the consequences for the product pipe is relatively complex. In particular, it is not clear that the model represented by Eq. (6.6) is appropriate for the asymmetric soil pressures applied to a partially constrained pipe in the cavity created by the bursting operation. However, for the present purpose of developing an easy-to-apply planning guide, it is again desired to make some simplifying assumptions. Thus, the effect of the estimated asymmetric earth pressure, assuming a locally collapsed cavity, is compared to the critical load as indicated in Eq. (6.6), recognizing that the procedure may be somewhat conservative due to a degree of lateral constraint that may be provided to the pipe.

The critical pressure, P_{cr}, as given in Eq. (6.6) may be expressed in terms of an equivalent head (ft) of water, for idealized conditions in which the ovality reduction factor, f_o, and tension reduction factor, f_R, are assumed equal to 1.0. Since the (effective) material stiffness, E, and Poisson's ratio, μ, are dependent upon the load duration, the critical pressure is also dependent upon duration. Table 6.3 is based on the corresponding information provided in ASTM F1962-05 or related industry information (Petroff, 2006), and is applicable to any diameter HDPE pipe. For MDPE pipe, the tabulated values must be adjusted by a factor of 0.75.

Since the soil is of significantly greater density than water, the indicated pressure head (ft) values of Table 6.3 must be reduced by a factor of $1/\gamma$, for which purposes the soil density is conservatively assumed to be double that of water. The values must be further adjusted (reduced) for the aforementioned ovality and tensile load considerations, as well as possible initial elevated temperature (ASTM F1962-05, App. 3).

In general, there are two phases to be considered with regard to possible collapse of the pipe during, or soon following, the installation. During the installation process, the short term, or possibly

Duration	Pipe Diameter to Thickness Ratio (DR)						
	7.3	**9**	**11**	**13.5**	**15.5**	**17**	**21**
Short term	2,896	1,414	724	371	238	117	91
10 hours	1,436	702	359	184	118	88	45
100 hours	1,205	588	301	154	99	74	38
1,000 hours	1,019	498	255	131	84	62	32
50 years	649	317	162	83	53	40	20

TABLE **6.3**   Ideal Critical Pressure (Water Head, ft) for Unconstrained HDPE (PE3608) Pipe (73°F)

10 hours, strength would be appropriate, in combination with anticipated values of f_o and f_R during this period. For the postinstallation phase, a 1000-hour collapse strength is conveniently assumed as which time the soil has locally redistributed itself around the pipe to provide some lateral constraint against collapse, in combination with the relatively tight clearance between the replacement pipe and the expanded borehole walls, characteristic of a pipe-bursting operation. For this postinstallation phase, the tension reduction factor, f_R, is equal to 1.0.

Based on the ovality reduction factor, dependency trend provided in ASTM F1962, and consistent with the present simplified approach, it is reasonable to assume a maximum overall value of f_o of approximately 0.5 to account for ring deformation due to initial ovality plus that induced by postinstallation loadings.

The greater short-term (or 10-hour) collapse strength relative to that at 1000 hours would tend to be offset somewhat by the tension reduction factor, f_R, during installation, as well as the degradation at possible elevated temperature as the HDPE pipe sits on the surface prior to installation. However, the latter effects would not be experienced simultaneously, as the pipe is placed at its maximum depth and distance. Therefore, the 1000-hour (postinstallation) collapse characteristics, as adjusted for the above soil density and ovality considerations, are conveniently employed to evaluate the vulnerability to collapse. Based upon the above discussion, the following planning guide is determined:

$$H_{all} \text{ (ft)} = 1000 \text{ hour water head} \times f_o \times f_R / \gamma$$

$$= 1000 \text{ hour water head} \times (0.5) \times (1.0)/(2.0)$$

or,

$$H_{all} \text{ (ft)} = 1000 \text{ hour water head}/4.0 \tag{6.7}$$

where H_{all} is the allowable (maximum) pipe depth.

An additional degree of margin (safety factor) may exist due to soil arching effects, reducing the effective load on the pipe), as well as the aforementioned degree of lateral constraint of the pipe provided within the cavity.

As an example, consider the feasibility of installing a 4-in. DR 17 pipe at a depth of 12 ft. Applying Eq. (6.7) results in

$$H_{all} \text{ (ft)} = 1000 \text{ hour water head}/4.0$$

$$= 62 \text{ ft}/4.0$$

$$= 15.5 \text{ ft}$$

Thus, the present planning guide indicates that the proposed 12-ft depth is reasonable.

It is noted that the above guidelines do not apply to applications with significant surface loads, such as vehicular traffic. For such cases, minimal depth requirements would also apply to avoid the effect of large surface loads. In addition, the ability of the pipe to successfully perform its long term function, including the ability to withstand external soil loads as well as internal operating pressure, should be confirmed, using conventional design procedures (PPI, 2008).

6.11   Summary

This chapter presented pipe replacement methods, primarily based on the use of HDPE pipe. Although other pipe materials may be used for pipe bursting, many references are made to HDPE since HDPE is most commonly used material for this application. In comparison to other renewal methods, including conventional open-cut construction, pipe bursting is often the method of choice when the existing pipe has few service lateral connections, is significantly deteriorated and, most importantly, *requires additional capacity*. This chapter discussed the predesign, design, and construction phases of pipe bursting. In addition, a relatively simple concept for estimating the pulling loads applied to the new pipe is provided to help determine the general feasibility of the HDPE pipe to withstand these forces. While PE pipe was used for pipe loading calculations, similar concept can be used for calculating loads on other types of pipe materials. The corresponding estimates allow the selection of an appropriate pipe wall thickness (DR value).

Polyethylene pipe has been most commonly used as a new replacement pipe, but other materials have also been successfully used; see Sec. 6.5. The reference to HDPE in this chapter by no means implies that this book endorses or recommends use of HDPE pipe over other types of pipe materials, a decision that needs to be made by the project owner and design engineer, considering specific project and site conditions. Chapter 4 presents a full discussion of different pipe materials and advantages and limitations of each.

CHAPTER 7

Construction and Inspection for Cured-in-Place Pipe

7.1 Overview of the CIPP Technology

7.1.1 Background

Cured-in-place pipe (CIPP) installation is one of the most widely used methods of trenchless pipeline renewal for both structural and non-structural purposes, and can be used for water, sewer (sanitary, storm, culvert), oil, and gas applications. The CIPP technology is used to renew deteriorated, leaking, and outdated pipelines for a new design life of 50 years.

The CIPP method involves a liquid thermoset resin-saturated material that is inserted into the existing pipeline by hydrostatic or air inversion, or mechanically pulled with a winch, a cable, and inflation. The material is heat-cured in place resulting in a CIPP product. Insituform Technologies introduced CIPP in the United Kingdom in 1971 and entered the United States market in 1976. Currently CIPP technology is offered by a large number of companies.

As for any other pipeline renewal method, CIPP design and installation is based on factors, such as the existing pipe condition, its defects, its applications (type of fluid and its properties) and its environment, nature of problem or problems involved, and future use of the pipeline. A existing pipeline system can exhibit a number of different defects. The defect can be structural in nature, operational or maintenance related, built into the system during construction, or can be a mix of different types of problems.

7.1.2 Method Description

Cured-in-place pipe (CIPP) is a thermoset resin system (polyester, vinyl ester, or epoxy) that is delivered via a felt, fiberglass, or carbon fiber tube of the design thickness specified. After cleaning and inspection, the existing pipe is used as a mold. The soft, resin impregnated and saturated, tube is inserted by inverting the tube into position using water or air. Another method is to first pulling the tube into place by a winch and then inflating it with air or water. Once in place and properly inflated, the resin is cured using hot water, steam, or ultraviolet light, resulting in a new pipe within a pipe. See Sec. 7.4 for more information.

Although originally developed for sanitary sewer applications, CIPP may be manufactured to suit many pipeline shapes, and can accommodate deformations and changes in direction of the pipe. Using this technology, pipe lengths up to half mile have been installed successfully in one single operation.

7.2   Site Compatibility and Applications

Cured-in-place pipe (CIPP) can be used effectively for a wide range of applications that include sanitary and storm sewers, gas pipelines, potable water pipelines, chemical and industrial pipelines, and pressure pipelines. There are a variety of tube materials and configurations, including reinforced or nonreinforced, that can be used with a variety of liquid resin systems, to address structural and corrosion resistance requirements. The physical properties of CIPP make it especially suitable for different types of pipe geometries including straight pipes, pipes with bends, pipes with different cross-sectional geometries, pipes with varying cross-sections, pipes with lateral connections, and deformed and misaligned pipelines. However, several factors must be evaluated before choosing CIPP as the method of renewal for an individual project. Factors such as space availability, chemical composition of the fluid carried by the pipeline, the number of laterals, the number of manholes, installation distance, renewal objectives, structural capabilities of the existing pipe, and the like, must be assessed before making a choice on the renewal system. CIPP is also used for localized repairs in a wide range of pipe applications.

The possibility of negotiating bends depends on the installation and curing process of the various systems. Pipe liners that are inverted during insertion can often negotiate bends up to 90°. There are, however, limitations to the degree of bending that ultraviolet (UV) cured liners can manage. Bends can also present problems for liners that are pulled into place, because the liner must remain on the underlying protective foil and must under no circumstance become twisted. It is important to be aware that vertical folds and wrinkles may occur in the liner when negotiating sharp bends. Table 7.1 presents an overview of CIPP applications and possible limitations.

Pipeline Type	Suitable
Sewers	Yes
Gas pipelines	Yes
Potable water pipelines	Yes[a]
Chemical/industrial pipelines	Yes[b]
Straight pipelines	Yes
Pipelines with bends	Yes
Circular pipes	Yes
Non-circular pipes	Yes[c]
Pipelines with varying cross-sections	Yes[d]
Pipelines with lateral connections	Yes[e]
Pipelines with deformations	Yes[f]
Pressure pipelines	Yes[g]

[a] Several CIPP lining systems are approved for lining potable water pipelines and can be combined with other lining systems to provide sufficient strength.

[b] As each manufacturer uses different types of resin, reinforcement, etc., the chemical resistance of liners varies. If special characteristics are required, individual manufacturers should be contacted in order to help finding the most suitable product for specific requirements with regard to chemical content, concentration, temperature, etc.

[c] All the described methods are suitable for lining main pipelines with circular, egg-shaped, or V-shaped cross-sections.

[d] Some proprietary liner systems allow CIPP techniques to be used in pipelines with varying cross-sectional area.

[e] Laterals can be reinstated using a robotic cutter. It is presently possible to reopen laterals on pipelines with a diameter of more than 4 in. (100 mm). In man-entry systems, laterals are reopened manually. With some CIPP systems, watertight connectors can be fitted between laterals and main lines.

[f] CIPP can be used where deformities are minor.

[g] A few liner manufacturers can supply laminates suitable for pressure pipelines.

TABLE 7.1 Overview of CIPP Applications and Possible Limitations

7.2.1 Effects of Pipe Defects

While cured-in-place pipe can be installed in a variety of shapes and configurations, levels of pipeline defect may dictate where the CIPP may not effectively renew the existing pipe or when additional remedial work is required before CIPP installation. With a thorough knowledge of pipeline defects, the designer can determine if a pipe can be renewed or must be replaced, what renewal or replacement method to use, or if the pipe must initially be spot repaired.

Defects	Codes	Defects	Codes
Brick gone	BG	Tile broken	TB
Brick loose	BL	Tile cracked	TC
Crown gone	CG	Tile gone	TG
Peeling T-lock	PTL	Tile loose	TL

TABLE 7.2   Sample Defect Codes used in Sewer Pipes

There are several ways to categorize and classify pipelines defects. Table 7.2 presents sample defect codes that may be used in sewer pipes. Similar codes can be used for different types of pipes. Table 7.3 presents a six category pipeline condition assessment system.

The defects listed in the "Condition Description" column of Table 7.4 include only the most common/critical defects affecting the structural integrity of a pipeline. Also, in this column, "light,"

Category	*Sample* Defects	Action/ Response	Response Period
I	• Severe corrosion throughout • Few areas with tile remaining • Exposed rebar • Soil voids visible	Renewal or replacement	Immediate
II	• Heavy to severe corrosion • Large areas with tile missing • Some rebar corrosion	Renewal or replacement	2–3 months
III	• Moderate to heavy corrosion • Many large areas with tile missing	Follow-up investigation	1 year
IV	• Light to moderate corrosion • Some large areas with all the tile missing	Follow-up investigation	3–5 years
V	• Slight corrosion in some areas • Patches of tile missing	Follow-up investigation	5 years
VI	• No corrosion visible • Some individual tiles missing	Follow-up investigation	5–10 years

[a] Overlapping of categories can occur as a result of the six categories lending themselves to subjective interpretations.

TABLE 7.3   Six Category Pipeline Condition Assessment System[a]

Category	Condition Description	Action Required	Defect Score (points/ft)
A	Very good • Condition is almost like new pipeline pipe	No repairs • Future routine inspection	Less than 10
B	Good to fair condition • Light cracks localized • Light corrosion localized • Light roots localized	No immediate repairs • Possible preventive measures such as chemical addition, other treatment, or maintenance to stabilize existing condition • Schedule next inspection in the order of pipeline system priority	10–25
C	Fair to poor condition • Moderate cracks/fractures • Moderate corrosion continuous • Moderate infiltration continuous • Moderate roots continuous	Routine repairs • Includes planning, environmental documentation, technical investigations, design, reviews, bid and award following established priorities. Possible preventive measures	25–65
D	Very poor condition • Severe cracks/fractures • Broken pipe with holes • Severe corrosion • Severe infiltration/roots	Expedited repairs • Includes fast track construction, accelerate planning/design, and preventive measures to avoid emergency	More than 65

TABLE 7.4 Pipeline Condition Categories (*Continued*)

281

Category	Condition Description	Action Required	Defect Score (points/ft)
E	Emergency condition • Collapsed pipe/street • Dirt pipe (CPD) • Crown of pipe gone (CPC, CG) • Void in backfill around pipe • Full flow obstruction/blockage	Emergency repair • Initiate urgent procedure	Ranking score not applicable

Table Notes:

A. All televised pipelines are ranked on a priority point per ft basis according to the severity of their defects. The ranking list is further subdivided into categories to identify and quantify the urgency of the corrective work needed.

B. The identification and ranking procedures are conservative. Certain defects may receive a high number of points to move them into the emergency category and to target them for immediate review. For example, upon discovery, a perceived emergency condition, although questionable, should be placed into Category E for immediate review by the P.E. in charge of emergencies.

C. Upon verification of an emergency condition, special order procedures for emergency pipeline repair work are initiated. Otherwise, the condition is upgraded to Category D. Once the emergency repair is completed, the repaired pipeline is ranked according to the remaining defects.

D. Category E condition pipelines must be repaired as quickly as possible (in accordance with emergency procedures) and therefore, should not be included on a priority list of pending projects.

TABLE 7.4 Pipeline Condition Categories

"moderate," and "severe" are meant to take into account the overall severity as well as frequency of defects affecting the entire pipe.

In another method, inspected pipelines can be assigned Categories A, B, C, and D on the basis of their unit scores, as listed in the "Defect Score" column of Table 7.4. Category E defects can be identified solely on the presence of critical defects noted during the closed-circuit television (CCTV), or other inspection methods. The defect score is not applicable for Category E pipelines.

The National Association of Sewer Services Companies (NASSCO)–Pipeline Assessment Certification Program (PACP), defines, in detail, each defect category and provides a program to train CCTV operators to identify these defects in a common descriptive language uniform for the United States and Canada. The condition data identified by the CCTV operator are then compiled and, through the use of a rating system, a priority level for renewal is assigned. This condition assessment will allow the designer to select the most applicable technology to accomplish the renewal. Based on NASSCO guidelines, Table 7.5 presents different types of defects that will impact CIPP installations. Figures 7.1, 7.2, and 7.3 represents the structural, operational and maintenance, and construction defects respectively.

7.3 Main CIPP Characteristics

As said previously, the primary components of CIPP are a flexible fabric tube and a thermosetting resin system. For typical CIPP applications, the resin is the primary structural component of the system. These resins generally fall into one of the unsaturated polyester, vinyl ester, and epoxy generic groups, each of which has distinct chemical resistance and structural properties.

Unsaturated polyester resins were originally selected for the first CIPP installations due to their chemical resistance to municipal sewage, good physical properties in CIPP composites, excellent working characteristics for CIPP installation procedures, and economic feasibility. Unsaturated polyester resins have remained the most widely used systems for the CIPP processes for over three decades.

Vinyl ester and epoxy resin systems are used in industrial and pressure pipeline applications, where their special corrosion and/or solvent resistance, tensile strength, and higher temperature performance are needed. These systems can also be used in sanitary sewer systems. In drinking water pipelines, epoxy resins are required.

The primary difference between heat cured and UV light cured resins is how the resin cure is activated and the workable life of the resin catalyst mixture (time from when the resin is first mixed, or catalyzed, until it cures at a specified temperature). Heat activated resins are designed to cure when a heat source is applied for

Defect Type	Defect Description	Possible CIPP Impact
Structural Defects (see Fig. 7.1)		
Cracks	A crack in the pipe is visible, but the pipe is still in place with no distortion.	Cracks typically will not reflect through the CIPP. This pipe condition is in an ideal time to renew the pipe. CIPP installation will result in a well-shaped, round pipe within a pipe.
Fractured pipe	A crack that has become visibly open although the sections of pipeline pipe wall are still in place and not moved or displaced.	Fractures typically will not reflect through the CIPP. This pipe condition is still in an ideal time to renew the existing pipeline, resulting in a well-shaped pipe within a pipe.
Broken pipe	Pipe where pieces are noticeably displaced and have moved from their original position.	The CIPP can be installed when the pipe is broken, but the displaced pipe sections will reflect in the final installed product.
Pipes with holes	Hole refers to a pipe which visibly has an unintentional hole in the pipe wall. Service connections are intentional holes in the pipe. A hole can be where the pipe has broken away and is missing.	The CIPP can be installed with a hole in the pipe. The hole configuration will reflect in the final installed product much like a service connection dimple. Stabilization of the surrounding soil structure should be evaluated.
Deformed pipe	The pipe is damaged to the point that the original cross-section of the pipeline is noticeably altered.	The CIPP can be installed when the pipe is deformed, but the deformation will reflect in the final installed product. If the crown of the pipe has developed a reverse curvature, the installed CIPP must be designed by a qualified engineer to determine the design requirements.
Collapsed pipe	A collapse is where deformation is so great there has been a complete loss of integrity of the pipeline with more than 40 percent of the cross-sectional area lost.	The CIPP should not be installed when the pipe collapse is greater than 40 percent of its cross-sectional area. A point excavation should be made to correct the collapsed section, and then the CIPP should be installed.
Joint problems	Joints can be offset, separated or angled.	The CIPP can be installed when the pipe is offset, separated, or angled less than 40 percent of the cross-sectional area. The existing condition, however, will reflect in the final installed product.

Pipe surface damage	Surface damage can appear in a number of ways. The surface can be rough, it can have aggregate missing or projecting, steel reinforcement visible and projecting, and more.	Most surface damage will not prevent the proper installation of CIPP.
Missing pipe sections	A section of pipe is missing with only the dirt left in place.	Typically, a preliner is installed first. Then the CIPP is installed inside the preliner. Grouting and stabilizing the surrounding soils should be evaluated.
Previous point repairs	An area of the existing pipeline system that was previously excavated and repaired with an undersized pipe section.	A CIPP liner can be installed, but will significantly wrinkle through the undersized pipe section. The better alternative is to replace the undersized pipe with the correct size and then install the CIPP.
Brick pipe defects	Include missing bricks, missing mortar, and loss of crown key and dropped invert.	A CIPP liner can be installed but will take the shape of the existing brick pipe. The alternative is to partially repair the existing pipe and restore its shape, then install the CIPP.
Operational and Maintenance Defects (see Fig. 7.2)		
Deposits & tuberculation	A multitude of different types of deposits may be found in an existing pipeline system.	All deposits should be removed from the existing pipe before installing the CIPP.
Tree roots	Roots are commonplace in various forms in a pipeline including fine roots, tap roots, medium roots, and root balls.	All roots must be removed from the existing pipe before installing the CIPP.
Infiltration/ inflow (I & I)	Infiltration into a pipeline can range from a weeper, a dripper, a runner to a gusher.	CIPP can be installed effectively, with minor infiltration such as weeper, drippers, and runners. Gushers should be sealed or a preliner used before installing the CIPP.
Obstructions in the pipe	Obstructions in the pipeline can vary from bricks and debris to pipes and cables installed through the pipeline.	All obstructions that will prevent the proper installation of the CIPP must be removed from the existing pipe.

TABLE 7.5 Existing Pipe Defects and Possible CIPP Impacts

Defect Type	Defect Description	Possible CIPP Impact
		Construction Defects (see Fig. 7.3)
Protruding service connections (hammer tap)	A protruding connection or hammer tap is a service connection installed, by a plumber into the main pipeline, usually by breaking a hole in the mainline and inserting the lateral pipe.	Protruding service connections should be cut back to within $\frac{1}{2}$-in. to 1-in. of the pipeline pipe wall before installing the CIPP. Lining over a long protruding connection may be possible however this will create a significant protrusion in the CIPP which may cause a blockage.
Intruding joint seal	Joint seal material between two pipe sections has become loose and is protruding into the pipeline pipe.	All significantly intruding joint seals should be removed, from the existing pipe, before installing the CIPP.
Pipe shape change	This occurs when a pipeline changes shape either from round to oval or from one size to another without the existence of a manhole at the change location.	Pipeline pipe size changes can be accommodated with a custom designed tube. Some CIPP wrinkling, and gaps behind the CIPP, will occur at the transition point of the size change.
Pipe material change	This occurs when the pipe material is changed in mid-section without a manhole being constructed.	Pipeline material changes should not affect the installation of a CIPP unless the size also changes. Then a custom tube should be fabricated.
Pipe alignment change	During installation, pipe joints may be deflected and brick pipe may be constructed to follow a curve in the road.	Small deflections and curves (see Section 7.2) in alignment of an existing pipe are typically not a concern when installing a CIPP.

Source: NASSCO

TABLE 7.5 Existing Pipe Defects and Possible CIPP Impacts (*Continued*)

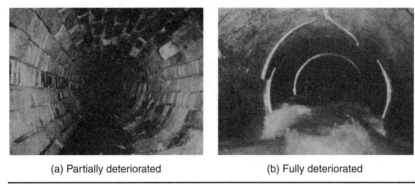

(a) Partially deteriorated   (b) Fully deteriorated

FIGURE 7.1   (a) and (b) Structural defects. (*Source: Insituform Technologies.*)

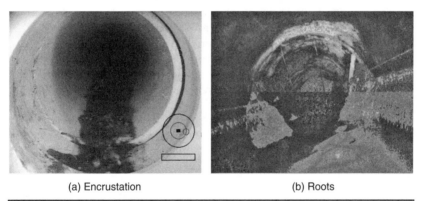

(a) Encrustation   (b) Roots

FIGURE 7.2   (a) and (b) Operational and maintenance defects. (*Source: Insituform Technologies.*)

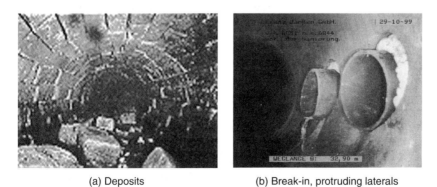

(a) Deposits   (b) Break-in, protruding laterals

FIGURE 7.3   (a) and (b) Construction defects. (*Source: Insituform Technologies.*)

a prescribed period of time. The workable life of a heat activated resin can be extended using ice or refrigeration to keep the resin temperature at a reduced level.

UV light activated resins are designed to cure when a UV light source is applied for a short period of time. The workable life of a UV light activated resin is typically equal to the resin life after being manufactured. If the wet-out tube remains isolated from UV light, then it can remain stored for weeks and months.

Ambient cured resins are designed to cure in a fixed period of time at a predetermined temperature after the resin and the catalysts are mixed together. Typically, if the ambient temperature is higher than the baseline temperature, the resin cure rate will be faster, conversely, if the temperature is lower, then the resin cure rate will be slower.

The primary function of the fabric tube is to carry and support the resins until it is installed and cured. This requires that the fabric tube withstand installation stresses with a controlled amount of stretch, but with enough flexibility to dimple at side connections and expand to fit existing pipeline irregularities. The fabric tube material can be woven or nonwoven with the most common material being a nonwoven, needled felt. Impermeable plastic coatings are commonly used on the exterior and/or interior of the fabric tube to protect the resin during installation. The layers of the fabric tube can be seamless, as with some woven material, or longitudinally joined with stitching or heat bonding.

In addition to the type of resin, the primary differences between the various CIPP systems are in the composition and structure of the tube, method of resin impregnation (by hand, by vacuum, or by pressure), installation procedure, and curing process. Typical CIPP design properties are shown in Table 7.6.

7.3.1 Tube Wet-Out

Wet-out is the process of introducing the catalyzed resin liquid into the fabric tube. Wet-out of a tube liner can be accomplished several ways. The traditional wet-out is termed vacuum impregnation. Other systems that are equally as efficient are resin immersion and resin bath wet-out.

Property	ASTM Test Method	Polyester (psi)	Vinyl Ester (psi)	Epoxy (psi)
Tensile strength	D638-08	2000–2500	2500–3500	3500–4000
Flexural strength	D790-07	4000–5000	4000–5000	4000–5000
Flexural modulus	D790-07	250,000–500,000	250,000–350,000	300,000–400,000

TABLE 7.6 Typical CIPP Design Properties

In the case of vacuum impregnation, once the resin and catalyst have been mixed, it is pumped into the fabric tube. Serial vacuum impregnation involves placing a vacuum suction attachment at predetermined intervals, evacuating the tube of air, allowing the resin to fill the evacuated liner area and patching the tube to seal the penetration of the vacuum device. Other impregnation techniques include resin bath immersion or utilizing gravity for the resin/catalyst mixture to fully impregnate the tube. Regardless of the method, *total saturation of the fabric tube is critical and mandatory*. Most tube coatings are transparent, which allows for a visual confirmation that the tube has been thoroughly wet-out with no apparent visible dry spots existing. During vacuum impregnation (wet-out) operations, the tube is fed through a set of rollers with a predesigned gap setting. The roller gap setting determines the volume of resin/catalyst mixture that is impregnated per unit length of wet-out tube. The roller gap setting will predetermine the final wall thickness of the cured pipe. If the gap is not set wide enough, the tube will have insufficient resin and may be below design standards. If the gap setting is too wide, excess resin will be placed in the tube. Figure 7.4 illustrates the resin impregnation (wet-out) process.

Factory wet-out is the most common method for vacuum impregnating the resin/catalyst system into the fabric tube. In the factory, the materials are stored in a controlled environment not subjected to weather or severe temperature changes. Typically, the resin is stored in large containers under controlled temperatures and the tube is allowed to reach required temperature under the factory environment.

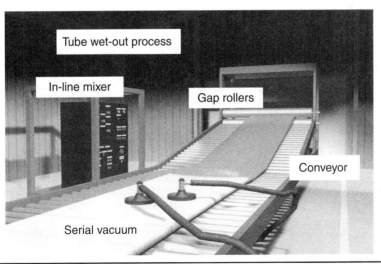

FIGURE 7.4   The resin impregnation (wet-out) process. (*Source: Insituform Technologies.*)

The wet-out equipment is generally fixed in place and can be maintained on as needed basis. Tubes can be wet-out using a static mixing system that regulates and combines the correct amounts of resin and catalyst for a proper wet-out of the tube. Once the tube is wetted-out, it can be placed, on ice, into a refrigerated unit and delivered to the job site. The majority of CIPP tube wet-out is performed in a factory setting. Specially designed truck trailers make it possible to haul large diameter and long length tubes.

7.4 CIPP Installation Methods

As mentioned earlier, there are two basic methods of CIPP installation: the pulled-in-place process and the inversion process. Different manufacturers provide specific variations of these techniques. This section provides more details for each method.

7.4.1 Pulled-in-Place Process

Using a power winch, the impregnated tube is typically pulled through an existing manhole or other approved access point, fully extending the tube to the next designated manhole or termination point. Water and/or air are used to inflate the liner after it is pulled in the existing pipe. Care needs to be exercised not to overstress and damage the tube during pull-in as a result of increased friction, especially where curve alignments, offsets, protruding services, and other friction-producing existing pipe conditions are present. The pulling force induced on the tube, as it is being pulled into the existing pipe, should be monitored and should not exceed the recommended maximum allowable pulling force for the size and thickness of tube being installed.

7.4.2 Inversion Process

In the inversion method, the resin saturated tube is inserted through an existing manhole or other approved access point and then installed into the existing pipe by means of a pressurized (water or air) head, with a volume sufficient to fully extend the tube to the next designated manhole or termination point. The speed of inversion typically should not exceed 30 ft per minute for small diameter tubes. Larger diameter tubes are installed significantly slower as recommended by the manufacturer. Figure 7.5 illustrates inversion technique of CIPP installation.

In order to monitor the curing temperature, a thermocouple or temperature measuring device is installed at access points between the existing pipe and the CIPP liner at the pipe invert or 6 o'clock position. These devices can be placed at each end and at all intermediate manhole access points along the length of the installation.

FIGURE 7.5   Inversion technique of CIPP installation. (*Source: Insituform Technologies.*)

The manufacturer should be consulted for a recommended use of thermocouples applicable to their liner system. The thermocouple wires can be used for both pull-in and inverted techniques, and are critical for monitoring the curing of the resin/catalyst system. These sensors must be installed before the tube is in place or before the tube is pressurized.

7.4.3   Preliner Options

In some cases it may be necessary to install a preliner into the existing pipeline to prevent washout or contamination of the resin, or to reduce friction during liner installation. A preliner may be required for heavily leaking pipelines, existing pipes located under groundwater, in contaminated soils, and existing pipes containing coatings that are not compatible with the liner and resin being installed. Preliners will remain in place and become part of the liner pipe installation. Depending on the nature of the leakage or friction, two other options are available: grouting of the existing pipeline to eliminate all infiltrating water, or installing a foil along the invert of the existing pipe to reduce friction during pull-in of the CIPP liner.

7.4.4   CIPP Liner Curing

As said earlier, there are three methods of resin curing. The most common method is curing by hot water. In this method, cure temperatures are monitored at access points and at each end of the installation. Figure 7.6 illustrates hot water curing. The second method is curing with hot air or steam. In this method, once the liner is in place and

Figure 7.6 Hot water curing. (*Source: Insituform Technologies.*)

properly inflated with air, the resin is cured using steam. Typically the CIPP components (such as tube material and resin) must be an integrated system specially designed for steam curing.

The third method is UV light curing. First, a fiberglass liner is pulled and inflated into the existing pipe. Then a CCTV camera with the UV light apparatus is inserted in one end and pulled to the opposite end recording the precure condition of the liner. The UV light apparatus is then pulled back to cure the liner, at a regulated speed, with the CCTV camera recording the curing process.

7.5 Inspecting CIPP Installation

A variety of project and site conditions can exist during the CIPP installations. Conditions can vary from wet to dry and from hot to cold environments. The inspector must document site and existing pipe conditions and make sure that the CIPP is installed according to the contract and specifications.

The inspector must become thoroughly familiar with the contract documents for the project and most importantly the technical specifications. The technical specifications will guide the inspector about inspection areas, documentation requirements, test samples (identification, type, number of test samples required, location, maintenance and delivery, testing agency, documentation, and so on), to ensure that the CIPP is installed with specified quality. The specifications determine degree of inspection a specific project requires. Table 7.7 presents the various actions to be taken by the inspector on various operations before, during, and after installation of CIPP.

Operation	Inspector Actions
Liner material delivered on site	• Inspect the liner condition and verify that the delivered material is not damaged nor has prematurely exothermed before it is installed. • The liner wet-out report should be reviewed and matched to the liner material and resin, submitted by the contractor for the project. Record all information on the inspector report.
Sample homeowner notification	• Verify that homeowner notifications have been distributed as required in the contract documents and document on inspection report.
Preinstallation cleaning and CCTV inspection of the pipe	• View CCTV inspection video of the existing pipe just prior to the installation of the liner by the contractor. Note condition and cleanliness of pipe.
Pipe plugging or bypassing	• Document the type of flow bypass provided by the contractor, including size and number of pumps and the size of the discharge piping installed by the contractor, as specified or required.
Traffic control	• Document the traffic control provided by the contractor including detours, flagmen, and police details as required by approved submittals.
Liner samples for testing	• Document that provisions for liner samples have been made by the contractor and record on inspection report.
Monitoring CIPP cure	• Maintain complete and accurate records of every step of the curing process and record all thermocouple temperatures from the start of the process to complete cool down of the liner.

TABLE 7.7 Inspector Action on Various Operations

293

Operation	Inspector Actions
Cure pipe samples	• Obtain samples of the cured liner directly from the contractor at the project site. • Send cured liner samples to an approved laboratory, maintaining a written chain of custody. • Review laboratory test results of the cured liner samples and compare to specified requirements in the contract documents. • Expedite testing and test results in a timely manner.
Service line reconnection	• Verify and document the condition of all service connections reinstated by the contractor.
Quality assurance & testing	• Visually inspect the installed CIPP and record all defects and any infiltration that is observed.
Defect documentation	• Record all defects encountered and have manufacturer recommend repair procedures.
Fit in existing pipe	• Check fit between the existing pipe and the CIPP at the manhole and record on the inspection report.
Installation lengths	• Check the length of the CIPP. It should protrude several inches into each manhole or at a minimum be cut flush with the manhole wall. Document on inspection report.
Physical property testing of cured CIPP	• Verify CIPP physical property compliance with the contract document requirements through implementation of testing methods. • Verify water tightness of CIPP in accordance with the specified requirements of the contract specifications and document on the inspection report.

Liner is not properly or sufficiently iced down at the folds in the liner during transport in the refrigerated container (premature exotherm)	• Cutout the exothermed section and, if possible, use this liner in another pipe. • Discard liner and furnish a new liner for installation.
Odors	• The resin component of the CIPP is typically a styrene-based material. The odor of styrene can typically be detected in extremely small quantities and may be detected at the manhole openings and sometimes it will migrate into a home if the home's trap is defective. The presence of styrene can be readily detected at levels of 1–2 parts per million (ppm) or less. The threshold for continuous exposure to styrene in a confined factory setting is 50 ppm. When exposed to the atmosphere, styrene will dissipate very quickly. When an odor is detected it should normally dissipate in a matter of several hours and should not present a harmful environment for personnel in the vicinity of the installation. If styrene odors are detected in a residence, ventilate to eliminate the odors.

Source: NASSCO

TABLE 7.7 Inspector Action on Various Operations (*Continued*)

7.6   Pipe Plugging or Bypass Pumping

Bypass pumping of the existing fluid is required for the installation of CIPP. With most small diameter pipelines, plugging may be adequate, but must be monitored on a regular basis to prevent flow backups. Service connections in the residential systems typically do not require bypass pumping. The homeowners are notified usually a week or two in advance and again 24 hours in advance to refrain from heavy use of water while the CIPP installation process is underway. A small amount of water from the service connection typically will not affect the installation of the CIPP. In commercial and industrial applications it may be necessary to bypass flow from each individual service connection, since flow interruption may not be an option. In these cases, individual cleanouts are identified and a small pump is set-up directing the flow to the mainline bypassing. In most cases, the project owner will require a pumping capacity that is equal to the anticipated peak flow (either recorded or anticipated) for the line section, plus 50 to 100 percent redundancy in the event the primary system fails to function. Most pump rental suppliers will determine bypass pumping suction and discharge configurations and setup requirements at the jobsite. However, pumping operations for the duration of the project is typically the responsibility of the general contractor. Figure 7.7 illustrates typical bypass pumping setup for a large volume of existing flow.

FIGURE 7.7   Typical bypass pump setup for a large volume of existing flow.

7.7   Quality Assurance and Testing

7.7.1   CIPP Inspection and Acceptance

The CIPP installation may be inspected visually if appropriate. Additionally, CCTV is commonly used for pre- and postinstallation inspection purposes. Variations from true line and grade may be inherent with CIPP because of the conditions of the existing pipe. However, certain conditions must be met, such as the new CIPP should not show any infiltration of groundwater. At the conclusion of the installation, all service connections must be accounted for and reconnected. Table 7.8 presents *sample* defect codes for CIPP lining evaluation.

7.7.2   Workmanship

The CIPP should be continuous over the entire length of an installation and be free of dry spots, lifts, and delaminations. If these conditions are present, the CIPP will be evaluated for its functionality to meet the applicable requirements of the contract documents. If the CIPP does not meet the requirements of the contract documents or the functions expected based on the technical specifications, the affected portions of CIPP may have to be repaired according to the contract provisions. Some project owners may apply a penalty in addition to the CIPP repair and/or replacement requirements (see following sections). See Sec. 2.8 for sample CIPP technical specifications.

7.7.3   Quality Control Issues

Design Issues

The key design consideration in CIPP liner installation is the ability of the cured-in-place pipe to withstand buckling due to ground water

Defects	Codes	Defects	Codes
Lining shrinkage	LS	Lining bubbles	LB
Lining wrinkles	LW	Lining fins or folds	LF
Foreign inclusions	FI	Lining delamination	LD
Lining cracks	LC	Hot dogs	HD
Dry spots	DS	Mis-cuts	MC
Lining wrinkles	LW	Annular space	AS
Lining elongation	LE	Lining thickness	LT
Lining dimples	LD	Lining lumps	LL
Under-cuts	UC	Overcuts	OC

TABLE **7.8**   Sample Defect Codes for Pipeline Lining Evaluation

Condition of Pipe	Design Parameters	Liner Thickness
Partially deteriorated	Groundwater level increases	Increases
	Ovality increases	Increases
Fully deteriorated	Groundwater level increases	Increases
	Soil cover increases	Increases
	Ovality increases	Increases
	Soil modulus increases	Decreases

TABLE 7.9   Factors Affecting Design of CIPP

and collapse due to soil and live loads. The resin and the lining tube make a composite material to determine the design thickness of the CIPP. The types of resins have been described in Sec. 7.3. There are generically two types of lining tubes—nonreinforced and reinforced liners. The reinforced liners can be divided into fiberglass and carbon fiber laminates.

Creep causes the CIPP to change shape, reducing its capability to resist external pressures. Factors influencing the design of a liner include physical properties of lining technology, and site and project specific conditions as shown in Table 7.9. The pipe wall thickness design is based on the anticipated effects on pipeline creep due to external loads over a 50-year period as specified by ASTM F1216-09, Eqs. X1.1 and X1.3. See Sec. 2.7 for sample CIPP design calculations.

Quality Issues

This section describes a *suggested* method for addressing CIPP quality issues. First, the project inspector will review CCTV post video within two weeks of submittal by the contractor. In the event the inspector finds that a corresponding tolerance level has been exceeded, the inspector will flag the defect and handover the CCTV post video to the project manager representing the owner or the city engineer for secondary review. If the project manager agrees that the defect is in excess of acceptable tolerance levels, he or she will determine acceptable remedies to correct the defects at no extra cost to the project owner. If the contractor chooses not to correct the defects, a deductive penalty can be imposed against the contractor. Once the corrective work has been performed by the contractor to the satisfaction of the project manager and the inspector, full payment for the CIPP must be issued. In the event the contractor chooses to accept the penalty deduction stipulated in the contract rather than to correct the defect, a change order will be issued for the deduction and the remaining balance for the subject CIPP installation will be paid to the contractor.

Testing Issues

CIPP samples must be tested based on the ASTM guidelines previously mentioned. If testing results fall short of required flexural modulus or thickness requirements, the required stiffness and actual stiffness of the lining samples will be compared. After comparison, if the average of actual stiffness is greater than 94 percent of the required stiffness, the pipe will be accepted, but marked as noncompliant and a deductive penalty proportional to the stiffness deficiency can be assessed against the contractor. A change order will be issued for the deductive penalty, in proportion to the deficiency, and the remaining balance for the CIPP section will be paid to the contractor. If the average of actual stiffness for the CIPP section is less than 94 percent of the required stiffness, the CIPP section will not be accepted and the contractor will be required to install a second liner over the first liner to compensate for the deficiency, at no extra cost to the project owner. Full payment for the CIPP section will be issued to the contractor after the second liner is properly installed and accepted. Assuming an acceptable second liner is properly installed, no deductive penalty will be imposed against the contractor.*

To identify every defective condition and to suggest every corrective measure or penalty is beyond scope of this book. This section is only intended as a guide to aid in identifying physical conditions of the installation, applying appropriate corrective measures, and assigning deductive penalties for defective installations, which should all be included in the bid and contract documents. This way the contractors know, before the bidding and start of the project, what expected quality measures are and possible outcomes if these quality measures are not met. This way, many quality problems, disputes, and potential litigations may be avoided. The final decision for application, enforcement, and final acceptance will ultimately be made by the owner's project/construction manager or the city engineer on a case by case basis and as stipulated in the contract documents. Using codes provided in Table 7.8, Table 7.10 presents a summary of *potential* CIPP defects, *suggested* acceptable tolerances, and *possible* repaired types that may be required. Figure 7.8 illustrates potential CIPP defects.

---

*The measured wall thickness has much more impact upon buckling resistance than the flexural modulus. Its value adjusts the buckling resistance by its cube power and is not directly proportional. Design engineers and project owners should be cautioned that contractors may submit higher bids, or choose not to submit a bid, if the contract contains unjustified quality requirements, low tolerance expectations, and possibilities of penalties or rework. Including or not including such language in the contract and justification of level of quality required needs to be evaluated based on project owner's needs, on the true project hydraulic, structural, and performance requirements, and must be evaluated in case by case basis using judgment of a knowledgeable professional engineer.

Defects	Probable Cause	Accepted Tolerance	Repair Requirements	Repair Types	Monetary Penalties
Wrinkles/fins/folds (LW)/(LF)	Undersized or deteriorated localized section of existing pipe.	Generally 5 percent of nominal diameter in top and 3% of nominal diameter in bottom of the existing pipe is tolerated, assuming such does not impede the normal flow characteristics.*	Repair if exceeding 5 percent of nominal diameter of existing pipe along top of the pipe (from 9 to 3 o'clock position) and 3 percent of nominal diameter of existing pipe along bottom of the pipe (from 3 to 9 o'clock position).	Removal of wrinkles (fins) and coat area with approved epoxy material. Trim to within 1/4 in. of pipe wall surface.	No deduction, assuming adequate thickness is achieved.
	The corrugated pipe may have been measured incorrectly at the inside diameter or high point of the corrugations, rather than the center of the corrugations.	The wrinkles are relatively minor and should not impede the flow characteristic of the pipe.	No action required.	N/A	No deduction, assuming adequate thickness is achieved.

Defect	Cause	Acceptance criteria	Action	Remedy	Deduction
Wrinkles (fins)—circumferential	1. Low head during inversion. 2. Pipe undersized due to deterioration in local area.	The wrinkle is acceptable if it does not disrupt or impede the normal flow characteristics of the pipeline.	The wrinkle should be cut out if it impedes the flow.	Removal of wrinkles (fins) and coat area with approved epoxy material. Trim to within 1/4 in. of pipe wall surface.	No deduction, assuming adequate thickness is achieved.
Blisters/bubbles	Inadequate bond between the fabric and the coating during manufacture.	Repair if exceeding 5 percent of nominal diameter of existing pipe.	Leaking pipe at defective blister.	1. Short liner patch repair. 2. No trimming or grinding permitted; point repair with lined section installed flush with original liner required.	No deduction, assuming a proper point repair or dig and replace is completed. If contractor chooses not to dig and replace a deduction can be applied for the reach.†
Lifts in liner	1. Premature pressure loss during installation. 2. Insufficient cure time calculated.	Repair if lift in liner is exceeding 5 percent of nominal diameter of existing pipe.	The lift should be cut out.	If the lift is hard, remove the liner section or remove the lift area and install a new liner.	N/A

TABLE 7.10 Addressing Potential CIPP Defects

301

Defects	Probable Cause	Accepted Tolerance	Repair Requirements	Repair Types	Monetary Penalties
Bulges/ lumps (LL)	1. Displaced broken or fractured pipe typically visible near a pipe joint. 2. Debris or rocks remaining in the pipe invert that was lined over.‡	Repair if bulges or lumps in liner is exceeding 5 percent of nominal diameter of existing pipe or if causing flow restrictions.	A new bulge in the invert, caused by residual debris left in the pipe that impedes the flow characteristics of the pipeline, should be cut out.	1. The area where the existing liner has been removed should be replaced by a short-liner patch repair. 2. Install point repair with smooth transition.	No deduction, assuming a proper point repair or dig and replace is completed. If contractor chooses not to dig and replace a deduction can be applied for the reach.†

Pinholes, cuts, and leaks through coating	1. Defects in the coating caused during manufacture and/or installation. 2. Liner may have been punctured during delivery and installation into the existing pipe. 3. When pulling the robotic cutter through the lined pipe, if the cutter bit is extended, damage to the liner coating may occur.	Pipe should have no visible leakage.	Should be repaired if leaking.	The area where the existing liner has pinholes should be patched with a short-liner repair.	N/A
Soft spot in liner	1. Cold spot at the bottom of the pipe. 2. No thermocouple installed to check curing temperatures. 3. Not heated long enough to cure the entire liner.	A soft liner is not acceptable and needs to be reheated and hardened or cut out.	All of the soft areas need to be cut out.	As recommended by the manufacturer.	N/A

TABLE 7.10 Addressing Potential CIPP Defects (*Continued*)

303

Defects	Probable Cause	Accepted Tolerance	Repair Requirements	Repair Types	Monetary Penalties
Dry tube or white spots	Liner not properly wet-out before installation.	Dry liner sections are not acceptable.	Dry liner section needs to be removed.	Short liner (patch repair) installed over dry area.	N/A
Hole in the liner	Pilot hole cut in liner or other defect.	Holes in the liner are not acceptable.	The hole needs to be repaired.	Holes can be repaired with epoxy as recommended by the manufacturer Short liner installed over the hole.	N/A
Loose or peeling vacuum patches	Patch not properly bonded to the liner after vacuum line is removed.	No action required unless pipe is leaking.	Should be repaired if leaking.	If leaking, install a short liner over the loose vacuum patch.	N/A
Crack in the liner	Shrinkage cracks may occur in smooth pipe where there is little or no locking between the existing pipe and the CIPP.	N/A	N/A	A short liner repair is installed over the cracked area.	N/A
Defective and leaking bung holes	Bung hole in liner (resin injection hole) was not patched correctly and is leaking.	No action required unless pipe is leaking.	Should be repaired if leaking.	A short liner repair is installed over the cracked area.	N/A

Seam tape, loose, blistered or leaking	Seam tape inadequately bonded during manufacture.	N/A	Tape should be removed to prevent potential flow blockage.	N/A
Annular space between existing pipe and liner at manhole/annular space (AS)	1. Insufficient pressure to fully expand the liner. 2. The end of the liner was not pulled sufficiently into the manhole for full expansion at the manhole face. 3. A factor of existing pipe geometry.	Should be repaired if leaking	1. If leaking between the existing pipe and the CIPP, inject a hydrophilic type grout to stop the leakage. 2. If no leakage, a cementitious grout can be used to fill the space.	No deduction, assuming a proper seal is achieved at each end and lateral. If contractor chooses not to dig and replace, a deduction can be applied for the reach.†
Ground water leakage between the liner and existing pipe at the manhole	Small gap between liner and existing pipe allowing water to penetrate.	N/A	In groundwater situations install a hydrophilic seal at the manhole interface between the existing pipe and the liner.	N/A

Adapted from NASSCO

*Wrinkles/fins/folds are diameter dependent, for example, 1/4-in. is a small wrinkle especially in medium and large CIPP.
†Contractors may choose to offer a deduction if the product does not meet the specification and is not acceptable to owner.
‡The debris and rocks should have been removed during cleaning operations.

TABLE 7.10 Addressing Potential CIPP Defects (*Continued*)

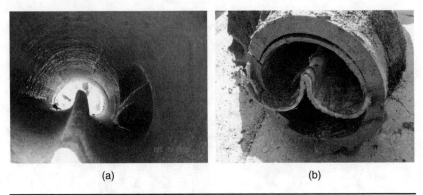

(a)   (b)

FIGURE **7.8** (a) and (b) Lifts (*CIPP defect*).

7.8 Summary

The CIPP technology can address a wide variety of pipeline problems. However, if pipe defects are not identified in the early stages of deterioration and are allowed to cause a pipe collapse, the CIPP technology may no longer be applicable and traditional methods of excavation and replacement may become the only solution. The inspection of a CIPP installation starts with a thorough knowledge of the technical specifications for the project and a full knowledge of CIPP technology. In order to properly inspect CIPP installations, inspectors need to understand quality requirements of specifications. Quality measures and possible outcomes of not meeting these quality measures should be clearly stated in the bid documents. Owners and engineers should have reasonable tolerances for CIPP installations based on requirements of their projects. Unreasonable quality measures may result for contractors to submit higher bids or discouraging contractors from submitting bids.

CHAPTER 8

Inspection and Quality Assurance/Quality Control for Trenchless Installation and Replacement Methods

---

8.1 Conventional Pipe Jacking

8.1.1 Introduction

Conventional pipe jacking is defined as a system of directly installing pipes behind a shield machine by hydraulic jacking from a drive shaft such that the pipes form a continuous string in the ground. Usually personnel are required inside the pipe to perform the excavation [such as operating a tunnel boring machine (TBM)] or spoil-removal process. The excavation can be performed manually or mechanically (see Fig. 8.1).

8.1.2 Materials

Pipe

The type of pipe used for the pipe-jacking method must be capable of transmitting the required jacking forces from the thrust plate in the jacking shaft to jacking field or TBM. Steel reinforced concrete (RCP), cenrifugally cast fiberglass-reinforced polymer mortar (CCFRPM),

Figure 8.1   Conventional pipe jacking. (*Source: Akkerman, Inc.*)

and polymer concrete (PCP) are the most common types of pipe used in pipe jacking.

Allowable Forces

Pipe to be jacked must be specifically designed by the pipe manufacturer with sufficient reinforcing and wall thickness to resist, without buckling or crushing, the horizontal, vertical, and longitudinal loads applied to it during the jacking operating.

Pipe Dimensions

1. The minimum diameter for a pipe installed by pipe jacking must be 42 in. OD or 36 in. ID, as the method requires people working inside the jacking pipe. Although no theoretical maximum size limit is specified, the most common diameter ranges from 48 to 72 in., with the largest being approximately 12 ft in diameter.

2. Steel pipe must have a minimum wall thickness of 0.25 in. or as specified in the current DOT standard specifications for construction, whichever is larger. Concrete pipe must have a minimum wall thickness as specified in current DOT standard specifications for construction. Likewise, CCFRPM and PCP must have similar minimum wall thickness specifications. See the specific pipe manufacturer's literature for more information.

3. Steel pipe must have a roundness tolerance, so that the difference between the major and minor outside diameters must not exceed 1 percent of the specified nominal outside diameter, or 0.25 in., whichever is less. Likewise, concrete pipe, CCFRPM, and PCP must have a similar roundness tolerance.

4. Pipe must have square and machine-beveled ends. The pipe end maximum out-of-square tolerance must be 0.04 in. (measured across the diameter).

5. Pipe must be straight.

6. Pipe must be without any significant dimensional or surface deformities. All pipes must be free of visible cracks, holes, foreign material, foreign inclusions, blisters, or other deleterious or injurious faults or defects. Any section of the pipe with a gash, blister, abrasion, nick, scar, or other deleterious fault greater in depth than 10 percent of the wall thickness must not be used and must be immediately removed from the site.

7. Any of the following defects warrants pipe rejection:
   - Concentrated ridges, discoloration, excessive spot roughness, and pitting
   - Insufficient or variable wall thickness
   - Pipe damage from bending, crushing, stretching, or other stress
   - Pipe damage that impacts the pipe strength, the intended use, the internal diameter of the pipe, and internal roughness characteristics
   - Any other defect of manufacturing or handling

Pipe Joint Cushion

A cushioning material must be used between pipe segments to assist in distributing the jacking loads evenly across the section of the pipe, and to prevent chipping or breaking of the pipe ends due to concentrated pressure caused by any slight irregularity of the pipe ends. The most common type of material used as a cushion material is particleboard.

8.1.3 Construction

Minimum Allowable Depths

The recommended minimum depth of cover must be 6 ft or 2 times outside diameter of the pipe. For slurry pipe jacking minimum depth cover must be 6 ft or 3 times outside diameter of the pipe. In locations where the road surface is superelevated, the minimum depth of the bore must be measured from the lowest side of the pavement surface. Conventional pipe jacking can work with least cover of all systems.

Equipment

1. *Jacking frame:* A jacking frame must be constructed of guide timbers, backstop, and pushing or jacking head. Guide timbers or rails must be constructed to the exact line and grade of the pipeline and must be anchored in such a manner as to be capable of maintaining the alignment and gradient throughout the jacking operations.

2. *Backstop:* The backstop must be constructed as to provide a bearing area capable of supporting no less than 200 percent of the estimated maximum jacking pressure and must be perpendicular to the alignment of the pipe. It must be anchored and braced in a manner to ensure that this position will be maintained throughout the jacking operation.

3. *Jacking head:* The pushing or jacking head must be constructed to fit the pipe to be jacked and to ensure that the pressure developed by the jacks will be evenly distributed on the periphery of the pipe. An opening large enough to permit the entrance of men and materials must be left and maintained in the jacking head.

Method Description

1. The contractor must excavate the boring and receiving pits at the approved locations. Any required sheeting, shoring, or bracing that is required to provide for safe working conditions must be provided by the contractor.

2. After the excavation is completed, the placing and jacking of the pipe must follow immediately to avoid unnecessarily disturbing the stability of the embankment and roadbed.

3. The contractor must dewater the pit excavation in a manner appropriate for the conditions.

4. The leading section of the pipe must be equipped with a jacking head securely anchored thereto to prevent any wobble or variation in alignment during the boring and jacking operation.

5. The driving end of the pipe must be properly protected against damage, and the intermediate joints must be similarly protected by the use of sufficient bearing shims to properly distribute the jacking stresses. Any section of pipe showing sign of damage must be removed and replaced, or repaired.

6. Efforts must be made to avoid loss of earth near the cutting head.

7. If jacking pressures are high, the contractor is to be allowed to work uninterrupted, (24 hours a day/7 days

a week), until the pipe has been jacked between the specified limits.

8. If appreciable loss of soil occurs during the boring and jacking operation, the voids must be packed promptly to the greatest extent practicable with flowable fill (see DOT flowable fill requirements).

9. The contractor must protect the excavation until the operation is complete and the excavation is backfilled.

Overcut Allowance
When using this method, the allowable overcut is 1 in. greater than the outside diameter of the pipe.

Soiltight Joints
Soil tight pipe joints are required to ensure the integrity of the roadbed. Pipe must be constructed to prevent earth infiltration throughout its entire length. Conventional pipe jacking is normally for use above the water table or for de-watered ground.

Lubrication Fluids
Lubrication fluids are required for this method of pipe installation to reduce jacking forces.

Pipe Locating and Tracking

1. During construction, monitoring, and plotting, boring progress must be undertaken to ensure compliance with the proposed installation alignment and allow for appropriate course corrections to be undertaken. Longer drives over 500 ft should have intermediate jacking shaft installed.

2. Pipe installed by this method must be located in plan as shown on the drawings, and must be no shallower than shown on the drawings unless otherwise approved.

Jacking Force Calculation

1. The estimated jacking force must not exceed one-third of specified allowable jacking capacity of the pipe to avoid pipe damage.

2. The estimated jacking force must exceed the resistant force required to install pipes with the microtunneling method.

3. Influential factors of the estimated jacking force include
   - Length of drive
   - Weight, grade, and diameter of the pipe
   - Height of the overburden
   - Soil characteristics

- Water table and dewatering operations
- Load on the face of TBM
- Operational interruptions
- Overcut size
- Lubrication
- Alignment and/or grade corrections
- Intermediate jacking stations

4. If the operation is interrupted by more than a few (8) hours, the jacking force required to install the remained of the pipe may increase up to 20 to 50 percent of the original jacking force estimate.

Figure 8.2 suggests a checklist for inspection of pipe-jacking operations.

8.2   Microtunneling

8.2.1   Introduction

The work for microtunneling methods generally includes jobsite planning and mobilization, construction of drive (jacking) and exit (receiving) shafts, and simultaneous tunneling and jacking of the pipe sections. After completion of the microtunneling operation, the work continues for service lateral connections, construction of manholes over the drive and reception shafts, and all other necessary related work items, such as site restorations.

Microtunneling is a remotely controlled, laser guided, pipe jacking operation that provides continuous support to the excavation face. Theoretically, pipes of 12 to approximately 90 in. or even more can be installed using microtunneling. The process starts with jacking of a microtunnel boring machine (MTBM) and pipe sections from a jacking shaft (Fig. 8.3) to a reception shaft. The machine has a closed faced shield. Figure 8.4 illustrates the front and back of a microtunneling machine.

Excavated soil is removed using auger removal system or hydraulic removal system with slurry fluid (more common), which also counterbalances groundwater and earth pressures. The MTBM is guided by a laser or other survey device mounted in the jacking shaft, which projects a beam onto a target in the articulated steering section of the MTBM. The MTBM is steered by extending or retracting remotely controlled steering jacks. Figure 8.5 illustrates the schematics of microtunneling operation.

Most of the microtunneling machines are equipped with a cone crusher. They are designed in such a way that cobbles, boulders, and

Inspection Guide for Pipe Jacking (PJ)

Preinspection Plan Review

☐ Review geotechnical and soil reports.

☐ Ensure DOT / agency facilities and nearby utility information are shown on the plans and profiles and that the proposed alignment does not interfere with them.

☐ Note the minimum cover above the top of the pipe and below the pavement surface, or ground elevation (for longitudinal installations outside the influence of the roadway) is _____ ft.

Pipe material class _____.
Pipe material yield strength _____ksi.
Allowable pipe jacking force _____ ton.
Pipe diameter _____ in.
Pipe wall thickness _____ in.
Overcut diameter _____ in.

☐ Review contingency plan.

☐ Review job-site layout including: distance from access pits to roadbed, proposed sheeting and bracing, materials storage and fabrication area, safety devices (barrels, guardrail, etc.), and dewatering pit locations.

☐ Note unique or special items /circumstances: _____

Construction Inspection

☐ Verify traffic control is consistent with the permit requirements, one call service has been contacted, and the DOT / agency permit is on-site.

☐ Verify job-site layout is consistent with the approved plans, especially the alignment of the pipe and machine.

☐ Verify continuous monitoring records indicate bearing and grade of the leading edge of the pipe is consistent with the approved plans, dewatering effort is satisfactory, soil volume removed is consistent with projection, and that workers understand the contingency plan.

☐ Verify pipe characteristics are consistent with permit requirements.

☐ Verify jacking pipe is certified for jacking, the pipe has a smooth interior and exterior surfaces, is used within the entire influence area of the roadbed, has clean and square ends, joints are watertight, defective pipe is not used, and damaged pipe is jacked through to the receiving access pit and removed.

☐ Verify the jacking head fits square with the pipe.

☐ Verify the back-thrust block is structurally adequate, jacking force is within the range of calculated jacking force, and that the jacking force is recalculated whenever the operation is stopped.

FIGURE 8.2 Sample checklist for inspection of pipe jacking operations.

❏ Verify sufficient lubrication fluid quantity is used, and a lubrication system properly injects lubricant on the inside and outside of the pipe.

❏ Verify any unsuccessful tunnel is back-filled immediately.

❏ Verify each end of the pipe is enclosed, restorations are completed, and attach Inspector's Daily Report (IDR), form 2228.

Permit No. _____
Inspector: _____
Date: _____

FIGURE 8.2   Sample checklist for inspection of pipe jacking operations. (*Continued*)

FIGURE 8.3   Jacking shaft.

pieces of rock up to a maximum size of around one-third of the external diameter of the cutterhead can be crushed to a particle size, which can be conveyed through the discharge pipe. Figure 8.6 illustrates common types of microtunneling cutterheads.

Quality Assurance/Quality Control

Quality assurance/quality control (QA/QC) is an important element in the prequalification and selection of microtunneling contractors, pipe selection, and using appropriate equipment that must be addressed in the bid documents and project specifications. Good quality control is also important in the inspection of the pipe, and is necessary during the installation process. Difficult jobs require contractors to demonstrate that they have sufficient work-related experience (or have a joint-venture agreement with a MTBM manufacturer) by meeting prequalification requirements. The MTBM must

FIGURE 8.4   Front and back of a microtunneling machine.

be compatible with expected ground conditions. Contract specifications must address (possibly through submittal requirements), methods for slurry disposal, lubrication, intermediate jacking stations, shaft construction and MTBM entry and exit seals, installation monitoring, possible dewatering, and ground restoration, among other subjects.

Preinstallation/Construction Planning

Prior to construction, underground utilities must be located and visually confirmed. A preconstruction survey must be made of the area to establish a damage and settlement baseline. A plan for addressing traffic control, handling and disposal of spoils, shaft construction and safety must be developed prior to beginning construction. Proper shaft construction is a key quality control issue. Groundwater and soil entry into the shaft must be controlled and minimized, especially

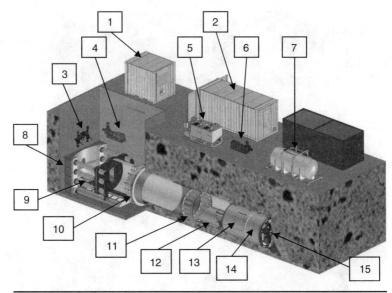

FIGURE 8.5 Microtunneling with hydraulic spoil removal in single-phase jacking. (1) Remote hydraulic power pack; (2) MTBM control container; (3) slurry pit valves; (4) crane way; (5) bentonite pump; (6) slurry feed pump; (7) cooling tank; (8) jacking frame back adapter; (9) keyhole jacking frame; (10) pipe clamp; (11) sealed intermediate jacking station; (12) MTBM trailing dolly with booster pump; (13) MTBM increase kit; (14) MTBM; (15) mixed face cutterhead with carbide discs and scrapers. (*Source: Akkerman, Inc.*)

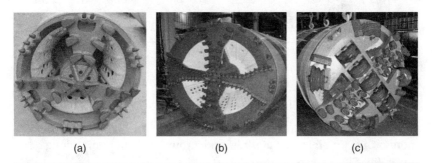

FIGURE 8.6 (*a*) Standard cutting heads in the form of tri-bar soft ground cutterhead. (*b*) Quad-bar soft ground cutterhead. (*c*) Mixed ground rock cutterhead with disc cutters. (*Source: Akkerman, Inc.*)

during breakout. The exit and entry seals must be monitored and maintained. The MTBM must be thoroughly inspected prior to launch. Recording and monitoring of key measurements such as distance, machine torque, thrust, steering jack pressure and position, inclination and position, roll, slurry charge and discharge pressures and flow rates, bag samples of soils from separation plant, and unusual problems encountered during excavation must be conducted.

Post–QA/QC Assessment

The visual inspection of microtunneling installations includes line and grade, joints, pipe damages (such as cracks, joints shoving, and so on), deformations, lateral connections, and linings and coatings (if available). The pipe's grade and alignment must be checked against plans/specifications and the pipe must be tested for joint integrity and design pressure.

Acceptance tests of nonaccessible pipes are carried out by means of closed circuit television (CCTV) technology according to NASSCO guidelines or laser inspection or other methods. For flexible jacking pipes (such as GRP and steel) the deformation is to be checked regarding its correspondence with acceptable tolerance limit set in the design specifications. The check on the diameter change can be carried out either optically or mechanically (e.g., calibration measuring and/or laser measuring apparatus; see Fig. 8.7). For measuring line and grade (horizontal and vertical positional deviation) in nonaccessible sewers, the following measuring devices and systems can be applied: inclinometer (vertical deviation), hose leveling unit measuring pressure (vertical deviation), and/or laser target beam (vertical and horizontal deviation).

8.2.2   Pipe Materials

Jacking pipe must be obtained from one manufacturer. Pipe must be specifically designed and certified for microtunneling by the pipe manufacturer and must comply with ASTM and ASCE (ASCE Standard Construction Guidelines for Microtunneling) specifications for use in microtunneling. The pipe joints must consist of an elastomeric sealing element, sleeve, and a compression cushion ring as required by applicable ASTM and ASCE standards. The pipe must be free from any imperfections that would impair the pipe's installation or use.

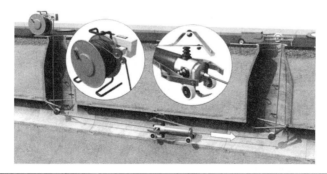

Figure 8.7   Measuring pipe deformation. (*Source: Stein and Partner.*)

Allowable Forces

The allowable jacking strength capacity of pipe must be capable of withstanding the maximum jacking forces imposed by the operation. The specified allowable jacking capacity of the pipe must be 2 times greater than the maximum jacking forces imposed by jacking operations as identified by theoretical calculations.

Pipe Requirements

1. The pipe diameters used for microtunneling method is dependent on the size of the microtunnel boring machine. The minimum size of the microtunnel machine is 12 in., and the maximum size is limited by the maximum size of pipe suitable for pipe jacking.

2. There are certain provisions for pipe thickness (or DR) based on the pipe material. For example, steel pipe may need to have a minimum wall thickness of 0.25 in. or as specified in the approved standard specifications for construction.

3. Pipe must be round. Steel pipe must have a roundness tolerance, so that the difference between the major and minor outside diameters must not exceed 1 percent of the specified nominal outside diameter, or 0.25 in., whichever is less. Likewise, concrete and other types of pipes must have similar roundness tolerances.

4. Pipe must have square and machine-beveled ends. The pipe end maximum out-of-square tolerance must be 0.04 in. (measured across the diameter).

5. Pipe must be straight. The maximum allowable straightness deviation over any 10-ft length of steel pipe is 0.125 in.

6. Pipe must be without any significant dimensional or surface deformities. All pipes must be free of visible cracks, holes, foreign material, foreign inclusions, blisters, or other deleterious or injurious faults or defects. Any section of the pipe with a gash, blister, abrasion, nick, scar, or other deleterious fault greater in depth than 10 percent of the wall thickness, must not be used and must be immediately removed from the site.

7. Any of the following defects warrants pipe rejection:
   - Concentrated ridges, discoloration, excessive spot roughness, and pitting
   - Insufficient or variable wall thickness
   - Pipe damage from bending, crushing, stretching, or other stress
   - Pipe damage that impacts the pipe strength, the intended use, the internal diameter of the pipe, and internal roughness characteristics
   - Any other defect of manufacturing or handling

Again, all pipe sections that do not comply with the specifications and/or show signs of defects or cracks must be immediately removed from the jobsite.

Pipe Transportation and Storage

Jacking pipes must be properly transported and stored at the construction site in such way that they are secured against rolling and sliding. Excessive stacking heights must be avoided so that pipes in the lower part of the stacks are not overloaded. Stacks of pipes should not be placed close to open trenches or shafts. Pipes with protective coatings must be stored on supports above ground to avoid damage to coatings and joints. All pipe sections must be stored on supports in very cold weather to avoid ground freezing. Elastomeric jointing components must be kept clean and be protected from sunlight, hydrocarbons, and extreme temperatures.

Careless handling can damage all pipe components. Any kind of high impact or point loadings create a hazard for sensitive elements of pipe joints, linings, or coating systems. Specific examples include

- Ends of concrete pipes are sensitive to impact damages due to the large weight of each pipe section.
- Pipe coatings made of relatively soft materials can be damaged during handling and backfilling. They may also become brittle in cold weather and must be handled with extra care.
- Rubber gaskets are subject to damage from sunlight or improper lubricants.

Pipes, pipeline components, and joint accessories must be inspected on delivery to ensure that they are appropriately marked and comply with the design requirements. Jacking pipes and pipe joints must be marked with manufacturer, production date, nominal size, appropriate ASTM standards, and production date. All pipe sections must be measured at each end for dimensional consistency (diameters and lengths), and those outside of tolerance limits must be marked and removed from the jobsite.

Extreme care must be taken when jointing unwelded, threaded, or infused pipe to ensure that the joints remain free of any foreign materials. Relatively small particles of foreign material can cause joint leaks, especially during high-pressure testing. Pipe joints using rubber gaskets are especially sensitive to contamination by any loose earth. These gaskets require lubricant and must be placed in accordance with the manufacturer's recommendations. Preventing gasket contamination must be part of the contractor's plan. The lubricant must conform to the manufacturer's criteria because joint failures have occurred when certain types of rubber were attacked by the chemistry of improper lubricants.

Protective Coatings

Pipeline coatings and linings that are specified for corrosion protection need to be protected against damage during all construction operations. Damaged coatings must be repaired in accordance with the pipe and/or coating manufacturer's guidelines and pipe sections with coating damage must be replaced. When necessary, cathodic protection systems must be supplied and installed in accordance with the project plans and specifications. A corrosion engineer must be retained to monitor the installation of all cathodic protection systems.

8.2.3   Construction

Minimum Allowable Depths

The recommended minimum depth of cover for microtunneling must be 6 ft or 3 times outside diameter of installed pipe, whichever is greater. In locations where the road surface is elevated, the minimum depth of the bore must be measured from the lowest side of the pavement surface.

Overcut Allowance

Overcut is the annular space between the excavated bore and the outside diameter of the pipe. When using microtunneling method, the allowable overcut must not exceed the outside pipe radius by more than 1 in. for soft ground, but overcut for large machines in rock will require 1.5 in.

Watertight Joints

1. Watertight pipe joints are required to ensure the integrity of the soil-pipe structure and street/roadbed. Pipe must be constructed to prevent water leakage or earth infiltration throughout its entire length.

2. Pipe manufacturers provide proprietary joints for microtunneling and other trenchless technology methods. More information on watertight specification for each type of pipe material can be obtained through specific pipe material association and/or specific pipe manufacturer.

Lubrication

Appropriate lubrication must be used to reduce jacking forces in different types of soil. The most common lubrication is bentonite.

Slurry Spoil Removal Syayem

The pumping rate, pressures, viscosity, and density of the slurry must be monitored to ensure adequate removal of spoil. The excess slurry must be contained until it is recycled or removed from the site. All slurry fluids must be disposed of or recycled in a manner acceptable to the appropriate local, state, and federal regulatory agencies.

Pipe Installation

Profile and alignment of microtunneling operation must be shown on the approved construction drawings. The contractor must grant the engineer/inspector access to all data and printouts obtained from MTBM, such as position of the MTBM, the fluid pressures, jacking loads, and spoil removal.

Jacking Force Calculation

1. As a minimum, the estimated jacking force must not exceed half or one-third of specified allowable jacking capacity of the pipe to avoid pipe damage. Pipe manufacturer will provide specific safety factor for the type of pipe used.

2. Influential factors for estimating jacking force include
   - Length of the drive
   - Weight, grade, and diameter of the pipe
   - Height of the overburden
   - Soil characteristics
   - Water table and dewatering operations
   - Calculated load on the shield (face pressure)
   - Operational interruptions (i.e., consideration for swelling of soil)
   - Overcut size
   - Lubrication
   - Any operational change in line and grade
   - Possible use of intermediate jacking stations
   - Pipe material, its dimensional consistency and squareness

If the operation is interrupted by more than 8 hours, the jacking force, required to install the remainder of the pipe sections, may increase up to 20 to 50 percent of the original jacking force estimate. Figures 8.8 through 8.11 presents *sample* jacking loads for different soil conditions

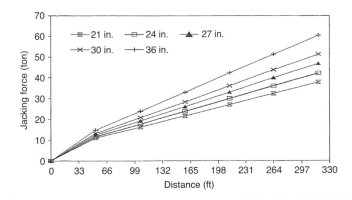

FIGURE **8.8**   Design graph for silty soil.

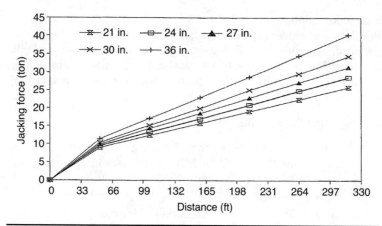

Figure 8.9 Design graph for clayey soil.

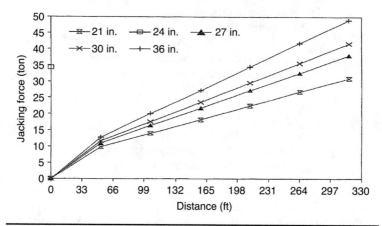

Figure 8.10 Design graph for sandy soil.

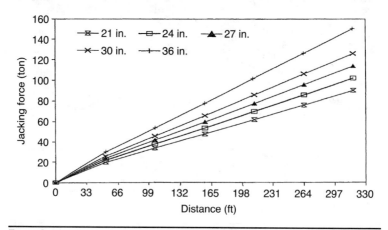

Figure 8.11 Design graph for clayey gravel soil.

and pipe diameters experienced in an evaluation program. These jacking load curves can only be used as a guide (rough order of magnitude) and comparison purposes for different pipe diameters and soil conditions, and cannot be used for estimating jacking loads on an actual project. It should be noted that the actual jacking loads may fluctuate during pipe installation but will increase as jacking distance increases.

Figure 8.12 suggests a checklist for inspection of microtunneling operations.

8.3 Pilot-Tube Microtunneling

8.3.1 Introduction

Pilot-tube microtunneling (PTMT) is a *hybrid* version of conventional microtunneling. Pilot-tube microtunneling combines the accuracy of microtunneling, the steering mechanism of a directional drill, and the spoil-removal system of an auger-boring machine. PTMT employs augers to transport spoil and a guidance system that includes a camera-mounted theodolite. The target uses electric light-emitting diodes (LEDs) to secure high accuracy in line and grade. When project conditions are suitable, pilot-tube microtunneling can be a cost-effective tool for the installation of small-diameter pipes of sewer lines or water lines. This technique can also be used for house connections direct from the main line sewers. Typically, pilot-tube machines can be used in soft soils and at relatively shallow depths. Jacking distances of 500 ft have been accomplished with newer guidance systems.

PTMT requires various components to function simultaneously for a successful installation. The integration of these components with each other plays a key role in productivity of a PTMT installation. Discussed below are the important components of PTMT.

8.3.2 Design of the Pipe

Design of the pipe must include the following:

1. *Service loading of the pipe:* The permanent (service) loading of the pipe for the specified designed life including, but are not necessarily limited to internal operating, transient and test pressures, soil overburden, surface loads, and external static water head. The installation loads that the pipes would be subjected to, which include but are not limited to, jacking forces, external pressure from groundwater, soil loads, surface loads, and annular space lubrication injection.

2. *Pipe diameter:* The pipe diameter is determined by the capacity of flow required. After the diameter and material of the

---

Inspection Guide for Microtunneling (MT)

Preinspection Plan Review

❑ Review geotechnical and soil reports.

❑ Ensure DOT/agency facilities and nearby utility information are shown on the plans and profiles and that the proposed alignment does not interfere with them.

❑ Note the minimum cover above the top of the pipe and below the pavement surface, or ground elevation (for longitudinal installations outside the influence of the roadway) is _____ ft.

❑ Note proposed pipe characteristics:
Pipe material class _____.
Pipe material yield strength _____ ksi.
Allowable pipe-jacking force _____ ton.
Pipe diameter _____ in.
Pipe wall thickness _____ in.
Overcut diameter _____ in.

❑ Review contingency plan.

❑ Review job-site layout including: distance from access pits to roadbed, proposed sheeting and bracing, materials storage and fabrication area, safety devices (barrels, guardrail, etc.), and dewatering pit locations.

❑ Note unique or special items /circumstances: _____

Construction Inspection

❑ Verify traffic control is consistent with the permit requirements, one call service has been contacted, and the DOT/agency permit is on-site.

❑ Verify job-site layout is consistent with the approved plans, especially the alignment of the pipe and machine.

❑ Verify continuous monitoring records indicate bearing and grade of the leading edge of the pipe is consistent with the approved plans, dewatering effort is satisfactory, soil volume removed is consistent with projection, and that workers understand the contingency plan.

❑ Verify pipe characteristics are consistent with permit requirements.

❑ Verify jacking pipe is certified for microtunneling, the pipe has a smooth interior and exterior surfaces, is used within the entire influence area of the roadbed, has clean and square ends, joints are watertight, defective pipe is not used, and damaged pipe is jacked through to the receiving access pit and removed.

❑ Verify the MTBM has the capability to steer, align the face, simultaneously remove spoil and install pipe, and remotely control the line and grade.

FIGURE 8.12   Sample checklist for inspection of microtunneling operations.

☐ Verify the back-thrust block is structurally adequate, jacking force is within the range of calculated jacking force, and that the jacking force is recalculated whenever the operation is stopped.

☐ Verify sufficient lubrication fluid quantity is used, and a lubrication system properly injects lubricant on the inside and outside of the pipe, and the volume or pressure of slurry flow in the supply and return side of the slurry loop is equal and steady.

☐ Verify any unsuccessful tunnel is back-filled immediately.

☐ Verify each end of the pipe is enclosed, restoration is completed, and attach Inspector's Daily Report (IDR), form 2228.

Permit No. _____
Inspector: _____
Date: _____

FIGURE 8.12   Sample checklist for inspection of microtunneling operations. (*Continued*)

pipe is fixed the next step is the selection of jacking frame. The jacking frames are chosen on the basis of specifications and actual site conditions and design requirements.

8.3.3   Construction Considerations

Every effort must be made to estimate soil conditions as accurately as possible. To design a PTMT project, step-wise planning is necessary. Following are the different aspects that must be considered while planning and designing a PTMT project.

Equipment

1. *Line and grade control system:* The line and grade control system includes, but not limited to, a theodolite, lighted target, camera, and monitor screen as shown in Fig. 8.13.

2. *Jacking frame:* The jacking frame *must* generate enough force to push the pipe from the drive shaft to receiving shaft. The design *must* be such that force is transferred to the pipes uniformly. It is very important to select a correct jacking frame for a given set of conditions. There are various factors to be considered while selecting a jacking frame, like pipe diameter, subsurface conditions, jacking force required, size of shaft, and the like. If a wrong jacking frame is selected, the project may prove to be very costly to the contractor. Therefore, a thorough investigation of the jacking frame specifications and project conditions is very essential. Figure 8.14 presents the jacking frame for the pilot-tube microtunneling.

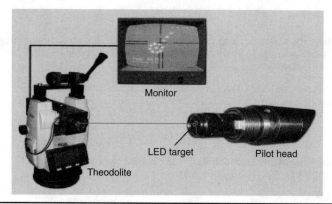

FIGURE 8.13 Line and grade control system. (*Source: Bohrtec, 2008.*)

FIGURE 8.14 Jacking frame. (*Source: Bohrtec, 2008.*)

3. *Pilot tubes:* The pilot tubes as shown in Fig. 8.15 are particularly made of small sections so that they can be accommodated in the shafts. The tubes shall rigidly connect to each other, the steering tip, and the enlargement casing, and have a clear inside diameter large enough to adequately view the lighted target. The tubes shall withstand the torque encountered in the steering process.

4. *Enlargement casing:* The enlargement casing used *must* be strong enough and *must have* a diameter slightly larger than the product pipe. The casing may match the product pipe but by use of the newest powered reaming heads and powered cutterheads only one casing is used for many larger product pipe size. It *must* connect properly with the pilot tubes.

5. *Soil-transportation system:* The soil is transported through the auger-system train, which is inside the casing.

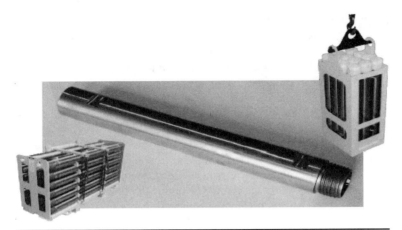

Figure 8.15   Pilot tubes. (*Source: Akkerman, 2008.*)

6. *Soil removal:* Suitable system *must* be provided to remove soil from the shafts to the surface.

7. *Hydraulic power unit:* The hydraulic power unit shall rest on the surface and be connected to the jacking frame by hoses. The unit shall meet all applicable noise standards.

8. *Lubricating system:* A good lubricating system is important so that the pipes can be transported from the drive shaft to the receiving shaft under safe load conditions.

9. *Monitoring system:* Monitoring system *must* be capable of continuously monitoring the jacking pressure, advancement of boring head and deviation of the PTMT machine. It is very important for a contractor to understand the PTMT operation, and functions of all the above components, to have maximum productivity and accurate installation.

Surface Survey

It is important to conduct a surface investigation of the PTMT *site* before the project begins to locate all the important structures that may affect the installation. The survey is typically done along the center line of the pipeline. Following are some of the important things to be considered while conducting a surface survey:

- The work area requirements at the pit locations
- Grade along the centerline
- Location of roads and other important structure along the centerline
- Locating the test pits
- Locating other waterways and wetlands along the centerline

Geotechnical Investigation

A geotechnical investigation involves identifying and classifying soil conditions along the center line of the installation. It is important to know if PTMT is a feasible method for the given conditions. For a simple trenchless project or an open-cut project, a geotechnical review may be sufficient. In a preliminary geotechnical review, the data is gathered from the surrounding constructions such as buildings, bridges, another pipeline project, and the like. This review would give an approximate estimation of the conditions to be encountered at the project site. For PTMT projects, however, it is important to know soil stratification at the project site. Boreholes must be made at various locations along the centerline of the pipeline to obtain the soil samples. These soil samples must be tested in laboratory to know the actual soil conditions. Also, the groundwater levels along the centerline of the pipeline must be measured. It must be verified that there are no large stones or cobbles along the centerline of the pipeline. PTMT is not a suitable method if there are a large number of boulders along the alignment. A thorough geotechnical investigation of the project site will lead to a good understanding of the subsurface conditions at the site.

Alignment Consideration

The next step is identifying a feasible pipeline alignment. This can be a difficult task, especially if the project is located in an urban area. For a trenchless method like PTMT, the alignment most of the time can be made straight between two points. The number and alignment of the utility lines already laid may play an important role in deciding the alignment. If there are too many utility lines, the new pipeline can be laid deeper. Similarly, the existence of the structures around the project alignment must be considered. The PTMT installation must be at a safe distance from any of the laid foundations. The distance between the shafts must be considered, which is normally not more than 400 ft. Also the space required for machinery around the jacking pits must be considered in choosing shaft locations.

Overcut Allowance

The overcut of the enlargement casing must be limited to no more than 1 in. of the diameter of the product pipe to be installed.

Lubrication Fluids

Lubricant must be selected by the contractor to achieve successful installation of the pipe system. Any additives used must meet NSF standard 060.

Figure 8.16 suggests a checklist for inspection of pilot-tube microtunneling operations.

Inspection Guide for Pilot Tube Microtunneling (PTMT)

Preinspection Plan Review

❑ Review geotechnical and soil reports.

❑ Ensure DOT / agency facilities and nearby utility information are shown on the plans and profiles and that the proposed alignment does not interfere with them.

❑ Note the minimum cover above the top of the pipe and below the pavement surface, or ground elevation (for longitudinal installations outside the influence of the roadway) is _____ ft.

❑ Note proposed pipe characteristics:
Pipe material class _____.
Pipe material yield strength _____ ksi.
Allowable pipe jacking force _____ ton.
Pipe diameter _____ in.
Pipe wall thickness _____ in.
Overcut diameter _____ in.

❑ Review contingency plan.

❑ Review job-site layout including: distance from access pits to roadbed, proposed sheeting and bracing, materials storage and fabrication area, safety devices (barrels, guardrail, etc.), and dewatering pit locations.

❑ Note unique or special items /circumstances: _____

Construction Inspection

❑ Verify traffic control is consistent with the permit requirements, one call service has been contacted, and the DOT / agency permit is on-site.

❑ Verify job-site layout is consistent with the approved plans, especially the alignment of the pipe and machine.

❑ Verify continuous monitoring records indicate bearing and grade of the leading edge of the pipe is consistent with the approved plans, dewatering effort is satisfactory, soil volume removed is consistent with projection, and that workers understand the contingency plan.

❑ Verify pipe characteristics are consistent with permit requirements.

❑ Verify jacking pipe is certified for pilot-tube microtunneling, the pipe has a smooth interior and exterior surfaces, is used within the entire influence area of the roadbed, has clean and square ends, joints are watertight, defective pipe is not used, and damaged pipe is jacked through to the receiving access pit and removed.

❑ Verify the thrust block is structurally adequate, jacking force is within the range of calculated jacking force, and that the jacking force is recalculated whenever the operation is stopped.

Figure 8.16 Sample checklist for inspection of pilot-tube microtunneling operations.

☐ Verify sufficient lubrication fluid quantity is used, and a lubrication system properly injects lubricant on the inside and outside of the pipe and the volume or pressure of slurry flow in the supply and return side of the slurry loop is equal and steady.

☐ Verify any unsuccessful tunnel is back-filled immediately.

☐ Verify each end of the pipe is enclosed, restoration is completed, and attach Inspector's Daily Report (IDR), form 2228.

Permit No. _____

Inspector: _____

Date: _____

FIGURE 8.16   Sample checklist for inspection of pilot-tube microtunneling operations. (*Continued*)

8.4   Horizontal Auger Boring

8.4.1   Introduction

The horizontal auger-boring (HAB) method involves the forming of a bore from a drive pit, by means of a rotating cutting head, and installing a steel pipe that serves as a casing for carriers (sewer, water, cable, and so on). It is a multistage process consisting of constructing a temporary horizontal jacking platform and a starting alignment track in a drive pit at a desired elevation (see Fig. 8.17). The casing pipe is

FIGURE 8.17   Horizontal auger-boring method. (*Source: Midwest Mole.*)

then jacked along the starting alignment track with simultaneous excavation of the soil being accomplished by a rotating cutting head in the leading edge of the casing pipe's annular space. The spoil is transported back to the entrance pit by a helically wound auger rotating inside the casing pipe. Horizontal auger-boring typically provides limited tracking and steering as well as limited support to the excavation face, unless a modified auger-boring machine is used. This method may also be referred to as bore and jack.

Horizontal auger-boring is a well-established trenchless method that is widely used for the installation of steel pipes and casings, especially under railways and road embankments. It is an economical trenchless pipe installation method that can be used under various soil conditions in diameters 8 to 72 in. This method can be used advantageously to reduce damage to pavements and disruptions to traffic, hence reducing the social costs associated with pipeline installations. Conventional pipe jacking (used for diameters 42 in. and up) and pipe ramming (used for diameters 4 to 140 in.) are other methods of casing installations.

The basic components of a horizontal auger-boring system include the base unit, casing pusher, power pack, auger sections, track and track extensions. Cutting bits are available for different soil conditions.

For successful execution of boring and jacking projects, presurvey of the site conditions for surface features and subsurface geotechnical conditions must be done, and utility data must be gathered and incorporated in the early stages of design process. It is required that the design engineer provide the contractor with sufficient information about the nature of the site and the possible obstacles. This will help in determining the suitability of utility installation by boring and jacking method.

The predesign survey includes the investigation of the general site condition, examination of existing underground utilities and potential for obstructions, geotechnical investigations (including groundwater), environmental conditions, required drive lengths, pipe diameters, site access, depth, grade, tolerances, impact to surface activities, location of existing/abandoned/proposed utilities, and rights-of-way requirements.

Geotechnical investigation of the site must be performed to identify the general and any special subsurface conditions. The extent of the investigation may vary depending on the knowledge of known local geological conditions.

Prior to the design of a boring project, all utilities must be located. This process involves the use of several tools, but is not limited to, utility locators, vacuum excavators, and use of utility maps. Each existing utility is then mapped or plotted, and the bore path established to avoid the existing utilities.

The boring pits must be located at a safe distance from existing structures to avoid any hazard to the structure or the public. The distance of the pit from the roadway must be adequate to allow sloping of the pit if necessary. If sufficient sloping cannot be accomplished due to space constraints, an earth support system of pit walls must be considered. Enough room for safe loading and unloading of equipment, and for spoil removal must be provided. Accidents are less likely to occur at sites that are open and kept clear of debris.

Typically steel pipes are used for casing in order to prevent the potential damage caused by the rotating augers. The casing must be of good quality and well prepared. Machine-cut beveled ends assure casing alignment, joint end squarness and exact lengths keep the head at the correct location relative to the casing, and smooth walls reduce the required jacking force and the tendency of the casing to rotate during the boring process.

After successful installation of casing pipe, the carrier pipe (also called product pipe) can be installed. Carrier pipe is installed by first attaching wooden skids or premanufactured casing spacers to the carrier pipe before assembly. The carrier pipe is then installed, one piece at a time, from either the entry or exit pit. It can be installed by pushing by hand or with a boring/jacking machine, or by pulling with a winch or other methods. The use of premanufactured casing spacers are not recommended for gravity sewers installations as they do not allow for differential blocking of the sewer inside the casing.

Postinstallation and installation monitoring includes the following:

- Preparing an as-built drawing.
- Amount of spoil removal must be checked for possible voids outside of the casing. Any voids needs to be grouted with approved grouting materials and methods.
- In sandy or unstable soil conditions, there is a possibility of void formation in the line of bore, and in this condition grouting of outside of casing is strongly recommended.

8.4.2 Materials

Pipe
Pipe used in this method includes an external *casing pipe* (also called *jacking pipe*) and may include an interior *carrier* or *product* pipe.

Allowable Forces
Considerable jacking forces may be required to install pipe using this method.

- Casing pipe must be obtained from one manufacturer.
- The allowable jacking strength capacity of casing pipe must be capable of withstanding the maximum jacking forces

imposed by the operation. The specified allowable jacking capacity of the casing pipe must be 3 times greater than the maximum jacking forces imposed by jacking operations as identified by theoretical calculations.

- Steel casing pipe must have minimum yield strength of 35,000 psi.

Casing Pipe

Steel casing pipe, also known as encasement pipe is most commonly used in horizontal auger boring.

1. Steel casing pipe must have a minimum wall thickness of 0.25 in. or as specified in the specifications for construction, whichever is larger.

2. Casing pipe must be round. Steel casing pipe must have a roundness tolerance, so that the difference between the major and minor outside diameters must not exceed 1 percent of the specified nominal outside diameter, or 0.25 in., whichever is less.

3. Casing pipe must have square and machine-beveled ends. The pipe end maximum out-of-square tolerance must be 0.04 in. (measured across the diameter).

4. Casing pipe must be straight. The maximum allowable straightness deviation over any 10-ft length of steel casing pipe is 0.125 in.

5. Pipe must be without any significant dimensional or surface deformities. All pipes must be free of visible cracks, holes, foreign material, foreign inclusions, blisters, or other deleterious or injurious faults or defects. Any section of the pipe with a gash, blister, abrasion, nick, scar, or other deleterious fault greater in depth than 10 percent of the wall thickness must not be used and must be immediately removed from the site.

6. Casing pipe must be used within the entire roadbed influence area. The *roadbed influence area* is defined as the subsurface area located under the road and shoulder surface, between each shoulder point or back of curb, and continues transversely outward and downward from each shoulder point or back of curb on a 1-on-1 slope.

7. Only new casing pipe must be used, unless otherwise approved by the engineer/inspector.

8. Casing pipe must have smooth interior and exterior walls to reduce jacking force and prevent casing rotation.

9. The inside diameter (ID) of the casing pipe must be at least 6 in. larger than the largest outside diameter (OD) of the carrier pipe to allow the carrier pipe to be inserted.

10. Any of the following defects warrants pipe rejection:
    - Concentrated ridges, discoloration, excessive spot roughness, and pitting
    - Insufficient or variable wall thickness
    - Pipe damage from bending, crushing, stretching, or other stress
    - Pipe damage that impacts the pipe strength, the intended use, the internal diameter of the pipe, and internal roughness characteristics
    - Any other defect of manufacturing or handling

Carrier Pipe

Carrier pipe material may be constructed with any material. The carrier pipe diameter must be small enough to insert into the casing pipe in conjunction with the casing spacers or wood blocking.

Casing Spacers

Casing spacers or wood blocking is required for all carrier pipes. Casing spacers can be wood, plastic, fiberglass, stainless steel, or carbon steel. For gravity sewer installations, the void between the casing and the carrier pipe must be filled utilizing the cellular grout, sand, pea gravel, or other approved product.

8.4.3 Construction

Equipment

Equipment used for this method varies greatly. However, the basic operations of boring, removing tailings, and jacking pipe are essential. Please refer to the specific operator's manual for more information.

Method

A full-size auger section must be used as the lead section of the casing.

Overcut Allowance

Overcut is the annular space between the excavated bore and the outside diameter of the casing pipe. The usual allowable overcut is 1 in. greater than the casing pipe radius.

Watertight Joints

Steel casing joints are to be welded to prevent earth infiltration through the joints and to withstand the required installation loads.

Lubrication Fluids

1. Lubrication fluids, consisting of a mixture of water and bentonite or bentonite/polymer, must be used in the annular space between the casing being installed and the native soil. This is done primarily to reduce friction on the casing. Lubrication may also be used inside the casing pipe to facilitate spoil removal.

2. Lubrication fluids are specifically recommended for this method regardless of the soil conditions.

3. Grease or hydrocarbons are not allowed for use as lubrication.

Pipe Locating and Tracking

Commonly, a waterlevel system or other approved method is used to monitor grade. A sonar transmitter or other approved method can be used for monitoring alignment.

Figure 8.18 suggests a checklist for inspection of horizontal auger-boring operations.

8.5   Pipe Ramming

8.5.1   Introduction

Pipe ramming is defined as a trenchless installation of steel casing pipes under roads and railroad tracks. A ramming tool attached to the rear of a steel pipe drives the pipe into the ground with repeated percussive blows. For casing pipe diameters less than 8 in., the pipe is installed closed face and soil is compacted around the pipe. For casing pipe diameters more than 8 in., the pipe is installed open face to allow the soil to enter the pipe during the installation (Figs. 8.19 and 8.20). The spoils inside the pipe can be removed after the installation by means of auger, compressed air, and/or water jetting. In larger diameters (e.g., more than 100 in.), a small backhoe or bobcat can be used for mechanical spoil removal.

Pipe ramming typically requires excavation of two pits. Before ramming, both the pipe and the ramming tool are placed into the insertion pit and lined up in the desired direction. Alternatively, the ramming can be launched without an insertion pit, if the operation is started in the slope of an embankment.

The roadbed influence area is defined as the subsurface area located under the road and shoulder surface, between each shoulder point or back of curb, and continues transversely outward and downward from each shoulder point or back of curb on a 1 on 1 slope.

Inspection Guide for Horizontal Auger Boring (HAB)

Preinspection Plan Review

❏ Review geotechnical and soil reports.

❏ Ensure DOT/agency facilities and nearby utility information are shown on the plans and profiles and that the proposed alignment does not interfere with them.

❏ Note the minimum cover above the top of the pipe and below the pavement surface, or ground elevation (for longitudinal installations outside the influence of the roadway) is _____ ft.

❏ Note proposed pipe characteristics:
Casing material _____.
Casing material yield strength _____ksi.
Allowable pipe-jacking force _____ ksi.
Casing diameter _____ in.
Overcut diameter _____ in.
Casing wall thickness _____ in.
Casing spacers material _____.
Spacer spacing _____, carrier material _____.
Carrier diameter _____ in.

❏ Note the inside diameter of the casing pipe, and the outside diameter of the carrier pipe. The difference is _____ in. (Typical minimum is 6 in., except for casing pipes smaller than 8 in.)

❏ Review contingency plan.

❏ Review job-site layout including: distance from access pits to roadbed, proposed sheeting and bracing, materials storage and fabrication area, safety devices (barrels, guardrail, etc.), and dewatering pit locations.

❏ Note unique or special items / circumstances: _____

Construction Inspection

❏ Verify traffic control is consistent with the permit requirements, one call service has been contacted, and the DOT/agency permit is on-site.

❏ Verify job-site layout is consistent with the approved plans, especially the alignment of the pipe and machine.

❏ Verify continuous monitoring records indicate bearing and grade of the leading edge of the pipe is consistent with the approved plans, dewatering effort is satisfactory, soil volume removed is consistent with projection, and that workers understand the contingency plan.

❏ Verify pipe characteristics are consistent with permit requirements.

❏ Verify the casing pipe is steel with smooth interior and exterior surfaces, is used within the entire influence area of the roadbed, has clean and square ends, joints are watertight, defective pipe is not used, and damaged pipe is removed.

FIGURE 8.18   Sample checklist for inspection of horizontal auger boring operations.

❑ Verify cutterhead does not protrude more than 3 in. beyond the leading edge of the casing pipe, and that full-size auger section is used in the lead section of casing. As a rule of thumb, for unstable soil, the cutterhead must be inside and for stable soil, the cutterhead must be outside.

❑ Verify sufficient lubrication fluid quantity is used if necessary, and a lubrication system properly injects lubricant on the outside of the pipe.

❑ Verify any unsuccessful boring is back-filled immediately.

❑ Verify each end of casing is bulkheaded, restoration is completed, and attach Inspector's Daily Report (IDR), form 2228.

Permit No. _____

Inspector: _____

Date: _____

FIGURE 8.18 Sample checklist for inspection of horizontal auger boring operations. (*Continued*)

FIGURE 8.19 Schematic view of pipe ramming. (*Source: TT Technologies.*)

FIGURE 8.20 Culvert installation under railroad using pipe ramming. (*Source: TT Technologies.*)

8.5.2 Materials

Pipe
Pipe used in this method includes an external *steel casing* pipe and may include an interior carrier pipe.

Pipe Dimensions and Requirements
Refer to pipe, casing, and spacer requirements for the horizontal auger-boring method mentioned in Sec. 8.3.2.

8.5.3 Construction

Equipment
Equipment used for this method varies greatly. However, the basic operations of hammering the pipe into position and removing the excess soil are essential. Please refer to the specific operator's manual for more information.

Method

1. Each pipe section must be rammed forward as the excavation progresses in such a way to provide complete and adequate ground support at all times. Lubrication must be applied to the external surface of the pipe to reduce skin friction. A hammer frame must be positioned to develop a uniform distribution of ramming forces around the periphery of the pipe. Special care must be taken by the contractor to ensure that the launch seal is properly designed and constructed. Special care must be taken when setting the pipe guard rails in the pit to ensure a correct alignment.

2. In open-end ramming, either a prefabricated soil-cutting shield must be attached to the front of the casing pipe leading edge, or a casing band must be welded around the outside or inside edge of the pipe. In closed-end ramming, a cone-shaped attachment must be welded or threaded to the front of the casing pipe.

3. Whenever possible, the casing pipe must be driven in a continuous, single run. If space is limited, the job can be completed in a series of short ramming sections. Length of pipe sections must be selected based on available space for an insertion pit setup. Casing pipe section length between 20 and 40 ft is frequently selected. Section lengths between 10 and 60 ft can be allowed considering the availability of pit space.

4. The contractor must dispose of all excavated material from the pipe ramming and access pit construction operations

off-site. Precautions must be taken to ensure worker safety if removal of soil is accomplished through the use of a foam pig and compressed air.

Overcut Allowance

The standard overcut is between $1/4$ and $3/4$ in. over radius both on the outside and inside of the casing pipe. Overcut up to 1 in. can be permitted depending on the diameter of the ramming tool, depth and ground conditions leaving the 6 o'clock open.

Watertight Joints

Soil tight pipe joints are required to ensure the integrity of the roadbed. Pipe must be constructed to prevent earth infiltration throughout its entire length.

Lubrication Fluids

1. Lubrication must be made of bentonite, water, and/or polymers and used for inside and sometimes outside of the pipe.

2. A lubrication system must be provided that injects lubricant on the inside to facilitate spoil removal and sometimes outside of the pipe to lower the friction developed on the sides of the pipe during ramming.

   Figure 8.21 suggests a checklist for inspection of pipe-ramming operations.

8.6 Horizontal Directional Drilling

8.6.1 Introduction

The elements of an horizontal directional drilling (HDD) installation are (see Fig. 8.22)

1. A rig, which provides the physical means—thrust and torque, to open the borehole and pull in the product.

2. A transmitter/receiver system for tracking the location of the bore.

3. The down-hole equipment—drill pipe, drill bits, and reamers, which converts the physical capabilities of the rig to open the borehole and pull in the product.

4. The drilling fluid, which serves to stabilize the borehole, cools the down-hole equipment, and removes the spoils from the borehole.

Inspection Guide for Pipe Ramming (PR)

Preinspection Plan Review

☐ Review geotechnical and soil reports.

☐ Ensure DOT / agency facilities and nearby utility information are shown on the plans and profiles and that the proposed alignment does not interfere with them.

☐ Note the minimum cover above the top of the pipe and below the pavement surface, or ground elevation (for longitudinal installations outside the influence of the roadway) is _____ ft.

☐ Note proposed steel casing pipe characteristics:
Casing material yield strength _____ ksi.
Casing diameter _____ in.
Overcut diameter _____ in.
Casing wall thickness _____ in.
Casing spacers material _____.
Spacer spacing _____.
Carrier material _____.
Carrier diameter _____ in.

☐ Note the inside diameter of the casing pipe, and the outside diameter of the carrier pipe. The difference is _____ in. (Typical minimum is 6 in.)

☐ Review contingency plan.

☐ Review job-site layout including: distance from access pits to roadbed, proposed sheeting and bracing, materials storage and fabrication area, safety devices (barrels, guardrail, etc.), and dewatering pit locations.

☐ Note unique or special items /circumstances: _____

Construction Inspection

☐ Verify traffic control is consistent with the permit requirements, one call service has been contacted, and the DOT / agency permit is on-site.

☐ Verify job-site layout is consistent with the approved plans, especially the alignment of the pipe and machine.

☐ Verify continuous monitoring records indicate bearing and grade of the leading edge of the pipe is consistent with the approved plans, dewatering effort is satisfactory, soil volume removed is consistent with projection, and that workers understand the contingency plan.

☐ Verify pipe characteristics are consistent with permit requirements.

☐ Verify the casing pipe is new steel with smooth interior and exterior surfaces, is used within the entire influence area of the roadbed, has clean and square ends, joints are watertight, defective pipe is not used, and damaged pipe is removed.

FIGURE 8.21 Sample checklist for inspection of pipe ramming operations.

☐ Verify hammer frame is used to distribute ramming forces around the pipe circumference, a launch seal is used at face of ramming operation, and a casing band is applied to the leading edge of pipe in open-end ramming operations, or a cone-shaped pipe head is attached to closed-end ramming operations.

☐ Verify sufficient lubrication fluid quantity is used, and a lubrication system properly injects lubricant on the inside and outside of the pipe.

☐ Verify any unsuccessful rammed hole is back-filled immediately.

☐ Verify each end of casing is bulkheaded, restoration is completed, and attach Inspector's Daily Report (IDR), form 2228.

Permit No. _____
Inspector: _____
Date: _____

FIGURE 8.21 Sample checklist for inspection of pipe ramming operations. (*Continued*)

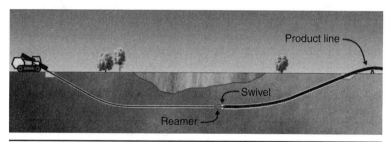

FIGURE 8.22 HDD method.

5. The drilling fluid delivery and recovery system made up of tanks, mixing systems, pumps; and, when recycling fluids, a system of screens, filters, shakers, cones, and the like to remove spoils brought to the surface from the fluid.

QA/QC Procedures

HDD QA/QC measures are provided in the installation contract, and by state and local regulations governing underground construction. In addition, industry guidelines such as references listed at the end of this book are helpful. Below is an overview of QA/QC requirements:

1. *Preconstruction:* Prior to construction, geological conditions need to be assessed to determine equipment and material needs. HDD operator must walk the site to identify potential hazards, sources of interference, and special conditions. The boreplan must be prepared with adequate setup area and separation from utilities.

2. *Determine geological conditions:* The type of soil in the path of the installation determines the type of equipments, cutterheads, and drilling fluids that are best suited for the project and whether HDD is a suitable tool for installing the pipe. Depending on the job size, a geological survey may or may not be conducted by an experienced geotechnical engineer. If no geotechnical survey has been preformed, the contractor must look for existing records associated with nearby construction sites, as well as other public records. Visual inspection of site geology and soil characteristics at the entry and exit pits also provide information on soil conditions.

3. *Hazards, obstructions, and utility location:* The installation of underground utilities using trenchless methods limits visual verification of conditions near and surrounding the installation. As with any underground construction, every means must be used to locate and verify existing conditions. These means include contacting the local one-call service and area utilities, reviewing records, having a locating service locate utilities, use locating equipment such as ground-penetrating radar (GPR). When operating near existing utilities or potential hazardous conditions, potholing must be used to verify the exact location of the existing utility and bore path.

4. *Drilling fluids:* As said earlier, drilling fluids stabilize the borehole, which mitigates hydro fracturing (inadvertent return of fluids or frac outs) and allows the product pipe to be pulled in. The proper mix of drilling fluids is determined by the soil conditions and characteristics of the water mixed with the drilling fluid. Geotechnical information must be gathered in advance of the project and soils extracted from the borehole must be tested from time-to-time during the installation to verify that the proper drilling fluid mix and additives are being used. Water must be checked and adjusted for pH and the presence of calcium.

5. *Downhole equipment:* The type of downhole equipment to be used in an HDD installation depends on the soil conditions, depth of the installation, and size and type of the product pipe being installed. Different bits and reamers work better in different soil conditions. Downhole transmitters come with different signal strengths, which must match the depth of the installation. In addition, certain installations may require that breakaway swivels be used to protect the product being installed. Also, pressure sensors can be used to measure fluid pressure at the swivel location as well as to measure pulling loads on the product pipe.

6. *Rig and mud circulation and recycling equipment:* The rig size, and mud system/recycling equipment capacity must match the job size. A rig with inadequate torque and trust capability will cause the installation to be more difficult, putting the installation at risk. Similarly, an under capacity mud circulation system, that is, inadequate tank or pump capacity, or recycling system, can have the same consequences.

7. *Installation:* A successful installation must follow the planned bore with as-built that approximates the planned path, maintains specified clearance from hazards and other utilities, stays within easement, minimizes and deals with inadvertent drilling nearby construction sites, as well as maintains integrity of the road embankment and other nearby structures.

8. *Record keeping:* Tracking systems can provide electronic records of the product pipe location. In the event that the tracking system does not record installation information electronically, accurate manual records must be maintained to verify location and compare against the boreplan. Newer systems provide planning software as well as recording of installation. Real-time records must be kept in a driller log of pitch and depth of each drill string, the drilling fluids used, and any special conditions encountered in installation. This information provides a record for verification of pipe location and drilling operations.

9. *Fluid monitoring:* Monitoring drilling fluid returns is also an important QA/QC procedure. Generally, drilling fluid, which carries soils from downhole, must exit the borehole at the entry or exit end of the installation. Drilling fluid flow provides visual verification that the borehole is open and that the fluids are not inadvertently escaping. Lost circulation may be an indication that something is wrong. Field tests that measure the drilling fluids viscosity and weight can help determine the need to adjust drilling fluid mix and the rate at which a product pipe can be safely installed. If the drilling fluid is being recycled, the recycling equipment must be inspected regularly to make sure that it is removing solids from the drilling fluid. If the percent solids in the fluid become excessive, adjustments in the system operation must be made to avoid damaging equipment.

10. *Bits and reamers safety:* Changing bits and reamers can be a potential safety hazard. Always use proper communications procedures and equipment such as breakout wrenches when changing bits and reamers.

11. *Pipe products:* HDD can be used to install a number of pipe products. The most common are plastic pipes (HDPE and PVC) and communications conduits, steel, and ductile iron.

Each of these products has specific QA/QC procedures for joining pipe sections.

12. *Postinstallation:* A successful installation using HDD is largely determined during the installation process. The product pipe condition at the exit end of the installation must be examined for gouges, cuts, or abrasions. Water and gas pipelines must be tested against leaks as specified specifications. The jobsite must be restored and all materials including drilling fluids must be disposed of, as required by local ordinance or as specified.

8.6.2 Pipe Material Standards

Pipe used for HDD must be smooth, flexible, and have sufficient strength to resist tension, bending, and external installation pressure loads. This method requires structurally strong joints that resist elongation or cross-section reduction.

HDPE pipes must conform to the current ASTM D1248-05, ASTM D3350-10, and ASTM F714-08. Steel pipe must conform to the current ASTM A 53-07 and ASTM 139-06. Ductile iron pipe must confirm to the current ASTM 716-95 and ASTM 746-95. PVC pipe must confirm to the current ASTM F1962-99 and ASTM D2321-09.

Allowable Forces

In case of HDPE pipe, an extra 6-ft section of the pipe must be pulled out of the borehole to check for any sign of stress or damage. Allowable pulling force for all diameters must be determined depending on the pipe size, wall thickness, manufacturer, field conditions, pull distance, bearing capacity of soils, adjacent infrastructure, and all other related considerations.

Pipe Dimensions

1. HDPE pipe may require a minimum DR of 11.

2. Pipe must be round. Steel pipe must have a roundness tolerance, so that the difference between the major and minor outside diameters must not exceed 1 percent of the specified nominal outside diameter, or 0.25 in., whichever is less. Likewise, HDPE, ductile iron, and PVC pipe must have similar roundness tolerances.

3. Pipe must have square and machine beveled ends. The pipe end maximum out-of-square tolerance must be 0.04 in. (measured across the diameter).

4. The maximum allowable straightness deviation over any 10-ft length of steel casing pipe is 0.125 in. Likewise, ductile iron and PVC pipe must have similar straightness tolerances. HDPE pipe does *not* need to be straight.

5. Pipe must be without any significant dimensional or surface deformities. All pipes must be free of visible cracks, holes, foreign material, foreign inclusions, blisters, or other deleterious or injurious faults or defects. Any section of the pipe with a gash, blister, abrasion, nick, scar, or other deleterious fault greater in depth than 10 percent of the wall thickness, must not be used and must be immediately removed from the site.

6. Any of the following defects warrants pipe rejection:
   - Concentrated ridges, discoloration, excessive spot roughness, and pitting
   - Insufficient or variable wall thickness
   - Pipe damage from bending, crushing, stretching, or other stress
   - Pipe damage that impacts the pipe strength, the intended use, the internal diameter of the pipe, and internal roughness characteristics
   - Any other defect of manufacturing or handling

Protective Coatings (Steel Pipe)

The product pipe may be exposed to significant abrasion during pullback. Therefore, a coating to provide a corrosion barrier as well as an abrasion barrier is required. The coating must be bonded well to the pipe and have a hard smooth surface to resist soil stresses and reduce friction. Usually a mill-applied fusion bonded epoxy (FBE) coating is required for steel pipes.

8.6.3   Construction

Minimum Allowable Depths

The minimum allowable installation depth of cover of a HDD installed pipe under the road and shoulder surface is correlated to the pipe diameter. Table 8.1 summarizes the minimum allowable depths.

Pipe Diameters (in.)	Depth of Cover (ft)
Small (< 4)	4
Mini (4–12)	8
Medium (13–24)	12
Large (> 24)	16

TABLE **8.1**   Minimum Allowable Depth

To help with future locating of installed pipes, installation of a trace wire on plastic pipes and submission of an as-built (both plan and profile) for all installations are required. In locations where the road surface is elevated, the minimum depth of the bore must be measured from the lowest side of the pavement surface.

Method

1. The ends of each section of HDPE pipe must be inspected and cleaned as necessary to be free of debris immediately prior to joining the pipes by means of thermal butt-fusion. The polyethylene pipe must be of the same type, grade, and class of the polyethylene compound used in the process. This process provides joint weld strength equal to or greater than the tensile strength of the pipe.

2. The handling of the joined pipeline must be in such a manner that the pipe is not damaged by dragging it over sharp or jagged objects. Sections of the pipes with cuts and gouges exceeding 10 percent of the pipe wall thickness or kinked sections must be removed and the ends rejoined.

3. HDPE pipes must be stored on level ground, free of sharp objects, which could damage the pipe. Stacking of the polyethylene pipe must be limited to a height that will not cause excessive deformation, bending, or warping of the bottom layers of pipes under anticipated temperature condition.

4. Sufficient space must be allocated to fabricate and layout the product pipeline into one continuous pipe length, thus enabling the pull back to be conducted during a single operation. If space considerations are discovered that make this impossible, the permit applicant must obtain specific alternative instructions from the design engineer.

5. Sufficient space is required on the rig side of the machine to safely set up and perform the operation.

6. The drill path alignment must be as straight as possible to minimize the frictional resistance during pullback and maximize the length of the pipe that can be installed during a single pull.

7. The minimum radius of curvature of HDD path must be 1200 times the nominal diameter of the pipe to be installed.

8. The required piping must be assembled in a manner that does not obstruct adjacent roadways or public activities. The contractor must erect temporary fencing around the entry and exit pipe staging areas.

9. Several prereams *may* be employed to gradually enlarge the borehole to the desired diameter and reduce road surface heaving potential. No backream diameter increase must exceed 1.5

times diameter of the product pipe. Furthermore, during the final pullback, the pullback rate must not exceed 10 ft/min.

10. The pipe must be sealed at both ends with a cap or a plug to prevent water, drilling fluids, and other foreign materials from entering the pipe as it is being pulled back.

11. Pipe rollers, skates, or other protective devices must be used to prevent damage to the pipe, eliminate ground drag, reduce pulling force, and reduce the stress on the pipes and joints.

12. The drilling fluid in the annular region outside of the pipe must not be removed after installation, and remain in place to provide support for the pipe and neighboring soil.

13. If the drilling operation is unsuccessful, the contractor must ensure to fill of any void(s) with flowable fill.

14. Entry penetration angles are limited by equipment capabilities. However, according to most HDD drilling rigs' design, the best entry angle must be between 10° and 12°.

15. Exit angles generally range from 5° (for large-diameter steel pipelines) to 12°. However, when high exit angles are encountered or designed, the pipe must be supported in an elevated position during the pullback operation to prohibit the pipe from bending, deforming, kinking, or even breaking.

Overcut Allowance

The overcut diameter must not exceed the outside diameter (OD) of the pipe by more than 1.5 times diameter of the product pipe, to ensure excessive voids are not created, resulting in postinstallation settlements.

Watertight Joints

Watertight pipe joints are required to ensure the integrity of the roadbed. Pipe must be constructed to prevent water leakage or earth infiltration throughout its entire length.

A watertight specification for each type of pipe material can be obtained for each pipe material. Refer to the appropriate industry specifications for more detailed information.

Drilling Fluids

1. Drilling fluid must be used during drilling and back-reaming operations. Using water exclusively may cause a collapse of the borehole while in unconsolidated soils, and may also cause soil swelling while in clayey soils. Either case may significantly impede the installation of the pipe.

2. Excess drilling fluids must be contained within a pit or containment pond, or trailer-mounted portable tanks, until removed from the site.

3. Drilling fluids must not enter the streets, manholes, sanitary and storm sewers, and other drainage systems, including streams and rivers.

4. Any damage to any DOT highway or non-highway facility caused by escaping drilling fluid, or the directional drilling operation, must be immediately restored by the contractor.

Pipe Locating and Tracking

1. During construction, continuous monitoring and plotting of pilot drill progress must be undertaken to ensure compliance with the proposed installation alignment and allow for appropriate course corrections to be undertaken. Monitoring must be accomplished by manual plotting based on location and depth readings provided by the locating/tracking system or by computer generated bore logs, which map the bore path based on information provided by the locating/tracking system. Readings or plot points must be undertaken on every drill rod.

2. Pipe installed by the HDD method must be located in plan as shown on the drawings, and must be no lower than shown on the drawings unless otherwise approved. The contractor must plot the actual horizontal and vertical alignment of the pilot bore at intervals not exceeding 30 ft. This "as-built" plan and profile must be updated as the pilot bore is advanced. The contractor must at all times provide and maintain instrumentation that will accurately locate the pilot hole and measure drilling fluid flow and pressure. The contractor must grant the engineer/inspector access to all data and readout pertaining to the position of the bore head, the fluid pressures, and flows.

Figure 8.23 suggests a checklist for inspection of horizontal directional drilling operations.

8.7   Pipe Replacement

8.7.1   Introduction

Trenchless pipe-replacement methods are defined as the renewal of existing pipelines by the simultaneous insertion of new pipe within the path of the existing pipe by the use of using a static pull, hydraulic expansion, or pneumatic bursting device. The existing pipe is fractured, and pushed into the surrounding soil or removed. At the same time, a new pipe is either pulled or pushed in the annulus left by the expanding operation. The new pipe size cannot significantly exceed

Inspection Guide for Horizontal Directional Drilling (HDD)

Preinspection Plan Review

❑ Review geotechnical and soil reports.

❑ Ensure DOT/agency facilities and nearby utility information are shown on the plans and profiles and that the proposed alignment does not interfere with them.

❑ Note the minimum cover above the top of the pipe and below the pavement surface, or ground elevation (for longitudinal installations outside the influence of the roadway) is _____ ft.

❑ Note proposed pipe characteristics:
Pipe material _____.
Pipe diameter _____ in.
Pipe wall thickness _____ in.
Overcut diameter _____ in.
Backream dia. increase _____ in.

❑ Ensure that the appropriate penetration angle and curvature rate are identified.

❑ Review contingency plan.

❑ Review job-site layout including: distance from access pits to roadbed, proposed sheeting and bracing, materials storage and fabrication area, safety devices (barrels, guardrail, etc.), and dewatering pit locations.

❑ Review steel pipe coating requirements.

❑ Note unique or special items / circumstances: _____

Construction Inspection

❑ Verify traffic control is consistent with the permit requirements, one call service has been contacted, and the DOT/agency permit is on-site.

❑ Verify job-site layout is consistent with the approved plans, especially the alignment of the pipe and machine.

❑ Verify continuous monitoring records indicate bearing and grade of the leading edge of the pipe is consistent with the approved plans, dewatering effort is satisfactory, soil volume removed is consistent with projection, and that workers understand the contingency plan.

❑ Verify pipe characteristics are consistent with permit requirements.

FIGURE 8.23   Sample checklist for inspection of horizontal directional drilling operations.

☐ Verify steel pipe is new with smooth interior and exterior surfaces, is used within the entire influence area of the roadbed, has clean and square ends, joints are watertight, defective pipe is not used, and damaged pipe is removed.

☐ Verify HDPE pipe is supported and secured reasonably with rollers, skates or other protective devices to prevent damage to the pipe, is new with smooth interior and exterior surfaces, is used within the entire influence area of the roadbed, has clean and square ends, joints are butt-fused and watertight, pipe with gouges exceeding 10 percent of the pipe wall thickness or kinked sections is not used, and damaged pipe is removed.

☐ Verify the pullback rate does not exceed 10 ft/min to avoid heaving.

☐ Verify drilling fluid departs entry and exit drilling points only, and that drilling fluid in the exterior annular region of installed pipe is not removed after installation.

☐ Verify any unsuccessful drill hole is back-filled immediately.

☐ Verify each end of the pipe is sealed with a cap, restoration is completed, and attach Inspector's Daily Report (IDR), form 2228.

Permit No. _____
Inspector: _____
Date: _____

FIGURE 8.23   Sample checklist for inspection of horizontal directional drilling operations. (*Continued*)

the size of the replaced pipe. The product originally transported by the replaced pipe can be temporarily rerouted (bypassed) to prevent overflows and provide uninterrupted service.

Pipe Bursting

In most of pipe-bursting operations, the new pipe is pulled into place. The rear of the bursting head is connected to the new pipe, and the front end of the bursting head to either a winching cable or a pulling rod assembly. The bursting head and the new pipe are launched from the insertion pit. The cable or rod assembly is pulled from the pulling or reception pit. Pipe bursting may use static, pneumatic, or hydraulic method of bursting the existing and host pipe. Pipe bursting must follow conditions set in this document. Figure 8.24 illustrates the pipe-bursting process.

8.7.2   Materials

Pipe

Pipe used in this method includes an existing *host* pipe and a *replacement* pipe.

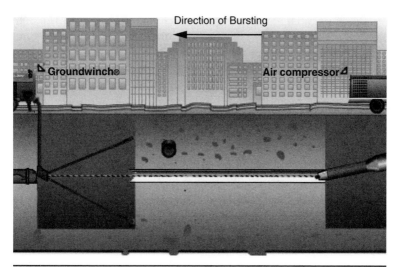

FIGURE 8.24   Pipe bursting. (*Source: TT Technologies.*)

Allowable Forces

See appropriate ASTM specification. As a guide, generally the pulling force must not exceed the following limitations:

- For less than 16-in. OD, 10 tons
- For 16-in. OD or more, 20 tons

Allowable pulling force for all diameters must be determined depending on the pipe material, size, wall thickness, manufacturer, field conditions, pull distance, manhole integrity, bearing capacity of soils, adjacent infrastructure, related equipment, cable strength, and all other related considerations.

Pipe Dimensions

1. The range in pipe diameters used in this method varies with the range in host pipe diameter. The upper size limit is based on the capabilities of the installation equipment. See equipment owner's manual for more specific information.

2. New pipe must have a minimum standard dimension ratio (SDR) of 11.

3. Pipe must be round. Steel pipe must have a roundness tolerance, so that the difference between the major and minor outside diameters must not exceed 1 percent of the specified nominal outside diameter, or 0.25 in., whichever is less. Likewise, HDPE, ductile iron, and PVC pipe must have similar roundness tolerances.

4. Pipe must have square and machine-beveled ends. The pipe end maximum out-of-square tolerance must be 0.04 in. (measured across the diameter).

5. The maximum allowable straightness deviation over any 10-ft length of steel pipe is 0.125 in. Likewise, ductile iron, and PVC pipe must have similar straightness tolerances. HDPE pipe does not to be straight.

6. Pipe must be without any significant dimensional or surface deformities. All pipes must be free of visible cracks, holes, foreign material, foreign inclusions, blisters, or other deleterious or injurious faults or defects. Any section of the pipe with a gash, blister, abrasion, nick, scar, or other deleterious fault greater in depth than 10 percent of the wall thickness, must not be used and must be immediately removed from the site.

7. Any of the following defects warrants pipe rejection:
   - Concentrated ridges, discoloration, excessive spot roughness, and pitting
   - Insufficient or variable wall thickness
   - Pipe damage from bending, crushing, stretching, or other stress
   - Pipe damage that impacts the pipe strength, the intended use, the internal diameter of the pipe, and internal roughness characteristics
   - Any other defect of manufacturing or handling

Host Pipe

1. Most brittle pipe materials make good candidates of host pipes for pipe bursting. Ductile pipe must be scored and then split as in the pipe-splitting operations described above.

2. Pipe made of nonductile abrasive material, but with ductile reinforcing, is the most difficult to replace using most pipe-replacement techniques.

3. Clay and cast-iron pipe are good candidates for pipe bursting. If PVC joints exist on clay pipe, or if ductile repair clamps, service saddles, and fittings exist on cast-iron pipe special application tools must be used in pipe bursting. Sacrificial external sleeve pipes must be used to ensure protection for plastic replacement pipes for high-pressure pipe applications against sharp fragments of host pipe.

4. Plain concrete pipe is a good candidate for pipe bursting. However, thick plain concrete pipe, reinforced encasements, or repair areas in the pipe need special considerations.

5. Replacing reinforced concrete pipe is difficult unless the concrete and reinforcing steel are deteriorated. Although powerful equipment may be used to burst the pipe, careful evaluation must be made on the reinforcement and deterioration.

6. Steel and ductile iron pipe are not suitable for pipe bursting. But in smaller diameters, they can be replaced by pipe-splitting technique.

7. PVC and other plastic pipes may be replaced using an appropriate combination of bursting and splitting techniques according to the strength and ductility of the pipe.

8. Asbestos cement pipes are generally good candidates for pipe bursting. Care must be taken to determine the class of the existing pipe. Thicker, higher tensile strength pipes require increased bursting forces. Modifications to standard bursting heads must include cutter blades to split the pipe.

Replacement Pipe

1. High-density polyethylene pipe (HDPE) is normally used as replacement pipe for pipe-replacement methods. HDPE must be solid wall and in conformance with ASTM F714. All pipe and fittings must be new high-density polyethylene pipe. The pipe material must be manufactured from a high-density, high-molecular-weight polyethylene compound which conforms to ASTM D 1248 and meets the requirements for Type III, Class C, Grade P-34, Category 5, and has a PPI rating of PE 3408. The pipe produced from this resin must have a minimum cell classification of 345434C (inner wall will be light in color) under ASTM D 3350. The value for the hydrostatic design basis must not be less than 1600 psi per ASTM D 2837. Pipe must have ultraviolet protection.

2. If there is insufficient space to fuse and layout the PE pipe, PVC pipe with a mechanical spline-locking design joint may be used for pipe-jacking operations.

3. Special flush-joint ductile iron pipe has recently been developed as a replacement pipe, which can be used as a carrier pipe, for either sewer or potable water.

4. Cast-iron pipe, vitrified clay pipe, and reinforced concrete pipe can be used as replacement pipe in pipe-bursting operations. However, the bursting head and pipe-installation technique needs to be modified to avoid significant tensile force on pipe joints.

Pipe Quality

1. All pipes must be free of visible cracks, holes, foreign material, foreign inclusions, blisters, or other deleterious or injurious faults or defects. Any section of the pipe with a gash, blister, abrasion, nick, scar, or other deleterious fault greater in depth than 10 percent of the wall thickness must not be used and must be immediately removed from the site. However, a defective area of the pipe may be cut out and the joint fused in accordance with the procedures stated above.

2. Any of the following defects warrants pipe rejection:
   • Concentrated ridges, discoloration, excessive spot roughness, and pitting
   • Insufficient or variable wall thickness
   • Pipe damage from bending, crushing, stretching, or other stress
   • Pipe damage that impacts the pipe strength, the intended use, the internal diameter of the pipe, and internal roughness characteristics
   • Any other defect of manufacturing or handling

8.7.3   Construction

Minimum Allowable Depths

Pipe-replacement systems follow closely the existing alignment and grade of the host pipe. Therefore the installation depth is relatively the same as the host pipe. However, if the replacement pipe diameter is larger than the host pipe diameter, the effective cover above the replacement pipe must be calculated as an increase of 10 in. for every inch of increased diameter.

Equipment

The contractor must utilize pipe-replacement equipment with adequate pulling/pushing force to complete pulls. The contractor must verify the pulling/pushing force exerted on the pipe does not exceed the manufacturer's recommendation for allowable pulling force to prevent damage to the pipe.

Method

1. Pipe must be assembled and fused on the ground in sections equivalent to the length of the anticipated pull.

2. During installation, all bending and loading of the pipe must be in conformance with the manufacturers recommendations and must not damage the pipe.

3. Manholes must be prepared to ease the pipe-installation process along the alignment and grade indicated on the plans.

The invert in the manholes must be removed to ease the pipe-installation process and in such a fashion to accommodate the invert replacement. Manhole inverts must be restored upon completion with 3000-psi concrete so as to establish a minimum 4-in.-thick manhole bottom.

4. Solid wall pipe must be produced with plain end construction for heat-joining (butt fusion) conforming to ASTM D 2657. The polyethylene pipe must be assembled and joined at the site using the thermal butt-fusion method to provide a leakproof and structurally sound joint. Threaded or solvent-cement joints and connections are not permitted. The butt-fused joint must have uniform roll back beads resulting from the proper use of temperature and pressure during the fusion process. The joint surfaces must be smooth. The joint must be watertight and must have tensile strength equal to that of the pipe. All defective joints must be cut out and replaced.

5. Bypass pumping is required for this method when needed; dewatering *may* be initiated prior to any pit excavation.

6. All laterals must be disconnected prior to pipe-bursting operation.

7. Upsizing is only allowed if soil and site conditions are suitable (new and existing pipe diameters considerations, minimum depth of existing pipe is allowable, and if a minimum distance to any existing utility or structure is followed).

Oversize Allowance

When using this method, the allowable oversize diameter is 1 in. greater than the pipe radius.

Watertight Joints

Watertight pipe joints are very important to the integrity of the roadbed. Every reasonable effort must be taken to ensure that watertight joints are installed. Pipe must be constructed to prevent water leakage or earth infiltration throughout its entire length.

A watertight specification for each type of pipe material can be obtained through each pipe material industry. Please refer to the appropriate industry specifications for more detailed information.

Lubrication Fluids

Lubrication fluids may be used to facilitate the installation of new pipe.

Pipe Locating and Tracking

Pipe locating and tracking is *not* required for this method of pipe installation. Figure 8.25 suggests a checklist for inspection of pipe-replacement operations.

Inspection Guide for Pipe-Replacement Systems (PRS)

Preinspection Plan Review

☐ Review geotechnical and soil reports.

☐ Ensure DOT / agency facilities and nearby utility information are shown on the plans and profiles and that the proposed alignment does not interfere with them.

☐ Verify site surveys before and after installation.

☐ Note the minimum cover above the top of the existing pipe and below the pavement surface, or ground elevation (for longitudinal installations outside the influence of the roadway) is _____ ft.

☐ Note new and host (existing) pipe characteristics:
Host pipe material _____.
Host pipe diameter _____in.
New pipe material _____.
New pipe diameter _____ in.
New pipe dimension ratio _____.
Allowable pulling force _____ ton.

☐ Review contingency plans.

☐ Review job-site layout including: distance from access pits to roadbed, proposed sheeting and bracing, materials storage and fabrication area, safety devices (barrels, guardrail, etc.), and dewatering pit locations.

☐ Verify all laterals are excavated and disconnected.

☐ Verify bypassing plans are adequate.

☐ Note unique or special items / circumstances: _____

Construction Inspection

☐ Verify traffic control is consistent with the permit requirements, one call service has been contacted, and the DOT / agency permit is on-site.

☐ Verify job-site layout is consistent with the approved plans, especially the alignment of the pipe and machine.

☐ Verify pit excavation and bypassing efforts are according to plans and satisfactory, and workers understand the contingency plan.

☐ Verify new pipe characteristics are consistent with permit requirements.

☐ Verify any upsizing is in accordance with permit.

FIGURE 8.25 Sample checklist for inspection of pipe-replacement operations.

☐ Verify HDPE pipe is supported and secured reasonably with rollers, skates, or other protective devices to prevent damage to the pipe, is new with smooth interior and exterior surfaces, is used within the entire influence area of the roadbed, has clean and square ends, joints are butt-fused and watertight, pipe with gouges exceeding 10 percent of the pipe wall thickness or kinked sections is not used, and damaged pipe is removed, and the host pipe is inspected with CCTV, and the bursting head diameter is comparable with the replacement pipe diameter.

☐ Verify sufficient lubrication fluid quantity is used, and a lubrication system properly injects lubricant on the inside and outside of the pipe.

☐ Verify any unsuccessful borehole is backfilled immediately.

☐ Verify each end of the pipe is enclosed, restoration is completed, and attach Inspector's Daily Report (IDR), form 2228.

Permit No. _____
Inspector: _____
Date: _____

Figure 8.25   Sample checklist for inspection of pipe-replacement operations. (*Continued*)

8.8   Access Pits/Driving and Receiving Shafts

1. *Location:* The location of drive and receiving shafts must be proposed (submitted) by the contractors and approved by the design engineer or the state DOT/agency/municipality based on approved guidelines and regulations and project and site requirements. For example, Michigan DOT requires a minimum distance of 20 ft, from the edge of the paved shoulder or curb to the face of any shaft, equipment, and supplies must be maintained in areas posted at 45 mph or less; otherwise, a minimum distance of 30 ft must be maintained.

2. *Sheeting and bracing:* Sheeting and bracing may be required whenever any part of the shaft excavation is located within the roadbed influence area. For example, an additional earth-retention structure may be required above and below the borehole on the boring face of all shafts to prevent loss of road embankment materials during construction.

3. *Surface and groundwater flow:* Ground surface around shafts and pits must be sloped away so surface runoff moves away from shafts and pits. To remove any storm or groundwater from shafts and pits, a sump pump must be installed at a pit at the bottom of the shaft.

4. *Protection:*
   - Traffic barriers must be installed adjacent to shaft locations according to the state DOT or an agency having jurisdiction or according to the standard specifications for construction. Temporary beam guardrail must also be installed according to the current DOT or agency having jurisdiction or construction specifications.
   - Fencing barriers must be installed adjacent to shafts, open excavations, equipment and supplies with suitable fencing and plastic drums to prohibit pedestrian access to the work site. Equipment must not be used as fencing to protect shafts and pits.
   - The contractor must construct and operate safe access shafts and pits according to OSHA guidelines and/or all applicable regulatory requirements.

8.9   Settlement/Heaving Monitoring

1. This method must be performed in a manner that will minimize the movement of the ground in front of, above, and surrounding the boring operation, and will minimize subsidence of the surface above and in the vicinity of the boring.

2. Potential settlement must be monitored at each edge of right-of-way, each shoulder point, each edge of pavement, the edge of each lane (or centerline for two lane roads), and otherwise at 50-ft intervals along the pipe centerline.

3. A survey must be performed 1 day prior to initiating this operation at each required monitoring location. A similar survey must then be performed at each location, on a daily basis, until the permitted activity has received a final inspection. This survey establishes the preexisting and postconstruction conditions, and the amount of settlement. All survey readings must be recorded to the nearest one-hundredth (0.01) of a foot. Whenever possible, trenchless pipe installations must *not* be installed directly under a pavement crack. Digital photographs of the pavement conditions must also be taken prior and after the pipe installation.

4. All operations must stop immediately whenever monitored points indicate a vertical change in elevation of 1 in. or more, or any surface disruption is observed. The contractor must then immediately report the amount of settlement to the engineer/inspector.

8.10   Groundwater Control

1. Dewatering must be conducted whenever there is a high groundwater table level to prevent flooding and facilitate the operation. The water table elevation must be maintained at least 2 ft below the bottom of the casing at all times. When needed, dewatering *may* be initiated prior to any excavation.

2. Minor water seepage or pockets of saturated soil may be effectively controlled through bailing or pumping. This control must be accomplished without removing any adjacent soil that could weaken or undermine any access pit, its supports, or other nearby structure.

3. Larger volumes of groundwater must be controlled with one or more well points or with staged deep wells. Well points and staged deep-well pumping systems must be installed and operated without damage to property or structures, and without interference with the rights of the public, owners of private property, pedestrians, vehicular traffic, or the work of other contractors. Any pumping methods used for dewatering and control of groundwater and seepage must have properly designated filters to ensure that the adjacent soil is not pumped along with the water. Well diameter, well spacing, and the pump's pumping rate must provide adequate draw down of the water level. Wells must be located to intercept groundwater that otherwise would enter the access pit excavation and interfere with the work. On removal of a well, the borehole must be filled and grouted according to the specifications identified in DOT's flowable fill special provision, and DOT's plugging drill holes special provision.

4. Existing storm sewers must only be used to discharge water from the dewatering operation in accordance with a permit obtained from the appropriate storm sewer owner. Filters or sediment control devices must be required to ensure that the existing system is not adversely affected by construction debris or sediment.

5. If grouting is used to prevent groundwater from entering the area of the access pit, the grouting must be installed without damage to property or structures and without interference with the rights of the public, owners of private property, pedestrians, vehicular traffic, or the work of other contractors. The material properties of the grout must conform to the specifications identified in DOT's flowable fill special provision.

6. Whenever a significant amount of unexpected groundwater enters an access pit, and a catastrophic pit failure is imminent, the pit must be backfilled immediately, until the groundwater level is at least 2 ft below the bottom of the pipe.

8.11   Boring/Ramming/Bursting Failure

1. If anything prevents completion of this operation, the remainder of the pipe must be constructed by methods approved by the engineer/inspector.

2. Abandonment of any component of the installation must only be allowed as approved by the engineer/inspector.

3. If an obstruction is encountered which prevents completion of the installation, the pipe must remain in place, taken out of service, and immediately filled with flowable fill.

4. The cost associated with encountering obstructions such as boulders shall be paid for as extra work.

8.12   Contamination

When an area of contaminated ground is encountered, all operations must stop immediately, and must not proceed until approved by the engineer/inspector. Any slurry must be tested for contamination and disposed of in a manner, which meets local, state, and/or federal requirements.

8.13   Bulkhead

In horizontal auger-boring and pipe-ramming operations, casing ends must be enclosed or bulkheaded with a commercial grade concrete, or approved alternate to seal the ends to prevent water leakage or earth infiltration. The concrete must extend longitudinally into the pipe end opening to create a minimum 1-ft-thick bulkhead barrier, or as required by permit. Engineer/inspector may allow rubber bulkheads in special situations.

8.14   Work-Site Restoration

1. Access pits and excavations must be backfilled with suitable material, and in a method approved by the engineer/inspector. The shafts must be backfilled and sealed upon completion of the microtunneling. The shaft and supports must be removed to 10 ft below the original ground surface. The disturbed

work-site area must be restored to existing grades and original material condition.

2. The disturbed grass-surface area must be topsoiled, seeded, fertilized, mulched, and anchored according the current DOT standard specifications for construction. Slopes steeper than 1-on-3, must be sodded according to the current DOT standard specification for construction. If a final site restoration is not completed within 5 days after completion of the operation, the installation of temporary soil erosion and sedimentation control measures must be required.

3. On completion of the work, the contractor must remove and properly dispose of all excess materials and equipment from the work site.

4. The permit, including the surety requirements, must remain in effect for a minimum of 1 year after completing the work to monitor for settlements of the pavement and/or slope.

8.15 Summary

This chapter presented an overview of quality assurance/quality control requirements for major trenchless installation and replacement methods. An overview of each method with sample guidelines for permitting and inspection were included. Pipeline owners and design and consulting engineers can customize these guidelines based for their specific project conditions.

CHAPTER 9

Planning and Safety Considerations for Trenchless Installation Methods*

9.1 Introduction

Every trenchless project is a unique undertaking. Although similar work may have been performed previously, no two projects, no matter how close they might be, will have identical job conditions. Planning is undertaken to understand potential project problems and to develop courses of actions. The goal of planning is to minimize resource expenditures and to successfully complete the project in a safe and productive manner. Planning is necessary in order to

1. Understand project objectives and requirements
2. Define work elements
3. Develop safe construction methods and avoid hazards
4. Improve efficiency
5. Coordinate and integrate activities
6. Develop accurate schedules
7. Respond to future changes
8. Provide a baseline for monitoring and controlling execution of project activities

---

*This chapter is not intended to substitute proper safety guidelines and training programs, as provided by OSHA, other regulatory agencies, the industry, or equipment manufacturers.

There is no single best way to conduct a construction activity. Performance of a work task depends on variables related to the work, such as the type of pipeline and utility application, type and diameter of pipe, the location of the project, the work hazards, the various contractual and legal constraints, and so on.

Early planning will provide more time to carefully consider the impacts of all project constraints and to devise efficient strategies for dealing with project requirements. For planning purposes, tasks should be divided into smaller independent executable subtasks. This enables project engineers to structure unwieldy and complex problems into smaller digestible tasks. The more time and resources allocated for planning at each stage of project, the greater the opportunity to develop optimal solutions rather than something that is "bare minimum."

In planning trenchless projects it is necessary to give considerable attention to safety. Engineers have both a moral and a legal responsibility to the public, to construction workers, and to the end-users of the projects to ensure that the workplaces, construction operations, and the work environment are safe. The machines that are used in trenchless installations can be very sophisticated, but they can also present a variety of hazards such as those associated with open-cut construction. Safety planning will identify hazards and develop ways to protect both the construction workers and the public.

All equipment operations entail risk. The level of safety planning that is needed should be based on the specific risks associated with work conditions. When risks are identified in advance, and steps are taken to control, reduce, or remove them, links are being removed from the chain of potential errors. Experience has shown that accidents are usually caused by a chain action of errors.

When the engineer prepares a plan and cost estimate for a trenchless project, the decision process is often not a list of sequential activities. The process takes the form of recurrent activities with feedback. As decisions are proposed, further investigation and collection of more information is usually necessary to reduce uncertainty.

Contractors have a right to rely on owner-provided information. Additionally, many contracts will contain a *differing site condition* clause. Material differences in conditions are applicable in either of two cases. A type I differing site condition exists when actual conditions differ materially from those "indicated in the contract." A type II differing site condition arises when actual conditions differ from reasonable expectations. These clauses provide the constructor some protection from geotechnical risks. They do not, however, eliminate the contractor's responsibility for performing a thorough examination of project conditions.

Field investigations, geologic and soil studies, and analysis of meteorological data, enable the contractor to better quantify what has been presented in the bid documents. The contract documents will usually include geotechnical data and information that was gathered during the design phase of the project. Supplemental site visits by the contractor are necessary to fully understand jobsite requirements and limitations.

Detailed studies of trenchless machine production can be performed only after the all the questions regarding surface and subsurface conditions and specific project conditions have been answered. At that point, the contractor will be able to undertake production analyses based on use of proposed trenchless equipment. This chapter is devoted to two most important aspects of a trenchless project: planning and safety.

9.2 Planning for a Trenchless Project

A successful trenchless construction project requires surface, subsurface investigations, and safety considerations. Trenchless installation methods require the design engineer to provide the contractor with sufficient information to reasonably anticipate the obstacles that might be encountered. During the design phase, surface and subsurface survey information will assist in determining the suitability of trenchless installation by specific methods. Obtaining and providing accurate surface and subsurface information will result in reducing the possibility of installation problems and change orders during the work as well as minimizing the possibility of litigations and disputes.

9.2.1 Surface Survey and Site Visit

A surface survey is required prior to designing of a trenchless construction project. Each trenchless construction project has specific site requirements. The surveys should be conducted along the center line of the proposed bore path for a width of 100 ft. As a minimum, surface surveys should include the following:

- Work area requirements
- Existing grade elevation data
- Surface features such as roadways, sidewalks, and utility poles
- Boring or test shaft/pit locations
- Waterways and wetlands
- Visible subsurface utility landmarks such as manholes, cleanouts, service laterals, or valve boxes
- Adjacent structures to bore path

Review of existing geological or geotechnical reports, maps, aerial photographs, and review of depositional history are important in developing a preliminary design survey. For example, if the area has been subjected to glaciations, then cobbles, boulders, and gravel can be expected. If the area has been subjected to large landslides, trees, and other natural or manmade objects may have been buried and could be encountered. If the area has been subjected to low-energy streams and rivers, then fine-grained deposits may be expected. Conversely, high-energy, steeply sloped stream beds may be covered with cobbles and boulders from nearby mountains. Reports on surface contours and elevations are also important so that any possible ground movements, settlements, and heaves can be monitored. During the design and bidding phase of the project, it may be necessary for engineers and contractors to visit the project site and visualize the degree of job difficulty that may be encountered during the project execution.

9.2.2   Subsurface Investigations

Subsurface investigation is the next step to surface survey. Subsurface features, which will impact trenchless installations, include presence of existing underground pipelines and utilities or other manmade structures, and method of their placement. Geotechnical conditions, such as existence of solid rock, heavy gravelly soil, and boulders and cobbles along the alignment of proposed trenchless installation must also be investigated.

Locating Existing Utilities

Trenchless installation projects require the contractor to install the pipe without seeing the excavation area. Therefore, the design engineers should give contractors a record of potential conflicts and utility crossings.

In the United States, the local one-call service like MISS DIG in the state of Michigan or the national number 811 should be contacted as a first step prior to design and construction of a trenchless project. In other countries, and in the absence of one-call service, municipalities and utility companies should be contacted individually to obtain the required information. Obtaining as-built and record drawings, geographic information system (GIS) data, utility maps, and locating pipeline markers are also important. Geophysical methods subsurface utility locating may include surface-applied pipe locators, ground-penetrating radar (GPR), electrical resistivity imaging, and seismic survey. In addition to the information provided by design engineers (usually in terms of soil-boring logs and possibly soil reports), and contacting utility companies and one-call centers, contractors must positively locate and expose existing utilities running parallel less than 10 ft (at specified intervals, such as every 50 ft, dependent on proximity and type of existing utility and trenchless method used) or where it proposed trenchless alignment crosses an existing utility or pipeline. Exposing utilities can be done by mechanical equipment (such as a backhoe), hand shoveling, and/or by vacuum excavation equipment. More information on locating existing utilities can be found from the references at the end of this book.

Subsurface Utility Engineering

This discussion of subsurface utility engineering (SUE) is based on the ASCE standard, *Standard Guidelines for the Collection and Depiction of Existing Subsurface Utility Data*, CI/ASCE 38-02, 2003. SUE can be defined as "a branch of engineering that involves managing certain risks associated with utility mapping at appropriate quality levels, utility coordination, utility relocation design and coordination, utility condition assessment, communication of utility data to concerned parties, utility relocation cost estimates, implementation of utility accommodation policies, and utility design." The use of SUE offers an opportunity for a more comprehensive and organized approach to the location of existing underground utilities. This method provides more in-depth information regarding existing utilities.

ASCE standards suggest use of four "utility quality levels," which are defined as "professional opinion of the quality and reliability of utility information." Each of the utility data quality levels is established by different methods of data collection and interpretation. The following sections present descriptions of different quality levels. Table 9.1 presents a summary of the four quality levels and their associated definitions.

Utility Quality Level	Definition
D	Information derived from existing records or oral recollections.
C	Information obtained by surveying and plotting visible above-ground utility features and using professional judgment in correlating this information to quality level D information.
B	Information obtained through the application of appropriate surface geophysical methods to determine the existence and approximate horizontal position of subsurface utilities. Quality level B data should be reproducible by surface geophysics at any point of their depiction.
A	Precise horizontal and vertical location of utilities obtained by the actual exposure (or verification of previously exposed and surveyed utilities) and subsequent measurement of subsurface utilities, usually at a specific point.

TABLE **9.1** Utility Quality Level

- *Quality Level D:* The minimum level of information is based on existing utility records. Such information is primarily useful for the purposes of project planning and route selection only.

- *Quality Level C:* In addition to the information from Quality Level D, this level includes information obtained from a site visit and a survey of ground surface features, such as manholes, valve boxes, posts, and the like, and correlation of this information with existing utility records. As a result, the presence of additional belowground utilities, or erroneously recorded location information of utility lines, may be determined. Although such information may be adequate for areas with minimal belowground facilities, or where possible repair is not a major issue, this quality level would typically not be sufficient for proceeding with trenchless construction in established areas.

- *Quality Level B:* In addition to the information from Quality Level C, for more useful and reliable information, Quality Level B surface locators are used to identify and mark the existing utility lines.

- *Quality Level A:* In addition to the information from Quality Level B, the highest quality level (Quality Level A) includes the use of minimal destructive equipment (such as vacuum potholing) at critical points to expose the utility to determine the precise horizontal and vertical position of underground utilities, as well as the type, size, condition, material, and other characteristics.

The reader is referred to the referenced ASCE standard for a more in-depth discussion of SUE. However, it is apparent that varying levels of existing utility location may be applicable depending on the specifics and objectives of the trenchless project. During the planning phase, the design engineer or owner should decide what quality level of information for underground utility is consistent with the project needs and relate that with their risk management strategy. The SUE process offers some potential benefits, such as avoiding conflicts with other existing utilities, reducing delays in the construction schedule, eliminating additional construction costs, and reducing inconveniences to the general public.

The level of utility quality information is an important decision that should be made and obtained as early as possible during the planning and design phase of the project.

9.2.3 Geotechnical Investigations

A second phase of subsurface investigation for trenchless installation projects involves determination of soil conditions. Once the proposed alignment has been identified, a geotechnical investigation should be performed. Investigations for complex installations should comprise two phases: a *general geotechnical review* and a *geotechnical survey*. A geotechnical survey alone may be sufficient for simpler installations.

General Geotechnical Review

A general geologic review involves examining existing geological data to determine what conditions might be encountered in the vicinity of the installation. Existing data may be available from past project records in the area of the trenchless project (buildings, piers, bridges, levees, and so on). Such an overall review will provide information that may not be otherwise developed from exploratory borings. This step allows the *geotechnical survey* to be tailored to the anticipated conditions at the site, thus enhancing the effectiveness of the geotechnical survey.

Geotechnical Survey

For trenchless installations, it is necessary to know the actual soil stratification at a given site, the laboratory test results of the soil samples obtained from various depths, and the observations made during drilling exploratory bores. The steps for subsoil investigation should include the following:

- Determining the nature of soil at the site and its stratification
- Obtaining disturbed and undisturbed soil samples for visual identification and appropriate laboratory tests
- Determining the depth and nature of bedrock, if encountered
- Performing in situ field tests, such as field density and standard penetration tests (SPT)
- Observing surface drainage conditions from and into the site
- Assessing any special construction problems with respect to the existing structures nearby
- Determining groundwater levels, sources of recharge, and drainage conditions

The main methods of geotechnical surveys are as follows:

- *Hand augers:* Suitable only for shallow depths, only disturbed or mixed samples of soil can be obtained in this method.
- *Test shaft/pits or trenches:* This method is suitable for shallow depths only, but allows visual observation over a larger area than is possible with samples from vertical soil borings.
- *Boring test holes and sampling with drill rigs:* This is the principal method for detailed soil investigations. Sampling interval and technique should be set to accurately describe the soil characteristics, taking into account the site-specific conditions.

Typically, split spoon samples will be taken in soft soil at 5-ft-deep intervals in accordance with ASTM D1586.

- *Ground penetration radar:* Useful in gravels and sands.
- *Acoustic (sonar):* Useful for determining depth of rock, interfaces between soft and hard deposits and buried objects.
- *Geophysical methods:* Variations in the speed of sound waves or in the electrical resistivity of various soils are useful indicators of the depth of water table and bedrock.

As said earlier, it is essential that a thorough subsurface investigation is carried out in the design phase to identify the geologic conditions along the pipeline alignment. The anticipated geologic conditions comprise the most important factor in the selection of an appropriate trenchless installation method for a specific project. Groundwater conditions will have an important influence on the behavior of the ground and constitute a major factor for loss of ground. Groundwater levels should be determined, and pumping tests or other field tests should be conducted to estimate the permeability of soil to see if dewatering is necessary and feasible. Contractors must investigate groundwater conditions just before start of trenchless installation, as water table conditions may have changed due to change in season when the design investigations were completed. In summary, the following is a list of recommended information to be obtained for a trenchless installation project:

- Soil information based on the unified soil classification system (USCS).
- Gradation curves on granular soils.
- Standard penetration test values where applicable (generally in unconsolidated ground).
- Particle-size distribution, including presence of cobbles and boulders.
- Shear strength of soil.
- Atterberg limits (liquid, plastic, shrinkage limits and plasticity index).
- Moisture content.
- Depth and movement of water table.
- Permeability.
- Cored samples of rock with description, rock quality designation, and percent recovery. It should be noted that rock hardness (Moh's hardness or Vickers test), texture, tenacity, and formation will determine penetration rate of drilling and tunneling equipment.

- Unconfined compressive strength for representative rock samples (frequency of testing should be proportionate to the degree of variation encountered in rock core samples). Rock should be cored in accordance with ASTM D2113 to the maximum depth of the proposed trenchless boring.
- Presence of contaminated soils (such as hydrocarbons and the like).
- Climatic data (such as temperature ranges).
- Information on quality of water [e.g., pH values (acidity or alkalinity), salinity, to determine possibility for pipe material corrosion].
- Special investigations for possible existence of swelling clays, or particular chemical conditions.

It is essential that subsurface investigations be conducted by geotechnical engineers who have knowledge of regional geology and hydrogeology as well as an experience in trenchless technology and boring/tunneling. If during the exploration of the underground a potentially difficult ground is found (such as mix-face conditions with sloped layers, presence of boulders and cobbles larger than one-third diameter of proposed borehole, swelling clays, running sands, gravelly soil, and the like), additional investigations must be carried out. The complete geotechnical report must be provided to potential bidders so they can submit realistic bids.

Settlement Potential

Surface settlement is mainly a result of loss of ground during tunneling and dewatering operations that cause subsidence. During a trenchless installation project, loss of ground may be associated with soil squeezing, running, or flowing into the tunnel boring machine (TBM) or the cutterhead; settlement due to large overcut size; and steering adjustments. Mixed ground conditions, unstable ground over stable ground (sand over clay), are the most common reasons for overexcavation. The actual magnitudes of these ground losses are largely dependent on the type and strength of the ground, groundwater conditions, size, and depth of the pipe, equipment capabilities, and the skill and experience of the operator in operating and steering the machine. If passive earth pressure is exceeded, heave of ground surface may occur, causing damage to nearby utilities, pavement, and other nearby structures.

Geotechnical Baseline Report

Subsurface conditions can vary, sometimes significantly, within a project site. Geotechnical engineers do not possess the ability to predict all possible ground variations in advance of trenchless technology construction. The design engineers can make educated guesses and sensible estimates as to the range of those conditions, and can account for potential variations in the design of the project. The challenge is for the project owner to secure the services of a contractor to construct the project for a fair (to both parties) price. The single greatest challenge for the owner is how to deal, contractually and financially, with the potential variability of the ground and groundwater conditions encountered during the trenchless work.

Some owners may try to contractually transfer and shift all the risk of the subsurface conditions to the contractor through "one-sided" contract language and a "you bid it, you build it!" mentality. However, this will persuade bidders to carry greater contingencies in their bids, which increase the cost of the project, and will not necessarily protect the owner from subsequent claims due to unforeseen subsurface conditions. Depending on the competitiveness of the marketplace, bidders may also choose to interpret the available information in the most optimistic manner possible, in an effort to present the most competitive bid they can. In both instances, there is a high likelihood that the contractor will win subsequent claims for additional compensation, because they may be able to prove that conditions encountered were *materially* more costly to bore, drill, excavate, support, stabilize, or otherwise control, than what they had assumed in their bid. There is an abundance of case law supporting the premise that the "owner owns the ground," and that the owner will ultimately pay the price of constructing his or her desired pipeline at the site along his preferred alignment.

Due to high possibility of changing ground conditions on which the bid was based, the presentation of the basis for design in the contract document allows the parties to understand the rationale behind the anticipated subsurface conditions. Baseline reports translate the facts and interpretations about subsurface conditions to be encountered into a set of relatively simple statements. The baseline statements may be preceded with a discussion of the potential variability and uncertainty of certain conditions; however, the baseline statements should be stated explicitly and in straight-forward, clear terms. The use of adverbs and undefined adjectives should be avoided. The baseline statements should be comprised of quantifiable terms that can be measured in the field during construction. By establishing clear baselines as part of the contract documents, the parties are more likely to agree on the conditions indicated in the contract, without time-consuming and costly arguments.

Baseline statements should be considered as contractual baselines, not necessarily geotechnical facts. Preparation of baselines may involve the interpolation between, and extrapolation beyond, the conditions depicted or inferred by the subsurface information.

Illustrative examples are provided in the guidelines document. The baselines establish a contractual statement of the conditions to be encountered during the work. In setting the baselines, the geotechnical baseline report (GBR) is allocating the risk of all conditions equal to or less adverse than the indicated baseline conditions to the contractor. Similarly, the financial risk of encountering conditions significantly more adverse than the baseline is allocated to the owner. While the baseline does not represent a warranty that the indicated conditions will, in fact, be encountered, it does represent a promise on behalf of the owner that the baselines will be considered in implementing the differing site conditions (DSC) clause, and will weigh in the determination of whether additional compensation for more adverse conditions is justified.

The GBR must be prepared by geotechnical engineers familiar with trenchless technology. Owners should retain design teams that include individuals experience in the design and construction of trenchless projects, and should ensure that those individuals will be closely involved with the preparation and review of the GBR document. The GBR should be developed after the design has been completed and the drawings and specifications have been advanced to a substantial level of completion. Ideally, the GBR should be reviewed by a knowledgeable engineer who has not been involved with the preparation of the drawings, specifications, or GBR. This independent "fresh look" is of critical importance in identifying and alleviating ambiguities that will inevitably exist within the GBR or between the GBR and other contract documents.

The owner should be advised of the range of conditions that could be encountered and the designer's best assessment of the most likely conditions to be encountered. The owner should also be included in the process of setting the baselines for construction. This should include an understanding of the reasonable range within which the contractual baseline(s) might be set, including the ramifications of where the baselines are set on the bid prices and potentials for change orders.

The *Geotechnical Baseline Reports for Construction,* published by the American Society of Civil Engineers (Essex, 2007) provides recommendations for what should be included in the GBR and what should not, and provides a checklist of items to consider when writing a GBR. This manual provides recommendations of the content and wording to be used in baseline statements to improve their clarity and precision, and presents illustrative examples of problematic and improved practice in stating baselines. While GBR is recommended for all types of trenchless technology projects, some owners and may not consider it to be applicable for smaller trenchless projects (such as Mini-HDD, horizontal auger boring, or pipe ramming operations).

9.2.4 Permits

Permits are typically permissions granted to the prospective project owner from a government agency [such as highway department (state Departments of Transportation or DOT), U.S. Army Corps of Engineers, a city (municipality) or county*] to construct or replace a pipeline under the existing facility. Permits may have construction and fee conditions for occupation of the space beneath the existing facility (such as annual fees, maintenance requirements, reporting requirements, insurance, etc.). Although the future pipeline owner is held ultimately responsible by the entity issuing the permit for any violations, the contractor will also be responsible (to the future pipeline owner) for compliance with permit requirements pertaining to the construction. Permits can require an extended processing time to obtain and therefore are typically obtained by the pipeline owner during the project design. The contract documents should list and contain a copy of the construction requirements for all permits obtained for the project. The contract should also require the contractor to adhere to the requirements of the permits. Some typical locations and types of permits that could be required for a given trenchless installation project include

- U.S. Army Corps of Engineers 404 discharge permit[†]
- Wetlands crossing permits
- Floodplain development permits
- Crossing permits for
  - City streets
  - County roads
  - U.S. and state highways
  - Interstate highways
  - Railroads
  - Waterways
- As mentioned earlier, construction permits can be issued by local governments, DOT, river authorities (such as U.S. Army Corps of Engineers), regional, state and federal regulatory agencies, and funding agencies

---

*City, municipality, county, and so on, is collectively called local government.
[†]Discharge of dredged and fill material into waters of the United States, including wetlands. Responsibility for administering and enforcing Section 404 is shared by the U.S. Army Corps of Engineers (USACE) and U.S. Environmental Protection Agency (EPA). EPA develops and interprets environmental criteria used in evaluating permit applications, identifies activities that are exempt from permitting, reviews/comments on individual permit applications, enforces Section 404 provisions, and has authority to veto USACE permit decisions. (See www.epa.gov/owow/wetlands/pdf/reg_authority.pdf.)

- To avoid project delays, it is important that the required permits be identified as early as possible during the project planning and measures are taken to secure the permits in a timely fashion

9.2.5 Job Site Logistics Requirements

The project designer and owner must recognize and address community and neighborhood needs in setting up the following details of the contract documents, which will direct the contractor in his work*:

- Traffic control

- Storage areas

- Equipment setup areas

- Construction staging areas

- Location of major supporting equipment

- Dust and noise restrictions

- Allowable working hours and working days

- Blackout periods, if any

- Project duration

- Project schedule

- Easement requirements

This information should be compiled in the bid documents so contractors understand the requirements and associated costs with avoiding violation of third party rights or unreasonable interference with any public or private property and quality of life. When using pipe materials that must be butt-fused or welded in advance of project start date [such as horizontal directional drilling (HDD), continuous sliplining, or pipe bursting] into long lengths, a barricade plan should be part of the planning process. It is important to recognize the logistical space requirements associated with the anticipated construction techniques during the planning phase of the project. In some instances, the availability or lack of available space for logistics will be the controlling factor in selecting a specific trenchless

---

*To avoid interference with "means and methods" of construction, project owners and design engineers can ask potential contractors for a series of "submittals" to be evaluated, modified and/or approved by the owner or engineer. This way, while contractors' expertise is taken into account, the final approval rests with the owner or the engineer on the project.

installation technique or a pipe material or a specific pipe joint (such as welding and butt fusion versus restraint joint).

9.2.6 Length of Installation

The length of the pipe segment that can be installed by trenchless technology is another detail that will be impacted by the material and selected trenchless method. For example, when segmental rigid pipe is chosen; the shaft/pit location can be placed wherever it will have the least impact on the area and the costs. If new manholes are being constructed or the old ones must be replaced, the shafts/pits will be located at these manhole locations. If the manholes are being saved, the shafts/pits could be located at service lateral excavations and downstream manholes are used as receiving shafts/pits.

When flexible pipe is chosen, the impact of fusing, welding, or using restraint joints for new pipe sections will dictate shaft/pit locations. The owner must recognize the limitation of tensile and jacking strength of any selected pipe system (material, joints, etc.). In addition, possible jacking (drive) distances are dependent on the pipe material, diameter, geometric tolerances, pipe material moisture absorption, steering of the cutterhead, trenchless installation method, friction of the surrounding soil on the pipe outer surface, size of overcut, and type and amount of lubrication used during the installation. There is presently no method to accurately estimate pulling or jacking forces and to quantify these effects. Therefore, allocation of a factor of safety (usually 2) for a given pipe system (specified material, diameter, wall thickness, joint, etc.) during the design phase is necessary.

9.2.7 Alignment Considerations

As in all pipeline projects, identifying feasible trenchless technology alignments involves evaluating available right-of-way (ROW) and easement acquisition issues, and determining the location of the existing utilities. Sometimes, alignments are not feasible or economical for open-cut method, but can be achieved by trenchless technology methods. Pipes installed using trenchless technology methods can be installed deeper by only increasing the depth of the shaft/pit. This can be of significant advantage if a deeper alignment can avoid potential conflicts with existing underground utilities, reduce pumping stations, and utility relocations.

Straight horizontal alignments are generally preferred for trenchless technology projects. Straight alignments provide for more accurate control of line-and-grade and for a more uniform stress distribution on the pipe and pipe joints, reducing the risk of concentrated loads that could damage the pipe.

To be feasible, a prospective alignment must have adequate jacking and receiving shaft/pit locations available. In addition, prospective jacking and receiving shaft/pit sites must be spaced at distances that are compatible with specific trenchless technology technique. The maximum distance a pipe can be installed with a proposed trenchless installation method is dependent on parameters such as pipe size, structural capacity of the pipe, possible use of intermediate jacking stations, thrust capacity of the thrust block and the main jacks, soil conditions, effectiveness of the lubrication system, and specific project conditions such as operator's skills in steering the cutterhead or the tunnel boring machine (TBM).

Providing adequate space for staging construction operations is important so that pipe installation can be completed in an efficient manner. Construction access to the jacking shaft/pit must be provided for hauling spoils, pipe sections, and tunneling/drilling equipment. In urban areas, traffic control requirements must be evaluated in selecting and laying out jacking shaft/pit sites. A typical jacking shaft/pit site needs enough space for the jacking shaft/pit, slurry tanks, a crane, pipe storage, and support facilities and equipment (e.g., as a generator, power pack, and bentonite lubrication unit), and operator's trailer. The jacking shaft/pit should be a sufficient distance from overhead electrical lines to avoid hazards in operating the crane, although in some areas a gantry system can be used instead of a crane for smaller pipe sizes.

The jacking equipment arrangement is quite flexible and space requirements can be reduced for smaller sites, if necessary. Frequently, jacking shafts/pits have been located in the parking strip along the edge of a street with the equipment setup in a linear arrangement. Similar linear arrangements have also been used to stage trenchless technology operations from the median of wider, more heavily traveled streets without significantly impacting traffic flow. Staging area requirements can be further reduced by using same machine location to install two drives, one in each direction (see Fig. 9.1). This approach further minimizes the environmental impacts of construction by reducing the number of jacking shaft/pit locations. Using pipe sections with smaller lengths, and indexing jacking equipment with several strokes, may allow smaller shafts/pits.

9.2.8   Accuracy and Tolerances Including Settlement and Heave

As with other details of the planning phase, the question of settlement and heave will be greatly impacted by many project-specific conditions and ground conditions. Trenchless technology has advanced so that pipe installation can successfully be performed in difficult ground conditions.

FIGURE 9.1   Staging HDD equipment to minimize environmental impacts.

The design engineer must fully understand the trenchless technology that has been chosen and have a set of realistic expectations. Setting realistic expectations can be difficult since the pipe is installed horizontally and may go through different ground conditions emphasizing the need for obtaining good subsurface information during the planning phase.

Accurate surface monitoring must be undertaken by the project owner, public agency, or the contractor as a proof that the trenchless operation did not affect the nearby structure. Sometimes the cost of repairing pavement heave or settlement may be less expensive than reinstatement of ground during open-cut construction which could impact local business and traffic in a much more disruptive way. Project engineers and owners need to consider that during a trenchless technology project, there is a possibility of ground movement that may show in the surface in terms of settlement or heave. The extent of settlement or heave on the surface is dependent on many factors, including type of trenchless method used, pipe diameter, experience of operator, equipment used, depth of installation, and soil conditions. With proper geotechnical investigations, planning, and contractor and method selection, the possibility and/or extent of surface/ground movements can be reduced tremendously.

9.3   Trenchless Safety Considerations

There is no doubt that a safe project starts in the planning and design phase. Considering trenchless projects are "engineered" jobs with more skilled and trained workers than conventional open-cut construction, they provide more opportunities for safe operations. Construction traditionally has been one of the most dangerous types of work, third after mining and agriculture.

In 1970, U.S. Congress passed the Occupational Safety and Health Act (OSHAct). This legislation became law on April 28, 1971, and it had a significant impact on the construction industry. According to OSHAct, the employer has an obligation to provide a place for employee that is free from recognized hazards and meets the OSHA standards. The employees must comply with the rules, regulations, and standards applicable to their type of work.

While the use of trenchless technology is steadily increasing, many municipalities, pipeline and utility owners, continue to award construction contracts to companies using open-trench method; a method that potentially can be hazardous for both workers and the general public. In fact, the OSHA excavation standard was revised in 1989 because excavating is one of the most dangerous of all construction operations. More workers are killed or seriously injured in and around excavations than in most other phases of construction work. Once a trench is opened, it requires the use of protective systems, such as shoring and trench boxes to prevent cave-ins and trench collapses. The trenches for sewer, water, and other pipelines represent the greatest concern for cave-ins because of the excavation in urban environments. Sewer lines are typically installed at depths of 8 to 15 ft, with some installations as deep as 40 ft. Water lines are typically installed at depths of 4 to 5 ft, and sometimes deeper installations are normal. Gas lines, electric, telephone, and other conduits and cables tend to be placed in shallow trenches of about 2 ft deep, but most of the times they require a minimum of 3 to 4 ft of cover, depending on the location and regulatory considerations.

According to data from the U.S. Bureau of Labor Statistics, between the years of 1992 and 2002, 384 construction workers were killed by trenching-related injuries, an average of 35 per year (see Table 9.2). In most cases, a cave-in was the main event leading to the death. Depending on the soil type and moisture content, one yd^3 of soil can weigh between 3000 and 4000 lb.

Compared to trench excavation, trenchless safety is negatively affected by several factors, such as

- *Lack of formal safety training:* Usually workers and operators switch back and forth between jobs that may include both traditional open-cut and trenchless work. Additionally, workers

Year(s)	Number of Deaths	Event Leading To Injury: Number (%)	Main Occupations Affected: Number (% of Deaths)
1992–2002	384 (35/yr)	Cave-in, 274 (71%) Struck by falling object, 18 (5%)	Constr. laborers, 201 (52%) Plumbers/pipefitters, 33 (9%)
2003	57	Cave-in, 38 (67%)	Constr. laborers, 33 (58%) Supervisors/mgrs, 9 (16%)

Source: BLS Census of Fatal Occupational Injuries Microdata provided to the Center to Protect Workers' Rights. (Plog et al., 2006.)

TABLE 9.2   Trenching-related Deaths from Injuries in Construction, United States, 1992–2002 and 2003

and operators may receive on-the-job training and not formal classroom/field training.

- *Lack of industry safety standards:* With trenchless methods, currently there are no specific safety standards which can serve as governing guidelines to owners, engineers, and contractors, to maintain safe working conditions. OSHA has standards for tunneling activities; however, they are more focused on mining and large-diameter tunnels.

- *Lack of accurate statistics:* As per OSHA requirements, contractors must keep track of accident occurrences, but there is no clear differentiation on what technology was used at the time of the accident.

Trenchless methods have many advantages, but at the same time they have some limitations, including risks to workers and the public. Most notably, operators cannot always see what lies ahead and, consequently, may strike the existing underground utilities. Such accidents may result in injuries or deaths depending on the circumstances, in addition to the added expense of repairing the damaged utilities and other properties. Conditions such as confined spaces inside the shaft/pit and tunnels, and activities such as movement of pipe sections and spoil-removal, working with hydraulic jacks, using machine power, and working with pneumatic and hydraulic hoses can potentially cause accidents. A dangerous situation may occur when workers are in the shafts/pits and several activities are conducted simultaneously in these tight spaces.

As with any construction project, it is important that safety be included in the design phase. For example, subsurface utility engineering will help designers find out what lies underneath and across the trenchless installation path. Furthermore, engineers are also required to perform extensive geotechnical investigations. The information provided by these investigations will help contractors to select proper equipment and tools to safely carry out trenchless operations.

9.3.1 Project Safety Planning

A partial checklist for safety planning of a trenchless project may include:

- *Proper safety training:* An effective training program can reduce the number of injuries and deaths, property damages, legal liabilities, workers' compensation claims, and missed time from work.

- *Appropriate work clothing (personal protection equipment):* Loose clothing never should be worn and allowed around machinery equipment.

- *Safe machine and tool operations:* It is important that equipment operators be qualified and certified for their tasks, and carefully follow equipment manufacturers' machine operations and safety guidelines. The employer (contractor) must ensure that operators and other workers have demonstrated proficiency in their duties, particularly safety issues. Primary personnel must have proper training, including classroom and field experience. Industry-based training and/or certification courses are available from equipment manufacturers, as well as professional organizations.

- *Safe handling of drilling and lubrication fluids:* While drilling fluids are usually considered to be environmentally safe, they are usually under high pressure and may potentially mix with contaminated soils. Workers must be trained in safe handling of these fluids.

- *Safe crossing and/or parallel installation with existing utilities:* Contractors must ensure that proposed alignment and profile of the boring/tunneling operation keeps a safe distance from existing utilities and carefully follows the planned alignment/profile during installation.

- *Safe repair of damaged utilities:* Usually the existing pipeline and utility operator must be contacted immediately and a call to an emergency phone number (911 for the United States) should be made.

- *Safe disposal of drilling fluids:* The bentonite or polymer material used must be certified by the National Sanitation Foundation (NSF). The additive materials should be chemically inert, biodegradable, and nontoxic. Petroleum-based or detergent additives should not be used. Although the bentonite-water, or commonly used polymer-water-slurry, is not inherently a hazardous material, special disposal may be required when drilling in an area known to contain toxic pollutants. In such cases, disposal must be in accordance with local laws and regulations. It may be necessary to dewater the spoils, transport the solids to an appropriate disposal site, and treat the water to meet disposal requirements.

- *Proper locating and markings of bore path:* Prior to the arrival of the locators, the contractor should mark the path of the proposed bore route, preferably using a white line or flags, and some means of identification. In general, belowground facilities within a minimum lateral distance of 10 ft of the proposed bore path should be marked, unless a greater distance is specified by state or other regulations. Other facilities known to be in the vicinity, but believed to be beyond 10 ft, or otherwise required minimum distance, should be confirmed by the corresponding owner. The actual paths and depths of identified utility lines are typically not provided during the planning process. They may subsequently be determined by the owner of the proposed pipe line or its representative.

- *Proper project startup:* Preconstruction meetings with the contractor are useful, and may be particularly important for unusual or difficult projects. These meetings are necessary when requesting temporary disruption of electric or gas service, to reduce the likelihood of associated safety hazards.

- *Risk Assessment Plan:* It is recommended that owners include provisions in the contract requiring contractors to submit a risk assessment plan customized for the specific trenchless project.* This plan should be reviewed and kept on file for implementation and future reference. Contractor risk assessment plan may include (see Table 9.3):
  1. All work tasks/steps (identification of sequence of work steps/tasks)
  2. Hazards, concerns, and potential accidents (identification of hazards for each tasks/step and site hazards that could affect workers)
  3. Controls, preventive measures, and boundaries (a list of controls for each hazard)
  4. Reference documents (a list of permits, operating manuals, and other reference procedures)

---

*It should be noted that extent of this risk assessment plan must be in accordance with the nature and size of the trenchless project.

Work Tasks/ Steps	Hazards, Concerns, and Potential Accidents	Controls, Preventive Measures, and Boundaries	Supplemental Documents	Training
Drilling operation	Existing utilities strikes	One-call system, subsurface utility engineering (SUE), potholing	Locator's manual	Operator training and certification programs
Pit excavation	Traffic accidents	Barricades/ flagging shall be erected around any open trenches and/or excavations and checked and maintained on a daily basis	OSHA regulations	• Company in-house training • Formal classroom training
	Equipment malfunction	Preoperational inspection of equipment by trained and qualified operators	Operator's manual	Equipment dealers and manufacturer
	Struck by, caught in between hazards of heavy equipment	Only trained and qualified personnel shall operate heavy equipment	Operator's manual	Toolbox meetings

TABLE 9.3 Sample Submittal: Contractor's Risk Assessment Plan

5. Required training (a list of training requirements for safe execution of each task)
6. Work plan outlining the procedure and the schedule to execute the work, including list of proposed equipment, proposed location of shafts/pits, and so on
7. Design of shaft or pit protection systems approved by a professional engineer

8. List of personnel, including backup personnel, and their qualifications and experience
9. Traffic control plan
10. Drilling fluid management plan, including potential environmental impacts, frac-out (inadvertent return of drilling fluids) control and mitigation, emergency procedures and associated contingency plans
11. Communication plan among field crew and operator of the equipment, which may include use of radios and hand signals

Working Shafts/Pits

Frequently, a starting pit will be required at the approach site of the bore. If the site conditions allow, the approach site is preferred to be the *downstream* side. The pit should be located far enough (according to DOT or local authority requirements) from existing road embankment or structures to allow adequate safety for the structure as well as the public. From experience, the pit size must provide the most convenient and safest working conditions. The pit should be properly constructed and prepared to be a safe working place. The success or failure of a bore or trenchless project depends on the preparation of the pit as well as the machine and operator. Any utilities located within the pit locations should be properly supported. If sloping of the pit wall is not feasible, sheeting of the pit should be considered. The OSHA Code of Federal Regulations 29 provides specific rules for pit construction, protection, barricades, traffic control, installation and type of ladders used in the pit, and personal safety equipment. The boring subcontractor should become familiar with these requirements. Such information can be obtained from the regional department of labor office.

Engineer's soil information will specify what to expect in the pit and along the bore path. For example, in a wet location, it may be required to ring the pit with well points for less than 20-ft-deep pits/shafts, or with deep wells and submersible pumps if deeper. If ground conditions permit, movement of ground water is free and no impervious stratum is close, deeper wells will provide beneficial influence further along bore path. The normal range of wellpoint spacing ranges from 3 to 12 ft. If the aquifer extends below the subgrade to more than 10 ft, spacing of the wellpoint will depend on the quantity of water to be pumped. Closer spacing may be necessary in case of stratified soil with a layer of clay. If horizontal wells seem appropriate, they may be installed when the radius of influence is greatest at the bore pit. The system will need close attention as long as it runs.

The use of horizontal drilling will permit a smaller, 4 or 6 in., solid header line under the highway (railroad, structures, etc.). This will also allow contractor to dewater the medium between the two-way roadways (e.g., in the cases of interstate highway crossings). These solid pipes may have to be filled with concrete and abandoned after dewatering. However, the middle section of the bore may allow for deeper zones of influence that benefit both lane-crossings.

Job Site Preparation

Heavy rainfalls, or the potential fall of a sudden flood, will be most costly unless jobsite is prepared before it happens. The site can be surrounded by a berm of sufficient height to prevent flooding. The berm needs to be kept high and solid at all times. The location of a pit sump pump should also be considered before the pit excavation takes place. It can be a matter of life and death should people be in the pit or in the bore line when a flush flood occurs. The berm may provide the additional time needed to recover people and some, if not all, of the tools and the equipment.

The backstop (thrust block) should be designed to withstand 1.5 to 2 times the expected maximum thrust for the planned bore. It is recommended that a steel plate be used between the track push-plate and the backstop. For larger-diameter and longer bores, drive sheeting, and/or a poured concrete pad should be considered. Experience and soil conditions will dictate the best method. A good base and a secure backstop are essential for all bores. Caution must be taken to ensure that the thrust pressures do not damage any existing utilities or structures in vicinity of the backstop.

When exposing existing utilities to verify depth, nonaggressive "potholing" techniques such as manual tools (with electrically insulated/nonconducting handles) or vacuum type excavators must be used. It is particularly important to visibly expose and verify the location of lines transporting electricity, gas, oil or petroleum products, or other flammable, toxic or corrosive fluids or gases. In general, utility lines must be routinely exposed at all anticipated crossings with the planned bore path, such as where the route along a right-of-way crosses laterals or service lines to residences or other structures.

The paths of existing belowground facilities should be marked, using paint, flags or equivalent, based on the uniform color code developed by the Utility Location and Coordination Council (ULCC) of the American Public Works Association (APWA), and/or ANSI standard Z535.1 (Safety Colors for Temporary Marking and Facility Identification). Table 9.4 presents the uniform color code for different utilities.

The tolerance zone (see Fig. 9.2) also defines the region within which the contractor must use nonaggressive methods of digging. The width of the zone is specified by local regulations and varies among the states. A minimum of 18 in. from the outer edges of the facility is recommended, unless a greater distance is specified by state or local regulations. For relatively close adjacent parallel utility lines

Color	Utilities
White	Proposed construction path
Red	Electric power
Orange	Communications
Yellow	Gas, oil, steam, petroleum
Green	Sewer, drain
Blue	Water, irrigation, slurry
Fluorescent pink	Temporary survey markings
Purple	Reclaimed water, irrigation, and slurry lines

TABLE 9.4 Uniform Color Code Developed by the ULCC

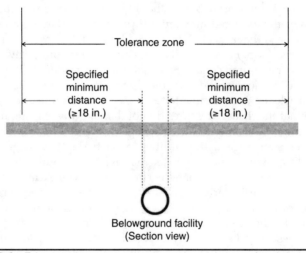

FIGURE 9.2 Tolerance zone.

(within twice the minimum specified distance laterally from each other), the tolerance zone is determined from the outer edge of the outermost utility line on each side. No portion of the cutting tool for the pilot bore, or the reamer used to expand the borehole, is allowed to enter the tolerance zone. Currently there are no general requirements for depth tolerance, so it must be decided on individual project conditions.

Shaft/Pit Space Planning

After all underground and overhead obstructions have been identified; the proposed drill path should be identified and documented. This allows for the evaluation of available right-of-way and easement acquisition issues. Proposed drill paths allow for the adequate placement of the jacking and receiving pit locations. Each trenchless method has their own requirements for setting spacing of the shafts/pits. Pit placement is dependent on such variables as type and size of pipe, the structural capacity of the pipe, soil conditions, and the type of lubrication systems of the pipe.

A typical pit site needs space to allow for the entrance and receiving pits, slurry separation or water tanks, and equipment and pipe storage. The entrance and receiving pits must be at a sufficient distance from overhead utilities. Careful planning of space and the drill path must also include traffic routes and the level of traffic on those streets.

Lack of space planning can lead to improperly placed pits, which can put improper stress on the tunneling technique. Cramped conditions can also lead to worker injury.

Regardless of the type of procedure, there are inherent job-site hazards. Securing the job site from curious bystanders, and slip and fall hazards are considerations that have to be made. Injuries of any type increase the liability costs for the construction company and can lead to OSHA penalties. Proper planning reduces the chance for many of these occurrences.

Planning must also include what-if scenarios on unanticipated construction problems. Water seepage into the drill hole or pit can lead to boring delays and to failure of the shoring materials that keep the pit walls from collapsing. Not only must the water removal procedure from the pit be considered, but also how the water is removed from the job site should be considered.

9.3.2 Hazard Assessment

A first critical step in developing a comprehensive safety and health program is to identify physical and health hazards in the workplace. This process is known as "hazard assessment."

Potential hazards may be physical or health-related and a comprehensive hazard assessment should identify hazards in both categories. Examples of physical hazards include moving objects; fluctuating temperatures; high-intensity lightening, rolling, or pining objects; electrocutions; and sharp edges. Examples of health hazards include overexposure to harmful dusts, chemicals, or radiation. The hazard assessment should begin with a walk-through survey of the jobsite to develop a list of potential hazards. In addition to reviewing any occupational illnesses or injuries, topics to look for during a walk-through survey include

- Sources of electricity
- Sources of motion such as machines or processes where movement may exist that could result in an impact between personnel and equipment

- Sources of high temperatures that could result in burns, eye injuries or fire
- Types of chemicals used in the workplace
- Sources of harmful dusts
- Sources of light radiation, such as welding, brazing, cutting, furnaces, heat treating, high-intensity lights, and the like
- The potential for falling or dropping objects
- Sharp objects that could poke, cut, stab, or puncture
- Biologic hazards such as blood or other potentially infected material.

When the walk-through is complete, the contractor should organize and analyze the data so that it may be efficiently used in determining the proper types of PPE required at the work site. The contractor should become aware of the different types of PPE available and the levels of protection offered. It is definitely a good idea to select PPE that will provide a level of protection greater than the minimum required to protect employees from hazards.

The workplace should be periodically reassessed for any changes in conditions, equipment, or operating procedures that could affect occupational hazards. This periodic reassessment should also include a review of injury and illness records to spot any trends or areas of concern and taking appropriate corrective action. The suitability of existing PPE, including an evaluation of its condition and age, should be included in the reassessment.

Documentation of the hazard assessment is required through a written certification that includes the following information:

- Identification of the workplace evaluated
- Name of the person conducting the assessment
- Date of the assessment
- Identification of the document certifying completion of the hazard assessment

9.3.3   Risk Assessment

The risks involved mainly concern the unforeseen circumstances that the construction may encounter during a trenchless project. Even if the soils have been identified and the trenching method outlined by the design firm in the construction contract, classifying and identifying risks by the construction company is the first step before undertaking any trenchless job. Secondly, the impact of each risk should be quantified and assessed. Finally, the mitigation and control of the risks identified should be finalized.

The risks that should be identified are those that affect performance, productivity, cost, schedule, quality, and safety. As with any project, not exceeding the budget and remaining on schedule are the primary goal. Risk management's purpose is to identify all associated risks to achieve this goal.

Many risks can be planned for if identified. There are still other risks, however, which cannot be planned for, but still must be considered. Among those are

- Unknown or unexpected ground conditions
- Weather delays
- Unexpected/unplanned alignment or grade changes of pipelines during pipe jacking
- Groundwater fluctuations and surface flooding
- Material shortages due to delay in shipping or supply

9.3.4   Utility Mapping

Any trenchless operation can have its own hazards. There are always hazards to construction workers in any type of underground construction; such as hazard of striking existing structures or utilities underground. Coordination is required among several parties and agencies prior to the start of the project. One of the first tasks is locating all the existing utilities. The one-call service (the 811 number for most of the United States) is essential in locating existing underground utilities; however, it would give the contractor an approximate alignment and no accuracy for the depth of utility (see Fig. 9.2). The cost associated with identifying the existing underground utilities is small when compared to the cost of striking a buried utility.

Additionally, visual checks and physical observations must also be made of the planned construction path. Just as important in identifying underground utilities is checking the site for evidence of substructures such as manhole covers, valve box covers, meter boxes, electrical transformers, conduits or drop lines from utility poles, and pavement patches to determine if any utilities may have been overlooked. Water and gas shutoffs attached to the houses or businesses in the area signify utility lines, even if they are not marked by the one-call service. Depressions in the ground may indicate that some type of excavation or trenching has taken place and, although there may not be an obstruction present, there is a possibility for some type of fill material that does not resemble the soil structure around it.

Written procedures have to be followed for the probability of striking an existing utility. Even if all of the checks and physical observations have been made, there is always a small probability that something could be overlooked. By having a procedure in place, the operating crew will have a set of written instructions to follow that can greatly reduce the downtime related to a strike. A list of all utilities, along with their phone numbers and contact personnel, should be kept at the jobsite.

9.3.5 Contingency Plans

The last step a company has to plan for in avoiding pitfalls during a trenchless project is to have contingency plans developed before the job starts to reduce delays. Regardless of the amount of investigation or research that may have been done to find existing utilities or structures, not everything can always be found or located. Even if the existing utilities are found, there may be other obstructions; both manmade and natural may exist below the surface. A construction site of a building long since torn down may have had a pile of dumped concrete left during its construction stage. Years ago many low areas became dumping areas of old building materials mixed with lumber or other organic materials. Only an open-cut method would be able to determine what types of materials have been buried.

Natural obstructions that may be present include the root system of a large tree that has since been cut down. In gravelly areas, large boulders may be present. A contingency plan has to be developed before the start of a project if an obstruction is found. Depending on what other obstructions may be in the area, changing location of the drill is not always possible. Even if an alternative location were possible, pulling a drill bit would leave a void beneath the surface, which needs to be filled with grout.

9.3.6 Communication

Communication guidelines should be established prior to the commencement of trenchless work. There should be good communication between the operator and other personnel (usually general laborers) to coordinate the drilling operation. Radios should be used for communication, but there should also be hand signals agreed to beforehand in the case the radios fail or reception is bad.

For example, communication breakdown can lead to delays through contact with existing utility companies if the drill-rig operator does not understand the direction being given to him or her. Because the general laborers may speak very little English, it is important that, if the job-site foreman does not speak the language of the laborers, at least hand signals are set up that all parties understand.

One last item that must be investigated during the planning/design stage is the geotechnical makeup of the soil along the length of the bore area. Because of the importance and complexity of this portion of the planning stage, it is discussed in Sec. 9.2.3.

Protection Against Utility Strike

Emergency procedures have to be developed and followed for when a buried utility is inadvertently hit. Electrical strikes, gas-line strikes, fiber-optic strikes, and water or sewer line strikes all have their own emergency procedures that should be initiated to reduce the likelihood of injury.

In an electrical strike, the most important thing to remember is for operators and workers to remain where they are and not move. If the electrical line contacts the drill rig, the drill operator should reverse the bore direction to see if the rig can break contact with the electrical line. Whichever worker on site is the safest distance from the crossed line should immediately contact the electric company. The drill operator should follow the manufacturer's procedure to determine if the drill is electrically charged before attempting to dismount the rig.

When a gas line is struck, the area must be immediately evacuated. The drill operator should shut down all engines and should not attempt to reverse the bore. The gas company should immediately be contacted.

Fiber optic lines are not as dangerous as an electrical or gas line strike, but workers have to remember not to look into the cut ends of the cable as severe eye damage can occur. Striking a fiber optic line usually occurs when digging the entrance or exit pit. Digging should be halted until the fiber optic cable company is contacted and the line repaired.

A water or sewer strike can be as dangerous to the construction crew and bystanders as that of a gas strike. Operations should immediately be stopped. The drill operator should not attempt to reverse the bore as sewer lines contain deadly pathogens. Pipe jacking, pipe bursting, or pipe ramming cannot reverse the bore and may have to dig an access pit to repair the damaged pipe. Medical attention should be sought for any workers that may have come in contact with the breakage. As with other strikes, the utility company should be contacted immediately.

9.3.7   Equipment Operator Training

Equipment operators not only must have general safety training, they also have to be thoroughly trained on the equipment they are operating. Understanding the equipment through knowledge of the manufacturer's operator's manual is important. Each trenchless method has its own unique training program that the operators need to understand. While HDD usually does not include the hazards associated with digging trenches, the drilling process does involve unique hazards that need to be followed by HDD operators. Some of the precautions are general to all trenchless techniques, others are not. Among those precautions are

- Operators should not exceed the manufacture's maximum torque and thrust.

- If operators notice a sudden slowness or stoppage in progress, they should stop the operation until possibility of hitting an obstruction is determined.

- Operators need to be aware of the hazards of the hydraulic action of the drilling fluid while passing through structures like roads, bridges, and buildings. There is potential for subsidence or elevation of these structures.

- The machine operator also has to realize that hydraulic oil under pressure can penetrate the skin and burn or cause blood poisoning. If the operator thinks there is a leak, precautions should immediately be taken.

- When problems do arise with the rig, properly trained personnel should be the only ones working on the machinery. Proper clothing and eye protection should be worn when working on equipment.

- Operators need to use caution when moving drill pipe with the crane. A tag line may work better when possible.

- Drilling and boring involve rotating equipment that can create hazards at the drilling rig, point of entry, and point of exit. Caution must be taken to protect these areas with guards and barricades. Workers must not wear loose clothing and should not be allowed to step over or stand above the rotating parts.

Experience

One of the main contributing factors to the success of a project is the experience and amount of training of the crew operating the equipment and steering the head. As stated earlier, even when utilities are marked, an inexperienced crew has a higher risk of striking buried utilities. An experienced crew is also more likely to salvage a project when unexpected conditions are encountered. The investment of time and money in training or hiring an experienced crew outweighs the cut of repairing damaged utility lines and inherent safety risks.

Manufacturers are committed to the safe operation of their equipment. They offer a variety of resources through their local dealers to educate operators in the safe use of the equipment through manuals, videos, and one-on-one training. These programs should be used to ensure operators are properly trained.

9.4 Summary

Trenchless methods have many advantages, but at the same time they have their own limitations, including risks to workers and the public.

This chapter presented an overview of safety issues and discussed various parameters that impact a safe trenchless installation. Many federal, state, and local regulations are to be followed in a construction project. Trenchless installations are executed blindly, a thorough utility locating is required to prevent strikes. Proper shaft and pit planning for spacing and sizing is important for proper working of the equipment. Good communication between the operator and other personnel should be established. Contractors must conduct a risk assessment study before start of the project and submit their safety plan to the project owners. This chapter presented a detailed description of planning and safety considerations in trenchless projects.

APPENDIX A

References

American Concrete Pipe Association (2000). "Concrete Pipe Design Manual," ACPA, Irving, Tex.

American National Standards Institute (2006). "National Electrical Safety Code," IEEE, New York.

American National Standards Institute and American Water Works Association. (2000). "AWWA Standard for Cement-Mortar Lining of Water Pipelines in Place–4 in. (100 mm) and larger," Denver, Colo.

American National Standards Institute and American Water Works Association. (2005). "Disinfecting Water Mains," AWWA, Denver, Colo.

American National Standards Institute and American Water Works Association. (2008). "AWWA Standard for Polyurethane Coatings for the Interior and Exterior of Steel Water Pipe and Fittings," AWWA, Denver, Colo.

American Society for Testing and Materials (2009). "Standard Practice for Rehabilitation of Existing Pipelines and Conduits by the Inversion and Curing of a Resin-Impregnated Tube," ASTM, F1216–09, West Conshohocken, Pa.

American Society for Testing and Materials (2008). "Standard Test Method for Standard Penetration Test (SPT) and Split-Barrel Sampling of Soils," ASTM, D1586-08a, West Conshohocken, Pa.

American Society for Testing and Materials (2006). "Standard Practice for Classification of Soils for Engineering Purposes (Unified Soil Classification System)," ASTM, D2487-06, West Conshohocken, Pa.

American Society for Testing and Materials (2005). "Standard Guide for Use of Maxi Horizontal Directional Drilling for Placement of Polyethylene Pipe or Conduit Under Obstacles, Including River Crossings," ASTM, F 1962–05, West Conshohocken, Pa.

American Society for Testing and Materials (1989). "Standard Practice for Underground Installation of Thermoplastic Pipe for Sewers and Other Gravity-Flow Applications," ASTM, D2321–89, West Conshohocken, Pa.

American Society of Civil Engineers (2009). "Inspecting Pipeline Installation," ASCE Manuals and Reports on Engineering Practice No. 117, Reston, Va.

American Society of Civil Engineers (2005). "Pipeline Design for Installation by Horizontal Directional Drilling," ASCE Manuals and Reports on Engineering Practice No. 108, Reston, Va.

American Society of Civil Engineers (2004). "Horizontal Auger Boring Projects," ASCE Manuals and Reports on Engineering Practice No. 106, Reston, Va.

American Society of Civil Engineers (2003). "Standard Guidelines for the Collection and Depiction of Existing Subsurface Utility Data," CI/ASCE Standard 38-02, Reston, Va.

American Society of Civil Engineers (2001). "Standard Construction Guidelines for Microtunneling," CI/ASCE Standard 36-01, Reston, Va.

American Water Works Association (2006). "Infrastructure Reliability: Service Life Analysis of Water Main Epoxy Lining," AWWA, Denver, Colo.

American Water Works Association (2002). "Decision Support System for Distribution System Piping Renewal," AWWA, Denver, Colo.

American Water Works Association (2001). "Manual of Water Supply Practices M28: Renewal of Water Mains," AWWA, Denver, Colo.

Ariaratnam, S. T., Harrison, S., and Milligan, J. (2010). " On-Grade Installation of Sanitary Sewers In Baton Rouge, Louisiana Using s Unique Trenchless Vacuum Boring System," Proceedings of No-Dig 2010, May 2–5, 2010, Chicago, Il, North American Society of Trenchless Technology, Liverpool, NY.

Atalah, A. (1998). "The Effect of Pipe Bursting on Nearby Utilities, Pavement, and Structures," Technical Report TTC-98-01, Trenchless Technology Center, Louisiana Tech University, Ruston, La.

Atalah, A. (2004). "Ground Movement in Hard Rock Conditions Related to Pipe Bursting," ASCE Pipeline Specialty Conference, San Diego, Calif., August 2004.

Atalah, A. (2004). "The Ground Vibration Associated with Pipe Bursting in Rock Conditions," North American NO-DIG 04, Annual Conference of the North American Society of Trenchless Technology, New Orleans, La., March 2004.

Atalah, A. (2004). "The Ground Movement Associated with Large Diameter Pipe Bursting in Rock Conditions and Its Impact on Nearby Utilities and Structures," Bowling Green State University, Bowling Green, Ohio.

Atalah, A. (2006). "The Safe Distance between Large-Diameter Rock Pipe Bursting and Nearby Buildings and Buried Structures," ASCE Journal of Transportation Engineering, Vol. 132, April 2006.

Atalah, A., Iseley, T., and Bennett, D. (1994). "Estimating the Required Jacking Force," NO-DIG 94, Annual Conference of the North American Society of Trenchless Technology, Dallas, Tex.

ATTERIS. Atteris Horizontal Directional Drilling. 2006. http://atteris.com.au/capabilities/hdd/. Accessed May 11, 2008.

Australian Society of Trenchless Technology (2004). "On-line Replacement," ISTT-ASTT, http://www.astt.com.au/On-line.pdf, pp. 1–5. Accessed September 12, 2009.

Bennett, R. D., Ariaratnam, S., and Khan, S. (2005). "Pipe Bursting Good Practices," North American Society for Trenchless Technology, Arlington, Va.

Bonds, R. W., Barnard, L. M., Horton, A. M., and Oliver, G. L. (2004). "Corrosion and Corrosion Control Research of Iron Pipe," Proceedings of ASCE International Conference on Pipeline Engineering and Construction, San Diego, Calif.

Boot, J., Woods, G., and Streatfield, R. (1987). "On–line Replacement of Sewer Using Vitrified Clayware Pipes," Proceedings of No-Dig International '87, ISTT, UK.

Bureau of Labor Statistics (2007). "National Census of Fatal Occupational Injuries in 2006," U.S. Department of Labor, Washington, D.C.

Clarke, N. W. (1968). "Buried Pipelines: A Manual of Structural Design and Installation," Maclaren and Sons Ltd., London.

Deb, A. K., Snyder, J. K., Chelius, J. J., and O'Day, D. K. (1990). "Assessment of Existing and Developing Water Main Rehabilitation Practices," AWWA, Denver, Colo.

Essex, R. J. (2007). "Geotechnical Baseline Reports for Construction Suggested Guidelines," American Society of Civil Engineers, Reston, Va.

Farshad, M. (2006). "Plastic Pipe Systems: Failure Investigation and Diagnosis," Elsevier, Maryland Heights, Mo.

Fraser, R., Howell, R. N., and Torielli, R. (1992). "Pipe Bursting: The Pipeline Insertion Method," Proceedings of No-Dig International '92, Washington, D.C.

Gabriel, L. H. (2006). "Corrugated Polyethylene Pipe Design Manual and Installation Guide," Plastics Pipe Institute, CPPA http://drainage.plasticpipe.org/Resources.Design_Manual.asp. Accessed July 14, 2006.

Gumbel, J. E. (1998). "Structural Design of Pipe Linings 1998—Review of Principles, Practice and Current Developments Worldwide," Paper presented at the 4th ASTT Conference, Brisbane, Australia , August (also available at http://www.insituform.com/resourceroom/resource_techpapersindex.html).

Gunsaulis, F.R. and Levings, R. (2008). "Installation of Gravity Sewers using Horizontal Directional Drilling (HDD)," Proceedings of Pipelines Congress 2008, July 22–27, 2008, Atlanta, GA, American Society of Civil Engineers, Reston, Va.

Hair, J. D., and Hair, C. W, III. (1988). "Considerations in the design and installation of horizontal drilled pipeline river crossings." Proceedings of Pipeline

Infrastructure Conference, Pipeline Division of the American Society of Civil Engineers, Boston.

Hair (1994). "Drilling Fluids in Pipeline Installation by Horizontal Directional Drilling," A Practical Applications Manual, J. D. Hair & Associates, Inc., Tulsa, Okla.

Howard, A. (1996). "Pipeline Installation: A Manual for Construction of Buried Pipe," Relativity Publishing, Lakewood, Colo.

Howell, N. (1995). "The Polyethylene Pipe Philosophy for Pipeline Renovation," Proceedings of No-Dig International '95, Dresden, Germany, ISTT, UK.

Iseley, D. T. and Najafi, M. (eds.) (1995). "Trenchless Pipeline Rehabilitation." The National Utility Contractors Association (NUCA), Arlington, Va.

Iseley, D. T., Najafi, M., and Tanwani, R. (1999). "Trenchless Construction Methods and Soil Compatibility Manual," National Utility Contractors Association (NUCA), Arlington, Va.

Janson, L. E. (1999). "Plastics Pipes for Water Supply and Sewage Disposal," Borealis, Stockholm.

Kroon, D. H., Lindemuth, D., Sampson, S., and Vincenzo, T. (2004). "Corrosion Protection of Ductile Iron Pipe," Proceedings of ASCE International Conference on Pipeline Engineering and Construction, San Diego, Calif.

Lindeburg, M. (1992). "Civil Engineering Reference Manual for the PE Exam," Professional Publications Inc., Belmont, Calif.

Makar, J. M. (2000). "A Preliminary Analysis of Failures in Grey Cast Iron Water Pipes," Engineering Failure Analysis 7, Manual of Procedures for Pipe Construction (1998), Office of Highway Management, Ohio Department of Transportation, 43–53.

Makar, J. M., Desnoyers, R., and McDonald, S. E. (2001). "Failure Modes and Mechanisms in Gray Cast Iron Pipe," International Conference on Underground Infrastructure Research: Municipal, Industrial and Environmental Applications, *Proceedings*, Kitchener, Ontario, June 10–13, 2001, 1–10.

Marne, D. J. (2006). "National Electrical Safety Code 2007 Handbook," McGraw-Hill Professional, New York.

Moser, A. P. and Folkman, S. (2008). *Buried Pipe Design*, 3d ed., McGraw-Hill, New York.

Najafi, M. (ed.) (2008). "Pipe Ramming Projects—ASCE Manuals and Reports on Engineering Practice No. 115." Reston, Va.

Najafi, M. (ed.) (2007). "Pipe Bursting Projects—ASCE Manuals and Reports on Engineering Practice No. 112." American Society of Civil Engineers, Reston, Va.

Najafi, M. and Gokhale, S. (2005). *Trenchless Technology: Pipeline and Utility Design, Construction, and Renewal*, McGraw-Hill, New York.

Najafi, M. and Kyoung, O. K. (2004). "Life-Cycle-Cost Comparison of Trenchless and Conventional Open-Cut Pipeline Construction Projects," Proceedings of ASCE International Conference on Pipeline Engineering and Construction, San Diego, Calif.

Najafi, M. (1994). "Trenchless Pipeline Rehabilitation: State-of-the-Art Review," Trenchless Technology Center, Ruston, La.

National Corrugated Steel Pipe Association (2000). "Corrugated Steel Pipe Design Manual," NCSPA, Dallas, Tex.

National Association of Sewer Service Companies (n.d.). "Guideline Specification for the Replacement of Mainline Sewer Pipes by Pipe Bursting," International Pipe Bursting Association Guidelines, NASSCO, Baltimore, Md.

National Association of Sewer Service Companies (2008). "Inspector Training Certification Program for the Inspector of Cured-In-Place Pipe installation, version 2.0," NASSCO, Baltimore, Md.

North American Society for Trenchless Technology (2008). Glossary of Terms, http://nastt.org/glossary.php?index=P. NASTT, Arlington, Va.

Occupational Safety and Health Administration (OSHA) (2005). "Code of Federal Regulations," U.S. Department of Labor, Washington, D.C.

O'Day, D. K. (1986). "Water Main Evaluation for Rehabilitation/Replacement," AWWA Research Foundation, Denver, Colo.

Petroff, L. J. (1995). "Installation Technique and Field Performance of HDPE Profile Pipe," Advances in Underground Pipeline Engineering II, 260–271.

Petroff, L. J. (2006). "Designing Polyethylene Water Pipe for Directional Drilling Applications Using ASTM F 1962," NO-DIG 2006.

Pipe Jacking Association (2005). "A Guide to Pipe Jacking and Microtunneling Design," PJA, London.

Plastics Pipe Institute (2008). "Handbook of Polyethylene Pipe," 2d ed., The Plastics Pipe Institute, Irving, Tex.

Plog, B. A., Materna, B., Vannoy, J., and Gillen, M. (2006). "Strategies to Prevent Trenching–Related Injuries and Deaths." The Center to Protect Workers' Rights, Madison.

Polyurea Development Association (PDA), www.pda-online.org. Accessed on August 15, 2009.

Poole, A., Rosbrook, R., and Reynolds, J. (1985). "Replacement of Small-Diameter Pipes by Pipe Bursting," Proceeding of 1st International Conference on Trenchless Construction for Utilities: No-Dig '85, April 16–18, London.

Rajani, B. and Kuraoka, S. (1995). "Field Performance of PVC Water Mains Buried in Different Backfills," Advances in Underground Pipeline Engineering, Second International Conference, 138–149.

Rajani, B. B. and Kleiner, Y. (2001). "Comprehensive Review of Structural Deterioration of Water Mains: Physically Based Models," Urban Water, Vol. 3, October, 151–164.

Reddy, D. V. (2002). "Long-Term Performance of Buried High Density Polyethylene Plastic Piping," Final Report, FDOT.

Schrock, B. J. and Gumbel, J. (1997). "Pipeline Renewal 1997," Paper presented at North American No-Dig '97, Seattle, Wash.

Selvakumar, A., Clark, R. M., and Sivaganesan, M. (2002). "Costs for Water Supply Distribution System Rehabilitation," ASCE Journal of Water Resources, Planning and Management, Vol. 128, Issue 4, pp. 303–306, Reston, Va.

Simicevic, J. and Sterling, R. (2001). "Guidelines for Pipe Ramming," U.S. Army Corps of Engineers, Engineering Research and Development Center, Vicksburg, Miss.

Spangler, M. G. and Handy, R. L. (1982). Soil Engineering, 4th ed., Harper & Row, New York.

SPI Composites Institute. (1989). "Fiberglass Pipe Handbook," Fiberglass Pipe Institute., New York.

Stein, D. (2001). "Rehabilitation and Maintenance of Drains and Sewers," 3d ed., Ernst & Sohn, Berlin, Germany.

Stein, D. and Braver, A. (2005). "Practical Guideline for the Application of Microtunnelling Methods," Rademann, Ludinghausen, Germany.

Telcordia (2007). "Blue Book-Manual of Construction Procedures," SR-1421, Telcordia Technologies (formerly Bellcore), November 2007.

Timoshenko, S. P. and Gere, J. M. (1961). Theory of Elastic Stability, 2d ed., McGraw-Hill, New York.

Vaslestad, J., Johansen, T. H., and Holm, W. (1994). "Load Reduction on Buried Rigid Pipes; Load Reduction on Rigid Culverts Beneath High Fills: Long Term Behaviour; Long-term Behaviour of Flexible Large-span Culverts," Directorate of Public Roads, Norwegian Road Research Laboratory, Oslo, Norway.

Water UK (2009). "In Situ Epoxy Resin Lining Operational Requirements and Code of Practice," WRc, London.

Watkins, R. K. and Anderson, L. R. (1999). "Structural mechanics of buried pipes," CRC Press, Boca Raton, Fla.

Watkins, R. K. and Spangler, M. G. (1958). "Some Characteristics of the Modulus of Passive Resistance of Soil—A Study in Similitude," Proceedings of Highway research, pp. 576–583, Logan, Utah.

Young, O. C. and Trott, J. J. (1984). "Buried Rigid Pipes—Structural Design of Pipelines," Elsevier Applied Science Publishers, New York.

APPENDIX B

Related Documents

American Concrete Pipe Association (2004). "Concrete Pipe Design Manual," ACPA, 2004, Irving, Tex.

American Iron and Steel Institute (2007). "Handbook of Steel Drainage & Highway Construction Products," 2d ed., AISI, Washington, D.C.

American Iron and Steel Institute (1995). "Modern Sewer Design," 3d ed., AISI, Washington, D.C.

American Public Works Association (1999). "Trenchless Technology Application in Public Works," Contributors U.S. Army Corp. of Engineer, TTC, APWA, Kansas City, Mo.

American Society of Civil Engineers (1998). "Pipeline Route Selection for Rural and Cross-Country Pipelines," ASCE Manuals and Reports on Engineering Practice No. 46, Reston, Va.

American Society of Civil Engineers. (2002). "Standard Guidelines for the Collection and Depiction of Utility Data," Standard CI/ASCE 38-02, ASCE, Reston, Va.

American Society of Civil Engineers (2004). "Horizontal Auger Boring Projects," ASCE Manuals and Reports on Engineering Practice No. 106, Reston, Va.

American Society of Civil Engineers (2005). "Pipeline Design for Installation by Horizontal Directional Drilling," ASCE Manuals and Reports on Engineering Practice No. 108, Reston, Va.

American Society of Civil Engineers (2007). "Geotechnical Baseline Reports for Underground Construction," ASCE, Reston, Va.

American Water Works Association (2001). "Manual of Water Supply Practices M28: Rehabilitation of Water Mains," AWWA, Denver, Colo.

American Water Works Association (2004). "Manual of Water Supply Practices M11: A Guide for Design and Installation," AWWA, Denver, Colo.

American Water Works Association (2006). "Manual of Water Supply Practices M55: PE Pipe—Design and Installation," AWWA, Denver, Colo.

Ariartnam, S. T., Bennett, D., and Khan, S. (2008). "Pipe Bursting Good Practices Manual," NASTT Publications, Arlington, Va.

Association of Metropolitan Sewerage Agencies (2002). "Managing Public Infrastructure Assets to Minimize Cost and Maximize Performance," AMSA, Washington, D.C.

Atalah, A. and Tremblay, A. (eds.) (2006). "Pipelines 2006: Service to the owner," Proceedings of the American Society of Civil Engineers Pipeline Division Specialty Conference, July 30 to August 2, Chicago, Ill. Reston, Va.

Benjamin Media Inc. (2009). "2009/2010 Directory of the North American Trenchless Technology Industry," Benjamin Media Inc., Peninsula, Ohio.

Bennett, D. and Ariaratnam, S. T. (2008). "Horizontal Directional Drilling: Good Practices Guidelines," 3d ed., HDD Consortium, NASTT Publications, Arlington, Va.

Bizier, P. (ed.) (2007). "Gravity Sanitary Sewer Design and Construction," American Society of Civil Engineers, Reston, Va.

Conner, R. C. (ed.) (1999). "Pipeline Safety, Reliability, and Rehabilitation," Proceedings of the group of ASCE technical sessions at the 1999 American Public Works Association International Public Works Congress & Exposition. September 19–22, Denver, Colo.

Das, B. M. (1994). *Principles of Geotechnical Engineering*, PWS Publishing Company, Boston, Mass.

Deb, A. K., Snyder, J. K., Hammell, J. O., Tyler, E., Gray, L. and Warren, I. (2006). "Service Life Analysis of Water Main Epoxy Lining," AWWA, Denver, Colo.

Federal Highway Administration (2002). "Manual for Controlling and Reducing the Frequency of Pavement Utility Cuts," Federal Highway Administration, Publication No. FHWA-IF-02-064, October.

Fisk, E. R. and Rapp, R. R. (2004). *Introduction to Engineering Construction Inspection*, John Wiley & Sons, Inc., Hoboken, N.J.

Galleher, J. J. and Kenny, M. K. (eds.) (2009). "Pipelines 2009: Infrastructure's hidden assets," *Proceedings* of the American Society of Civil Engineers 2009 Pipeline Division Specialty Conference: August 15–19, 2009, San Diego, California, Reston, Va.

Galleher, J. J. and Stift, M. T. (eds.) (2004). "Pipeline Engineering and Construction: What's on the Horizon?" Proceedings of the American Society of Civil Engineers Pipelines 2004 International Conference: August 1–4, San Diego, California, Reston, Va.

GASB Statement 34 (1999). "Basic Financial Statements and Management's Discussion and Analysis—for State and Local Governments," Government Accounting Standard Series, No. 171-A, June 1999, Norwalk, Conn.

Gokhale, S. B. and Rahman, S. (eds.) (2008). "Pipeline Asset Management Maximizing Performance of our Pipeline Infrastructure," Proceedings of the American Society of Civil Engineers Pipelines Congress 2008, July 22–27, Atlanta, Ga., Reston, Va.

Goodman, A. S. and Hastak, M. (2006). "Infrastructure Planning Handbook: Planning, Engineering, and Economics," American Society of Civil Engineers, Reston, Va.

Grant, D. M. (1991). "ISCO Open Channel Flow Measurement Handbook," 3d ed., ISCO, Inc., Lincoln, Neb.

Heald, C. C. (1994). "Cameron Hydraulic Data, 18th ed," Ingersoll-Dresser Pump Co., Liberty Corner, N.J.

Hovland, T. J., and Najafi, M. (eds.) (2009). "Inspecting Pipeline Installation," American Society of Civil Engineers, Reston, Va.

Howard, A. (1996). "Pipeline Installation," Relativity Publishing, Lakewood, Colo.

International Infrastructure Management Manual (2006).

Hughes, D. (2002). "Assessing the Future: Water Utility Infrastructure Management," American Water Works Association, Denver, Colo. 2002.

Hughes, J. B. (ed.) (2009). "Manhole inspection and rehabilitation," American Society of Civil Engineers, Reston, Va.

Iseley, D. T. and Gokhale, B. S. (1997). "Synthesis of Highway Practice 242: Trenchless Installations of Conduits beneath Roadways," Transportation Research Board, National Research Council, Washington, D.C.

Iseley, D. T. and Najafi, M., (eds.) (1995). "Trenchless Pipeline Rehabilitation," A Manual Published by the National Utility Contractors Association, Arlington, Va.

Iseley, D. T., Najafi, M., and Tanwani, R. (eds.) (1999). "Trenchless Construction Methods and Soil Compatibility Manual," the National Utility Contractors Association, Arlington, Va.

Janson, L. E. (2003). "Plastic Pipes for Water Supply and Sewage Disposal, 4th ed." Stockholm, Sweden.

Jeyapalan, J. K. and Jeyapalan, M. (eds.) (1995). "Second International Conference: Advances in Underground Pipeline Engineering," ASCE, June 25–28, 1995, Bellevue, Wash.

Kramer, R. S., McDonald, J. W., and Thomson, C. J. (1992). *An Introduction to Trenchless Technology*, Chapman & Hall, New York.

Liu, H. (2003). *Pipeline Engineering*, CRC Press, New York.

Meyer, E. G. (2005). "Polyethylene Pipes in Applied Engineering," Brussels, Belgium.

Mohitpour, M. (2008). "Energy Supply and Pipeline Transportation: Challenges & Opportunities," ASME, New York.

Moser, A. P. and Folkman, S. (2008). *Buried Pipe Design*, 3d ed., McGraw-Hill, New York.

Najafi, M. (1994). "Trenchless Pipeline Rehabilitation: State-of-the-Art Review," Results of a Two-Year Research Conducted Jointly by the U.S. Army Engineer—Waterways Experiment Station and the Trenchless Technology Center—Louisiana Tech University.

Najafi, M. (ed.) (2003). "New Pipeline Technologies, Security and Safety, Two-Volume," Proceedings of International Pipeline Conference July 13–16, Baltimore, Md.

Najafi, M. (ed.) (2007). "Pipe bursting projects," ASCE manuals and reports on engineering practice, no. 112. Reston, Va.

Najafi, M. (ed.) (2008). "Pipe Ramming Projects," ASCE manuals and reports on engineering practice, no. 115. Reston, Va.

Najafi, M. and Ma, B. (eds.) (2009). "ICPTT 2009 Advances and Experiences with Pipelines and Trenchless Pipeline Technology for Water, Sewer, Gas and Oil Applications," Proceedings of the American Society of Civil Engineers International Conference on Pipelines and Trenchless Technology October 18-21, Shanghai, China.

Najafi, M., Gunnink, B., and Davis, G. (2005). "Preparation of Construction Specifications, Contract Documents, Field Testing, Educational Materials, and Course Offerings for Trenchless construction," University of Missouri-Columbia and Missouri Department of Transportation, Jefferson City, Mo.

Najafi, M. and Rush J. (eds.) (2008). "Guide to water & wastewater asset management, Benjamin Media Inc," Underground Infrastructure Management(UIM), Peninsula, Ohio.

National Asset Management Steering Group (NAMS) (2002). "International Infrastructure Management Manual Version 2.0," New Zealand.

National Association of Sewer Service Companies (2008). "Inspector Training Certification Program for the Inspector of Cured-In-Place Pipe installation, version 2.0," NASSCO, Baltimore, Md.

National Association of Sewer Service Companies (2006). "Jetter Code of Practice: A Guide for Selection and Operation of Sewer Jetter Equipment and the Selection of Nozzles for Different Applications," NASSCO, Baltimore, Md.

National Association of Sewer Service Companies (2001). "Pipeline Assessment and Certification Program," NASSCO, Baltimore, Md.

National Clay Pipe Institute (2003). "Clay Pipe Engineering Manual," NCPI, Lake Geneva, Wis.

National Utility Contractors Association (2010). "NUCA Seeing is Believing: Safety Exposing Buried Utilities," NUCA (DVD/CD-ROM), Arlington, Va.

National Utility Contractors Association (2002). "NUCA Safety Manual, Fifth Revision," NUCA, Arlington, Va.

North American Conference and Exhibition of Trenchless Technology (2010). "2010 No-Dig Show Conference Proceedings," Chicago (Schaumburg), Illinois, May 2–7, Liverpool, New York.

North American Conference and Exhibition of Trenchless Technology (2009). "No-Dig, 2009 NASTT/DCCA, 29 March–3 April," North American Society for Trenchless Technology, Arlington, Va.

North American Conference and Exhibition of Trenchless Technology (2008). "No-Dig, 2008 NASTT/DCCA, April 27–May 2," North American Society for Trenchless Technology, Arlington, Va.

North American Conference and Exhibition of Trenchless Technology (2007). "Annual Technical Conference of the North American Society for Trenchless Technology: No-Dig 2007: Riding the Wave of Trenchless Technology, April 16–19, San Diego, California, USA," North American Society for Trenchless Technology, Arlington, Va.

North American Conference and Exhibition of Trenchless Technology (2006). "No-Dig, 2006 NASTT/DCCA, March 26–31," North American Society for Trenchless Technology, Arlington, Va.

North American Conference and Exhibition of Trenchless Technology (2005). "The Amazing Trenchless Experience: North American No-Dig 2005, Orlando, Florida, April 24–29," North American Society for Trenchless Technology, Arlington, Va.*

Orman N. R. and Lambert J. E. (2004). "Manual of Sewer Condition Classification," 4th ed, WRc, Swindon, Wiltshire.

Osborn, L. E. and Najafi, M. (eds.) (2010). "Trenchless Renewal of Culverts and Storm Sewers," American Society of Civil Engineers, Reston, Va.

Osborn, L. E. and Najafi, M. (eds.) (2007). "Pipelines 2007: Advances and Experiences with Trenchless Pipeline Projects," Proceedings of the American Society of Civil Engineers International Conference on Pipeline Engineering and Construction July 8-11, Boston, Mass.

Pipe Jacking Association (1995). "Guide to Best Practice for the Installation of Pipe Jacks and Microtunnels," Pipe Jacking Association, London.

Pipe Jacking Association (1995). "A Guide to Pipe Jacking & Microtunneling Design," PJA, National Utility Contractors Association, London.

Plastics Pipe Institute (2008). "Handbook of Polyethylene Pipe," 2d ed., The Plastics Pipe Institute, Irving, Tex.

Public Works Standards, Inc. (2010). "BNI 2010 Supplement to the Greenbook Standard Specifications for Public Works," BNI Publications, Vista, Calif.

Public Works Standards, Inc. (2009). "Greenbook: Standard Specifications for Public Works Construction," 2009 ed., BNI Publications, Inc., Vista, Calif.

Rameil, M. (2007). "Handbook of Pipe Bursting Practice," Vulkan-Verlag Gmbh, Germany.

Reinholtz, J. D. (ed.) (2006). "OSHA Standards for the Construction Industry," CCH, Chicago, Ill.

Sarrgand, S. M., Mitchell, G. F., and Hurd, J. O. (eds.) (1993). "Structural Performance of Pipes 93," A. A. Balkema, Brookfield, Vt.

Slavin, L. M. (ed.) (2009). "Belowground Pipeline Networks for Utility Cables," American Society of Civil Engineers, Reston, Va.

Stein, D. (2005). "Trenchless Technology for Installation of Cables and Pipelines," Becker Druck, Arnsberg, Germany.

Stein and Partner. (2005). "Practical Guideline for the Application of Microtunnelling Methods for the Ecological, Cost-Minimised Installation of Drains and Sewers," Stein and Partner, Bochum.

Sterling, R. L. (2000). "Utility Locating Technologies: A Summary of Responses to a Statement of Need," A Report prepared by the Trenchless Technology Center (TTC) for the Federal Laboratory Consortium for Technology Transfer, Washington, D.C.

Summit on the Future of Civil Engineering. (2007). "The Vision for Civil Engineering in 2025: Based on the Summit on the Future of Civil Engineering 2025, American Society of Civil Engineers, June 21–22, 2006." ASCE, Reston, Va.

Tanwani, R. (1991). "An Evaluation of Trenchless Excavation Construction Methods and Equipment in the United States," Unpublished MS Thesis, Louisiana Tech University, Ruston, La.

Vermeer Manufacturing Company (2006). "Horizontal Directional Drilling Guide," Pella, Iowa.

Vipulanandan, C. and Ortega, R. (eds.) (2005). "Pipelines 2005: Optimizing Design, Operations, and Maintenance," Proceedings of the American Society of Civil Engineers Pipeline Division Specialty Conference: August 21–24, Houston, Tex.

Watkins, R. K. and Anderson, L. R. (2000). *Structural Mechanics of Buried Pipes*, CRC Press, Boca Raton, Fla.

Whidden, W. R. (ed.) (2009). "Buried Flexible Steel Pipe: Design and Structural Analysis," American Society of Civil Engineers, Reston, Va.

*For more NASTT No-Dig Proceedings please go to http://www.nastt.org/store/conference_proceedings.html

Journal

American Society of Civil Engineers (ASCE). *Journal of Pipeline Systems Engineering and Practice (JPS)*. http://www.editorialmanager.com/jrnpseng/.

Magazines

Damage Prevention Professional, published by Infrastructure Resources, LLC. 10740 Lyndale Avenue South, Suite 15W, Bloomington, MN 55420. Website: www.damagepreventionprofessional.com.

Municipal Sewer & Water, published by COLE Publishing, Inc. 1720 Maple Lake Dam Rd. PO Box 220, Three Lakes, WI 54562-0220. Website: www.mswmag.com.

NASSCO Times, published by NASSCO Inc., 1314 Bedfrd Avenue, Suite 201, Baltimore, MD 21208.

North American Pipelines, published by Benjamin Media, Inc. 1770 Main Street, P.O. Box 190, Peninsula, OH 44264 USA. Website: www.napipelines.com.

Pipeline & Gas Journal, published by Oildom Publishing Company of Texas, Inc. P.O. Box 941669, Houston, TX 77094-8669. Website: www.pipelineandgasjournal.com.

Pipeline News, published by Oildom Publishing Company of Texas, Inc. P.O. Box 941669, Houston, TX 77094-8669. Website: www.pipelineandgasjournal.com.

TBM: Tunnel Business Magazine, published by Benjamin Media, Inc. 1770 Main Street, P.O. Box 190, Peninsula, OH 44264 USA. Website: www.tunnelingonline.com.

Trenchless International, published by Great Southern Press GPO Box 4967, Melbourne Victoria 3001 Australia. Website: http://trenchlessinternational.com.

Trenchless Technology, published by Benjamin Media, Inc. 1770 Main Street, P.O. Box 190, Peninsula, OH 44264. Website: www.trenchlessonline.com.

Trenchless World, published by Aspermont UK (Mining Communications Ltd). Albert House, 1 Singer Street, London, EC2A 4BQ. Website: www.trenchlessworld.com.

Underground Construction, published by Oildom Publishing Company of Texas, Inc. P.O. Box 941669, Houston, TX 77094 USA. Website: www.undergroundconstructionmagazine.com.

Utility Contractor, published by Benjamin Media, Inc. 1770 Main Street, P.O. Box 190, Peninsula, OH 44264 USA. Website: www.utilitycontractoronline.com.

Water Utility Infrastructure Management (UIM), published by Benjamin Media, Inc. 1770 Main Street, P.O. Box 190, Peninsula, OH 44264 USA. Website: www.uimonline.com.

APPENDIX C

Acronyms and Abbreviations

	Description
ABS	Acrylonitrile-butadiene-styrene
AC	Asbestos cement
ACP	Asbestos cement pipe
CCFRPM	Centrifugally cast fiberglass reinforced polymer mortar
CCTV	Close-circuit television
CFP	Close-fit pipe
CI	Cast iron
CIP	Cast-iron pipe
CIPP	Cured-in-place pipe
CIPP/SL	Cured-in-place pipe/short liners
CL	Coatings and linings
CM	Compaction methods
CML	Cement-mortar lining
CP	Closed profile
CP	Concrete pipe
CPJ	Conventional pipe jacking
DDM	Design decision model
DIP	Ductile iron pipe
DIPS	Ductile iron pipe sized
DOT	Department of Transportation
DR	Deformed and reformed pipe
DR	Dimension ratio

	Description
DSC	Differing site condition
DSS	Decision support system
DWCP	Dual wall corrugated profile
EPBM	Earth pressure balance machine
EPDM	Ethylene polypelene diene monomer
ERW	Electric resistance welded
FBE	Fusion bonded epoxy
FF	Fold and formed pipe
FR	Frictional resistance
FRPP	Fiberglass reinforced polyester panel
FWC	Fiber wound collar
GBR	Geotechnical baseline report
GFRP	Glass fiber reinforced plastic pipe
GFRPM	Glass fiber reinforced polyester mortar
GIS	Geographical information system
GPR	Ground penetrating radar
GRP	Glass reinforced pipe
GRP	Glass fiber reinforced polyester
GWT	Ground water table
HAB	Horizontal auger boring
HDB	Hydrostatic design basis
HDD	Horizontal directional drilling
HDPE	High density polyethylene
HDS	Hydrostatic design stress
HEB	Horizontal earth boring
I/I	Infiltration/inflow
ID	Inside diameter
IJS	Intermediate jacking station
ILR	In-line replacement
IPS	Iron pipe size
JF	Jacking force
L-CIPP	Lateral—cured-in-place pipe
LCP	Lining with continuous pipe

	Description
LDPE	Low density polyethylene
LOR	Localized repair
L-PB	Lateral pipe bursting
LR	Lateral renewal
MAOP	Maximum allowable operating pressure
MDPE	Medium density polyethylene
MH	Manhole
MJ	Mechanical joint
MOP	Manual of practice
MSL	Modified sliplining
MT	Microtunneling
MTBM	Microtunnel boring machine
MTM	Manhole-to-manhole
O&M	Operation and maintenance
OC	Open-cut
OD	Outside diameter
OP	Open profile
PACP	Pipeline assessment certification program
PB	Pipe bursting
PB-HY	Pipe bursting–hydraulic
PB-PE	Pipe bursting–pneumatic
PB-PS	Pipe bursting–pipe splitting
PB-ST	Pipe bursting–static
PCCP	Prestressed concrete cylinder pipe
PCP	Polymer concrete pipe
PE	Polyethylene
PJ	Pipe jacking
PL	Panel lining
PM	Piercing method
PP	Polypropylene
PPE	Personal protection equipment
PR	Pipe ramming
PRS	Pipe replacement systems
PRM	Pipe reaming

	Description
PSR	Point source repair (same as localized repair)
PTMT	Pilot tube microtunneling
PVC	Poly-vinyl-chloride
PVC-M	Modified poly-vinyl-chloride
PVC-O	Molecularly oriented poly-vinyl-chloride
PVC-U	Unplasticized poly-vinyl-chloride
PVDF	Poly-vinylidene chloride
PVDM	Polyvinylidene difluoride membranes
QA/QC	Quality assurance/quality control
RCCP	Reinforced concrete cylinder pipe
RCP	Reinforced concrete pipe
RCP	Rapid crack propagation
ROW	Right-of-way
RQD	Rock quality designation
SBR	Styrene-butadiene rubber
SCS	Stress corrosion cracking
SDR	Standard dimension ratio
SH	Shotcrete
SIPP	Spray-in-place pipe
SL	Sliplining
SMR	Sewer manhole renewal
SMYS	Specified minimum yield stress
SP	Steel pipe
SPT	Standard penetration test
SR	Spot repair (same as localized repair)
SR	Steel ribs
SUE	Subsurface utility engineering
SWP	Spiral wound pipe
TBM	Tunnel boring machine
TCM	Trenchless construction methods
TCP	Traffic control plan
TEB	Three-edge-bearing load
ThP	Thermoformed pipe

	Description
TLP	Tunnel liner plates
TRM	Trenchless renewal methods
TRS	Trenchless replacement systems
TT	Trenchless technology
UCL	Underground coatings and linings
USCS	Unified soil classification system
UT	Utility tunneling
UV	Ultra-violet
VCP	Vitrified clay pipe
VOC	Volatile organic compounds
WL	Wooden lagging

Acronyms for Organizations Related to Trenchless Technology

Acronym	Description	Website
AASHTO	American Association of State Highway Transportation Officials	www.transportation1.org/aashtonew
ACIPCO	American Cast Iron Pipe Company	www.acipco.org
ACPA	American Concrete Pipe Association	www.concrete-pipe.org
ACPPA	American Concrete Pressure Pipe Association	www.acppa.org
AEM	Association of Equipment Manufacturers	www.aem.org
AGA	American Gas Association	www.aga.org
AISI	American Iron and Steel Institute	www.steel.org
ANSI	American National Standards Institute	www.ansi.org
APC	American Plastics Council	www.americanplasticscouncil.org

Acronym	Description	Website
APGA	American Public Gas Association	www.apga.org
API	The American Petroleum Industry	www.api.org
APWA	American Public Works Association	www.apwa.net
ASA	American Shotcrete Association	www.shotcrete.org
ASCE	American Society of Civil Engineers	www.asce.org
ASDWA	Association of State Drinking Water Administrators	www.asdwa.org
ASME	American Society of Mechanical Engineers	www.asme.org
ASTM	American Society of Testing and Materials	www.astm.org
AWWA	American Water Works Association	www.awwa.org
AWWRF	American Water Works Research Foundation	www.awwarf.org
BAMI-I	Buried Asset Management Institute–International	www.bami-i.com
BSI	British Standards	www.bsi-global.com
CUIRE	Center for Underground Infrastructure Research and Education	www.cuire.org
DIN	Deutsches Institut fur	www.din.de
DIPRA	Ductile Iron Pipe Research Association	www.dipra.org
EPA	Environmental Protection Agency	www.epa.gov
FHWA	Federal Highway Administration	www.fhwa.dot.gov
ICRI	International Concrete Repair Institute	www.icri.org
ISO	International Organization of Standardization	www.iso.org
ISTT	International Society for Trenchless Technology	www.istt.com

Acronym	Description	Website
JIS	Japanese Industrial Standards	www.jsa.or.jp
MSS	Manufacturer's Standardization Society	www.mss-hq.org
NACE	National Association of Corrosion Engineers	www.nace.org
NACWA	National Association of Clean Water Agencies	www.nacwa.org
NAPCA	National Association of Pipe Coating Applicators	www.napca.com
NASSCO	National Association of Sewer Service Companies	www.nassco.org
NASTT	North American Society for Trenchless Technology	www.nastt.org
NCPI	National Clay Pipe Institute	www.ncpi.org
NCSPA	National Corrugated Steel Pipe Association	www.ncspa.org
NFPA	National Fire Protection Association	www.nfpa.org
NIOSH	National Institute of Occupational Safety and Health	http://origin.cdc.gov/niosh/
NRCS	Natural Resources Conservation Service	www.nrcs.usda.gov
NSF	National Sanitation Foundation	www.nsf.org
NSF	National Science Foundation	www.nsf.gov
NTSB	National Transportation Safety Board	www.ntsb.gov
NUCA	National Utility Contractors Association	www.nuca.com
NWRA	National Water Resources Association	www.nwra.org
OSHA	Occupational Safety & Health Administration	www.osha.gov
PCCA	Power & Communication Contractors Association	www.pccaweb.org
PPFA	Plastic Pipe and Fittings Association	www.ppfahome.org

Acronym	Description	Website
PPI	Plastics Pipe Institute	www.plasticpipe.org
PRC	Pipeline Research Council	www.prci.com
SPE	Society of Plastics Engineers	www.4spe.org
SPI	Society of Plastic Industry, Inc.	www.plasticsindustry.org
SSPC	The Society of Protective Coatings	www.sspc.org
STI	Steel Tank Institute	www.spfa.org
TRB	Transportation Research Board	http://trb.org
TSI	Transportation Safety Institute	www.tsi.dot.gov
TTC	Trenchless Technology Center, Louisiana Tech University	www.latech.edu/tech/engr/ttc
Uni-Bell	Uni-Bell PVC Pipe Association	www.uni-bell.org
USACE	U.S. Army Corps of Engineer	www.usace.army.mil
USDOT	U.S. Department of Transportation	www.dot.gov
WEF	Water Environment Federation	www.wef.org
WRc	Water Research Center	www.wrcplc.co.uk

APPENDIX D

Glossary of Terms

ABS: Acrylonitrile-butadiene-styrene; a form of thermoplastic.

AC: Asbestos cement; a composite material used in pipe construction.

AMP: Asset management plan; a structured approach for utilities to achieve long-term defined service standards or an external bearing used to isolate the final drive from the thrusting force of the machine.

ASTM: American Society of Testing Materials.

Auger: A flighted drive tube having hex couplings at each end to transmit torque to the cutting head and transfer the spoil back to the machine.

Auger boring: Also horizontal auger boring, a technique for forming a bore from a drive pit to a reception pit, by means of a rotating cutting head. Spoil is removed back to the drive shaft by helically wound auger flights rotating in a steel casing. The equipment may have limited steering capability. See guided auger boring.

Auger machine: A machine to drill earth horizontally by means of a cutting head and auger or other functionally similar device. The machine may be either cradle or track type.

Auger MTBM: A type of microtunnel boring machine, which uses auger flights to remove the spoil through a separate casing placed through the product pipeline.

Auger TBM: Tunnel boring machine in which the excavated soil is removed to the drive shaft by auger flights passing through the product pipeline pushed in behind the TBM.

Back reamer: A cutting head attached to the leading end of a drill string to enlarge the pilot bore during a pullback operation to enable the carrier, sleeve, or casing to be installed.

Backstop: Also thrust block, a reinforced area of the entrance pit wall directly behind the track or where the jacking loads will be resisted.

Band: A ring of steel welded at or near the front of the lead section of casing to cut relief and strengthen the casing (used in horizontal auger boring).

Bedding: A prepared layer of material below a pipeline to ensure uniform support.

Bent sub: An offset section of drill stem close behind the drill head that allows steering corrections to be made by rotation of the drill string to orientate the cutting head (used in horizontal directional drilling).

Bentonite: Colloidal clay sold under various trade names that form a slick slurry or gel when water is added; also known as driller's mud. See drilling fluids.

Bits: Replaceable cutting tools on the cutting head or drill string.

Bore: A generally horizontal hole produced underground, primarily for the purpose of installing services.

Boring: (1) The dislodging or displacement of spoil by a TBM; a rotating auger or drill string to produce a hole called a bore. (2) An earth-drilling process used for installing conduits or pipelines. (3) Obtaining soil samples for evaluation and testing.

Boring machine: An automated mechanism to drill earth.

Boring pit: Also entrance pit, launch or drive pit; an excavation in the earth of specified length, depth, and width for placing the boring machine on required line and grade.

Bottom inversion: The CIPP tube is inverted through a specially designed elbow located at the elevation of the pipe, typically in a manhole or excavated pit.

Breakout: Controls the joint make and/or break mechanism (in HDD operations).

Burst strength: The internal pressure required to cause a pipe or fitting to fail within a specified time period.

Butt-fusion: A method of joining polyethylene and PVC pipe where two pipe ends and are rapidly brought together under pressure to form a homogeneous bond.

Bypass: An arrangement of pipes and valves whereby the flow may be passed around a hydraulic structure or appurtenance. Also, a temporary setup to route flows around a part of a sewer system.

Bypass pumping: Taking all existing flow in a pipe and routing around the section of pipe to be renewed, replaced, or repaired.

Calibration hose: A flexible hose inverted into a pulled-in liner and used to hold the resin saturated fabric tube up tight against the existing, old, or host pipe until final cure has been achieved.

Can: A principal module, which is part of a shield machine as in microtunneling or tunnel boring machines (TBMs). Two or more may be used, depending on the installation dimensions required and the presence of an articulated joint to facilitate steering. May also be referred to as a trailing tube.

Carbon fiber: A reinforcing material used to strengthen cured-in-place pipe.

Carriage: The mechanical part of a nonsplit boring machine that includes the engine or drives motor, the drive train, thrust block, and hydraulic cylinders.

Carrier pipe: The tube, which carries the product being transported, and which may go through casings at highway and railroad crossings. It may be made of steel, concrete, clay, plastic, ductile iron, or other materials. On occasion it may be bored direct under the highways and railroads.

Cased bore: A bore in which a pipe, usually a steel sleeve, is inserted simultaneously with the boring operation. Usually associated with horizontal auger boring or pipe ramming.

Casing: See casing pipe.

Casing adapter: A circular mechanism to provide axial and lateral support of a smaller-diameter casing than that of the casing pusher.

Casing pipe: A steel pipe usually installed by horizontal auger boring or pipe ramming methods to support boreholes under roadways or railroad tracks through which a carrier (or product) pipes or ducts are installed.

Casing pipe method: Method in which a casing, generally steel, is pipe jacked into place, within which a product pipe is inserted later.

Casing pusher: The front section of a boring machine that distributes the thrusting force of the hydraulic cylinders to the casing and forms the outside of the spoil ejector system.

Catalyst: The component of a resin system that induces a reaction to form heat and subsequently cure the liner.

Cathodic protection: Preventing corrosion of a pipeline by using special cathodes (and anodes) to circumvent corrosive damage by electric current. Also a function of zinc coatings on iron and steel drainage products is galvanic action.

Caulking: General term which, in trenchless technology, refers to methods by which joints may be closed within a pipeline.

Cave-in: The separation of a mass of soil or rock material from the side of an excavation, or the loss of soil from under a trench shield or support system, and its sudden movement into the excavation, either by falling or sliding, in sufficient quantity so that it could entrap, bury, or otherwise injure and immobilize a person.

CCTV: See closed-circuit television inspection.

Cell classification: Method of identifying plastic materials, such as polyethylene, as specified by ASTM D3350, where the cell classification is based on these six properties: (1) density of base resin (2) melt index (3) flexural modulus (4) tensile strength at yield (5) ESCR (6) hydrostatic design basis and color.

Cellar drain: A pipe or a series of pipes that collect wastewater which leaks, seeps, or flows into subgrade parts of structures and discharges them into a building sewers; by other means, dispose of such wastewaters into sanitary, combined, or storm sewers. Also referred to as *basement drain*.

Chemical grouting: Method for the treatment of the ground around a shaft or pipeline, using noncementitious compounds, to facilitate or make possible the installation of an underground structure.

Chemical stabilization: A repair method in which a length of pipeline between two access points is sealed by the introduction of one or more compounds in solution into the pipe and the surrounding ground and, where appropriate, producing a chemical reaction. Such systems may perform a variety of functions such as the sealing of cracks and cavities, the provision of a new wall surface with improved hydraulic characteristics or ground stabilization.

Chimney: The small vertical section between a manhole frame and cone, which is built from brick, masonry, or concrete adjusting rings.

Chippers: See bits.

CIPP: Cured-in-place pipe; a renewal technique whereby a flexible resin-impregnated tube is installed into an existing pipe and then cured to a hard finish, usually assuming the shape of the existing pipe.

Circumferential: The perimeter around the inner surface of a circular pipe cross section.

Circumferential coefficient of expansion and contraction: The fractional change in circumference of a material for a unit changes in temperature; expressed as inches of expansion or contraction per inch of original circumference.

Closed face: The ability of a tunnel boring machine to close or seal the facial opening of the machine to prevent, control, or slow the entering of soils into the machine. Also may be the bulk heading of a hand-dug tunnel to slow or stop the inflow of material.

Closed-circuit television inspection (CCTV):   Inspection method using a closed-circuit television camera system with appropriate transport and lighting mechanisms to view the interior surface of sewer pipes and structures.

Close-fit: Description of a lining system in which the new pipe makes close contact with the existing defective pipe at normal or minimum diameter. An annulus may occur in sections where the diameter of the defective pipe is in excess of this.

Coefficient of thermal expansion and contraction: The fractional change in length of a material for a unit change in temperature.

Collapse: Critical failure of a pipeline when its structural fabric disintegrates.

Collaring: The initial entry location of casing or a cutting head into the earth.

Collection system: A network of sewers that serves one or more catchment areas.

Collector sewer: A sewer located in the public way collects the wastewaters discharged through building sewers and conducts such flows into larger interceptor sewers, pumping and treatment works (referred to also as *street sewer*).

Combined sewer system: A single network of sewers designed to convey stormwater as well as sanitary flows.

Competent person: One who is capable of identifying existing and predictable hazards in the surroundings, or working conditions that are unsanitary, hazardous, or dangerous to employees, and who has authorization to take prompt corrective measures to eliminate them.

Compressed air method: In trenchless technology, refers to the use of compressed air within a tunnel or shaft to balance ground water and prevent ingress into an open excavation.

Compression gasket: A device that can be made of several materials in a variety of cross sections and that serves to secure a tight seal between two pipe sections (e.g., "O" rings).

Compression ring: A ring fitted between the end bearing area of the bell and spigot to help distribute applied loads more uniformly. The compression ring is attached to the trailing end of each pipe and is compressed between the pipe sections during jacking. The compression rings compensate for slight misalignment, pipe ends that are not perfectly square, gradual steering corrections, and other pipe irregularities. Compression rings are also referred to as spacers.

Conduit: A broad term that can include pipe, casing, tunnels, ducts, or channels. The term is so broad that it should not be used as a technical term in boring or tunneling.

Cone: The section between the top of a manhole wall and chimney or the frame. The diameter of the manhole is reduced over the cone section to receive the frame. The cone section may be concentric or eccentric.

Continuous sliplining: See sliplining.

Control console: An electronic unit inside a container located on the ground surface, which controls the operation of the microtunneling machine. The machine operator drives the tunnel from the control console. Electronic information is transmitted to the control console from the heading of the machine. This information includes head position, steering angle, jacking force, progression rates, machine face torque, slurry and feed line pressures, and laser position. Some control consoles are equipped with a computer that tracks the data for a real-time analysis of the tunnel drive.

Control lever: A handle that activates or deactivates a boring machine function.

Conventional pipe jacking: Jacking pipe sections simultaneously as tunnel excavation proceeds using various forms of TBMs or hand mining (not microtunneling or pilot tube microtunneling).

Conventional trenching: See open-cut.

Corrugated pipe: Pipe with ridges (corrugations) going around it to make it stiffer and stronger. The corrugations are usually in the form of a sine wave a+B181nd are usually made of galvanized steel or aluminum.

Cradle machine: A boring machine typically carried by another machine that uses winches to advance the casing.

Creep: The dimensional change, with time, of a material, such as plastic, under continuously applied stress after the initial elastic deformation.

Crossing: Pipeline installation in which the primary purpose is to provide a passage beneath a surface obstruction, such as a road, railroad track, lake or river.

Crown: (1) Top of pipe segment, or (2) The highest elevation within a pipe.

Cured-in-place pipe (CIPP):   A lining system in which a thin flexible tube of polymer or glass fiber fabric is impregnated with thermoset resin and expanded by means of fluid pressure into position on the inner wall of a defective pipeline before curing the resin to harden the material. The uncured material may be installed by winch or inverted by water or air pressure, with or without the aid of a turning belt.

Cut and cover: See open-cut.

Cutterhead: Any rotating tool or system of tools on a common support that excavates at the face of a bore; usually applies to the mechanical methods of excavation.

Cutter bit (cutter head):   The actual teeth and supporting structure that is attached to the front of the lead auger, drill stem, or front face of the tunnel-boring machine. It is used to reduce the material that is being drilled or bored to sand or loose dirt so that it can be conveyed out of the hole. Usually applies to mechanical methods of excavation, but may also include fluid jet cutting.

Dead man: A fixed anchor point used in advancing a saddle or cradle-type boring machine.

Deformed and reshaped: See modified sliplining.

Diameter of reamer: Largest diameter of reamer (in horizontal directional drilling).

Dimension ratio (DR):   See standard dimension ratio (SDR).

Dimple: A term used in tight fitting pipeline renewal, where the new plastic pipe forms an external departure or a point of expansion slightly beyond the underlying pipe wall where unsupported at side connections. The dimples are used for location and reinstatement of laterals.

Directional drilling: A steerable system for the installation of pipes, conduits, and cables in a shallow arc using a surface launched drilling rig. Traditionally the term applies to large-scale crossings in which a fluid-filled pilot bore is drilled using a fluid-driven motor at the end of a bend-sub, and is then enlarged by a washover pipe and back reamer to the size required for the product pipe. The positioning of a bent sub provides the required deviation during pilot boring. Tracking of the drill string is achieved by the use of a downhole survey tool.

Dog plate: See backstop.

Drill bit: A tool that cuts the ground at the head of a drill string, usually by mechanical means, but may include fluid jet cutting.

Drill string: (1) The total length of drill rods or pipe, bit, swivel joint and so on in a drill borehole. (2) System of rods used with cutting bit or compaction bit attached to the drive chuck.

Drilling fluid or mud: A mixture of water and usually bentonite and/or polymer continuously pumped to the cutting head to facilitate cutting, reduce required torque, facilitate the removal of cuttings, stabilize the borehole, cool the head, and lubricate the installation of the product pipe. In suitable soil conditions water alone may be used.

Drive or entry or launch or jacking shaft or pit: Excavation from which trenchless technology equipment is launched for the installation of a pipeline. In pipe jacking, it incorporates a thrust wall to spread reaction loads to the soil.

Dry bore: Any drilling or rod pushing system not employing drilling fluid in the process. Usually associated with guided impact moling, but also some rotary methods.

Earth piercing: (1) Term commonly used in North America as an alternative to impact moling. (2) The use of a tool, which comprises a percussive hammer within a suitable casing, generally of torpedo shape. The hammer may be pneumatic or hydraulic. The term is usually associated with nonsteered devices without rigid attachment to the launch pit, relying upon the resistance (friction) of the ground for forward movement. During operation, the soil is displaced not removed. An unsupported bore may be formed in suitable ground, or a pipe drawn in, or pushed in, behind the tool. Cables may also be drawn in.

Earth pressure balance (EPB) machine: Type of microtunneling or tunneling machine in which mechanical pressure is applied to the material at the face and controlled to provide the correct counterbalance to earth pressures to prevent heave or subsidence. The term is usually not applied to those machines where the pressure originates from the main pipe jacking rig in the drive shaft/pit or to systems in which the primary counterbalance of earth pressures is supplied by pressurized drilling fluid.

Earth pressure balance shield: Mechanical tunneling shield that uses a full face to support the ground in front of the shield and usually employs an auger flight to extract the material in a controlled manner.

Emergency repair: An unscheduled repair that must be made during a pipe failure or collapse. This type of repairs may cost many times (usually 10 times more, not including social costs) of planned repair costs and may not be as effective and/or permanent.

Emergency stop: A red, manually operated push button that, when activated, stops all functions of the machine.

Entrance pit: (1) See boring pit or drive shaft.

Entry/exit angle: Angle to horizontal (the ground surface) at which the drill string enters or exits in forming the pilot bore in a horizontal directional drilling operation.

EPDM (ethylene-propylene-diene monomer): Type of rubber that has excellent resistance to ozone, sunlight, and oxygen. It also has excellent resistance to acids, alkalis, and ketones. Plus, it has excellent heat resistance and aging. However it has poor resistance to fuels and oils.

Epoxy: Resin formed by the reaction of bisphenol and epichlorohydrin.

Epoxy lining: A curable resin system based on epoxy resins.

Exfiltration: The leakage or discharge of flows being carried by pipes or sewers out into the ground through leaks in pipes, joints, manholes, or other sewer system structures; the reverse of *infiltrations*.

Exit pit: See reception shaft.

Exit shaft: See reception shaft.

Expander: A tool, which enlarges a bore during a pullback operation by compression of the surrounding ground rather than by excavation. Sometimes used during a pipe bursting process as well as during horizontal directional drilling

Face stability: Stability of the excavated face of a tunnel or pipe jack operation.

Felt: A material specially designed to soak up, hold, and transport resins in place to produce a hard cured-in-place pipe.

Fiberglass: A high strength material that is commonly layered with a felt material to add reinforcement for the resin cured-in-place pipe.

Fillers: Materials used to enhance the capabilities of a resin system.

Film: Either an outer or inner material to protect the resin impregnated tube from contaminants.

Flexural modulus: The slope of the curve defined by flexural load versus resultant strain. A high flexural modulus indicates a stiffer material.

Flexural strength: The strength of a material in bending expressed as the tensile stress of the outermost fibers at the instant of failure.

Flight: The spiral plates surrounding the tube of an auger.

Fluid cutting: (1) An old trenchless method where pressurized fluid jets are mainly used to provide the soil cutting action. (2) A process using high-pressure fluid to wash out the face of a utility crossing without any mechanical or hand excavation of the soils in the face. This method is no longer allowed.

Fluid-assisted boring or drilling: A type of horizontal directional drilling technique using a combination of mechanical drilling and pressurized fluid jets to provide the soil cutting action.

Fold and form lining: Method of pipeline renewal in which a liner is folded to reduce its size before insertion and reversion to its original shape by the application of pressure, or heat, or both.

Fold and form pipe: A pipe renewal method where a plastic pipe manufactured in a folded shape of reduced cross-sectional area is pulled into an existing conduit and subsequently expanded with pressure and heat. The reformed plastic pipe fits snugly and takes the shape of the ID of the existing, old, or host pipe.

Force-main: A pipeline that conveys sanitary, combined, or stormwater flow under pressure from a pumping (or lift) station to a discharge point (treatment plant).

Forward rotation: The clockwise rotation of the auger as viewed from the machine end.

Gel Time: The time, at which a catalyzed resin system will begin to cure, creating internal heat and thus beginning to harden.

Geographical information system (GIS): A computer software system designed to store, manipulate, analyze, and print geographically referenced information.

Gravity sewer: A sewer that is designed to operate under open channel conditions (below pipe full capacity) up to a maximum design flow at which point it will become surcharged.

Ground mat: Usually used in horizontal directional drilling, metal mats rolled out on either side of drill rack for operators and crew to stand on during operation to give grounding protection in case of electrical strike.

Ground rod: This is a copper or brass rod that is hand driven into the ground and is connected to the drill rack and mats to provide adequate grounding of an HDD rig.

Groundwater table (or level):   Upper surface of the zone of saturation in permeable rock or soil (when the upper surface is confined by impermeable rock, the water table is absent).

Grout: (1) Material used to seal pipeline and manhole cracks; also used to seal connections within pipe or sewer structures. (2) A material, usually cement or polymer based, used to fill the annulus between the existing pipe and the lining; and also to fill voids outside the existing pipeline. (3) A material such as cement slurry, sand, or pea gravel that is pumped into voids.

Grouting: (1) Filling of the annular space between the existing, old, or host pipe and the carrier pipe. Grouting is also used to fill the space around laterals and between the new pipe and manholes. Other uses of grouting are for localized repairs of defective pipes and ground improvement prior to excavation during new installations. (2) The process of filling voids, or modifying or improving ground conditions. Grouting materials may be cementitious, chemical, or other mixtures. In trenchless technology, grouting may be used for filling voids around the pipe or shaft, or for improving ground conditions. (3) A method of filling voids with cementitous or polymer grout.

GRP: Glass reinforced plastic, a family of renewal linings. Often generically known as reinforced plastic mortar (RPM) and reinforced thermosetting resin (RTR).

Guidance system: The guidance system continuously confirms the position of the TBM.

Guide rail: Device used to support or guide, first the shield and then the pipe within the drive shaft during a pipe jacking or utility tunneling operation.

Guided auger boring: Pilot tube microtunneling. A modified version of horizontal auger boring method.

Guided boring: This term is used in Europe for small diameter horizontal directional drilling method.

Guided drilling: See guided boring.

Gunite: A renewal technique that employs steel reinforcement fixed to the inside surface of an existing sewer line, which is sprayed with dry concrete.

HDPE: High density polyethylene, see polyethylene.

Head (static): The height of water above any plane or point of references. (The energy possessed by each unit of weight of a liquid, expressed as the vertical height through which a unit of weight would have to fall to release the average energy posed.) Standard unit of measure shall be the foot. Head in feet for water is 1 psi = 2.310 ft.

Heat cure: The application of either steam or hot water to cure a resin saturated tube.

Heaving: A process in which the ground in front of a tunneling or pipe jacking operation may be displaced forward and upward, causing an uplifting of the ground surface.

Height of cover (HC): Distance from crown of a pipe or conduit to the finished road surface, or ground surface, or the base of the railway.

High density polyethylene: A plastic resin made by the copolymerization of ethylene and a small amount of another hydrocarbon. The resulting base resin density, before additives or pigments, is greater than 0.941 g/cc.

Hoop stress: The circumferential force per unit areas, psi, in the pipe wall owing to internal pressure.

Horizontal directional drilling (HDD): See directional drilling.

Horizontal earth boring machine: A machine used to bore horizontally through the earth by means of a MTBM, cutterhead, rotating tool, or ramming tool.

Horizontal auger boring: The use of auger boring machines to prepare holes by the installation of a casing whereby the spoil is removed by the use of augers.

Host pipe: In trenchless renewal and replacement methods. Existing, old, or deteriorated pipe.

Hydraulic gradient line (HGL): An imaginary line through the points to which water would rise in a series of vertical tubes connected to the pipe. In an open channel, the water surface itself is the hydraulic grade line.

Hydrogen sulfide: An odorous gas found in sewer systems with chemical formula of H_2S.

I/I: Infiltration/inflow; infiltration stands for ground water seepage and inflow stands for surface water flow into the gravity collection system.

Impact moling: Method of creating a bore using a pneumatic or hydraulic hammer within a casing, generally of torpedo shape. The term is usually associated with nonsteered or limited steering devices without rigid attachment to the launch pit, relying upon the resistance of the ground for forward movement. During the operation the soil is displaced, not removed. An unsupported bore may be formed in suitable ground, or a pipe drawn in, or pushed in, behind the impact moling tool.

Impact ramming: See pipe ramming.

Impervious coating: The outer layer of a CIPP tube that will prevent the installation water or steam from mixing with the resin system in the tube.

Impregnated tube: A felt tube fully saturated with a catalyzed resin system.

In-line replacement: The process of breaking out of an existing, old, or host pipeline and the installation of a new pipeline at the same location. With this method, the existing pipeline will serve as a "pilot bore" for the new pipe installation, which might be a different pipe material with the same or larger diameter, and will be installed with the same alignment of the existing pipe.

Infiltration: Penetration of groundwater into the sewer system through cracks and defective joints in the pipeline, or through lateral connections, or manholes.

Infiltration or inflow (I/I):   The total quantity of water from both infiltration and inflow without distinguishing the source.

Inflow: Storm water discharged into a sewer system and service connections from sources on the surface.

Interjack pipes: Pipes specially designed for use with an intermediate jacking station used in pipe jacking and microtunneling operations.

Interjack station: See intermediate jacking stations.

Intermediate jacking method: Pipe jacking or microtunneling method to redistribute the jacking force by the use of intermediate jacking stations.

Intermediate jacking stations: A fabricated steel cylinder fitted with hydraulic jacks that are incorporated into jacking pipes strung between two pipe segments. Its function is to distribute the jacking load over the pipe string on long drives thereby decreasing the total jacking forces exerted on the thrust block and pipe sections near the shaft.

Internal seal: Internal seals are used for structural repair pipe joints and missing pipe sections. It can be used in both worker entry and non worker entry pipes.

Inversion: The process of turning a fabric tube inside out with water or air pressure as is done at installation of a cured-in-place pipe (CIPP).

Invert: (1) The lowest point on the pipe circumference; also the defined channel in the manhole platform that directs flow from inlet pipe to outlet pipe. (2) The inside bottom, lowest elevation, of a pipe.

Jacking: The actual pushing of pipe or casing in an excavated hole. This is usually done with hydraulic cylinders (jacks), but has been done with mechanical jacks and air jacks.

Jacking force: Force applied to pipes in a pipe jacking operation.

Jacking frame: A structural component that houses the hydraulic cylinders used to propel the microtunneling machine and pipeline. The jacking frame serves to distribute the thrust load to the pipeline and the reaction load to the shaft wall or thrust wall.

Jacking pipes: Pipes sections with smooth outside joints designed to be installed using pipe jacking techniques.

Jacking pit: See jacking shaft.

Jacking shaft (also launch or entry shaft):   Excavation from which trenchless technology equipment is launched for the installation or renewal of a pipeline.

Jacking shield: A steel cylinder from within which the excavation is carried out either by hand or machine. Incorporated within the shield are facilities to allow it to be adjusted to control line and grade.

Joint sealing: Method in which an inflatable packer is inserted into a pipeline to span a leaking joint, resin, or grout being injected until the joint is sealed and the packer then removed.

Lateral: A service line that transports wastewater from individual buildings to a main sewer line.

Lateral connection: The point at which the downstream end of a building drain or sewer connects into a larger-diameter sewer.

Launch pit: See drive shaft or pit.

Lead pipe: The leading pipe designed to fit the rear of a jacking shield and over which the trailing end of the shield is fitted.

Light cure: The curing of a CIPP liner using UV light energy in lieu of either water or steam.

Liner: A fabric tube that has been saturated with a liquid resin.

Liner plate: A proprietary product, used to line tunnels instead of casing, and comes in formed steel segments. When these segments are bolted together they form a structural tube to protect the tunnel from collapsing. The segments are made so that they may be bolted together from inside the tunnel.

Lining: A renewal process where a new pipe or coating material is inserted or cured in place to give an existing pipe a new design life.

Localized (spot) repair: Repair work on an existing pipe, to an extent less than the run between two access points or manholes.

Locator: An electronic instrument used to determine the position and strength of electro-magnetic signals emitted from a transmitter (sonde) in a directional drilling operation, or from existing underground services, which have been energized, thereby identifying its location. In horizontal directional drilling it is referred to as a walkover system.

Machine upset: The inadvertent action of a horizontal auger boring machine that rotates the machine and track from its normal and upright position to another position.

Man-entry: Also worker-entry, describes any inspection, construction, renewal or repair process, which requires an operator to enter a pipe, duct, or bore. OSHA currently has no minimum size limit for worker-entry operations; however, they address a much broader concept of *confined space* in Title 29 Code of Federal Regulations Part 1910.146. The minimum size for which this is currently permissible in the United Kingdom is 900 mm (approximately, 36 in.). Many trenchless technologies do not require worker-entry inside the pipe.

Manhole: A structure that allows access to the sewer system.

Manual inspection: Method of sewer inspection that usually involves physical entry and hands-on examination.

Microtunneling: A trenchless construction method for installing pipelines. Microtunneling uses all of the following features during construction: (1) Remote controlled—The microtunneling-boring machine (MTBM) is operated from a control panel, normally located on the surface. The system simultaneously installs pipe as spoil is excavated and removed. Personnel entry is not required for routine operation. (2) Guided—The guidance system usually references a laser beam projected onto a target in the MTBM, capable of installing gravity sewers or other types of pipelines to the required tolerance, for line and grade. (3) Pipe jacked— The process of constructing a pipeline by consecutively pushing pipes and MTBM through the ground using a jacking system for thrust. (4) Continuously supported—Continuous pressure is provided to the face of the excavation to balance groundwater and earth pressures.

Microtunnel boring machine (MTBM): See microtunneling.

Midi-HDD: Steerable surface-launched horizontal directional drilling equipment for installation of pipes, conduits, and cables. Applied to intermediate sized drilling rigs used as either a small directional drilling machine or a large guided boring machine. Tracking of the drill string may be achieved by either a downhole survey tool or a walk-over locator.

Mini-horizontal directional drilling (Mini-HDD): Small diameter horizontal directional drilling. In Europe it is called guided boring.

Mixed face: A soil condition that presents two or more different types of material in the path of the bore.

Modified sliplining: A range of techniques in which the liner is reduced in cross-sectional diameter before insertion into the carrier pipe. It is subsequently restored close to its original diameter, generally forming a close-fit with the original pipe. There are different methods of cross sectional area reduction.

Needled felt: A highly absorbent felt material specially designed for the CIPP tube fabrication.

Non-worker entry: Size of pipe, duct, or bore, less than that for worker-entry.

Occupational illness: Any abnormal condition or disorder caused by exposure to environmental factors associated with employment. For excavations this might include illnesses caused by the inhilation of toxic vapors.

Open-cut: The method by which access is gained to the required level underground for the installation, repair, or replacement of a pipe, conduit, or cable. The excavation is then backfilled and the surface restored.

Open face shield: Shield in which manual excavation is carried out from within a steel tube at the front of a pipe jacking operation.

On-site wet-out: A non-factory tube wet-out that is performed over the hole in the field.

Ovality: There are two definitions: (1) the difference between the maximum and mean diameter divided by the mean diameter, and (2) the difference of the mean and minimum divided by the mean, at any one cross section of a pipe, generally expressed as a percentage.

Overcut: The annular space between the excavated borehole and the outside diameter of the pipe.

PACP: Pipeline Assessment Certification Program by NASSCO.

Panel lining: Panel lining is a modified sliplining method. The shape of the culvert is covered by preparing panels and fitting them to the culvert. It can be used to structurally renew large diameter pipes. This method can accommodate different shapes.

PE: Polyethylene; a form of thermoplastic pipe.

pH: A measure of the acidity or alkalinity of a solution. A value of seven is neutral; lower numbers indicate more acidity.

Physical pipe inspection: The crawling or walking through manually accessible pipe lines. The logs for physical pipe inspection record information of the kind detailed under television inspection. Manual inspection is only undertaken when field conditions permit this to be done safely. Precautions are necessary.

Piercing tool: Similar to closed-face pipe ramming, but for small diameter (2 to 6 in.) boring; used for cable installations under roadways.

Pilot bore: The action of creating the first (usually steerable) pass of any boring process, which later requires back-reaming or similar enlarging process to install the product pipe. Most commonly applied to horizontal directional drilling but also used in pilot tube microtunneling and guided auger boring systems.

Pilot tube method: See pilot tube microtunneling.

Pinch rollers: Used to control the thickness of the resin impregnated tube during wet-out.

Pipe bursting: A pipe replacement method for breaking the existing pipe by brittle fracture, using force from within, applied mechanically, the remains being forced into the surrounding ground. At the same time a new pipe, of the same or larger diameter, is drawn in behind the bursting tool. The pipe-bursting device may be based on an impact moling tool to exert diverted forward thrust to the radial bursting effect required, or by a hydraulic device inserted into the pipe and expanded to exert direct radial force or a static hammer. For new pipe, generally a HDPE pipe is used, but currently PVC, ductile iron, clay and GRP is also used. Also known as pipe cracking and pipe splitting.

Pipe displacement: See pipe bursting.

Pipe eating: A pipe replacement technique, usually conducted by use of a horizontal directional drilling rig, in which a defective pipe is pulverized during the backreaming operation. Also microtunneling machines can be used where the existing pipe is excavated together with the surrounding soil as for a new installation. The microtunneling shield machine will usually need some crushing capability to perform effectively. The defective pipe may be filled with grout to improve steering performance.

Pipe jacking: A system of directly installing pipes behind a shield machine by hydraulic jacking from a drive shaft, such that the pipes form a continuous string in the ground.

Pipe ramming: A nonsteerable system of forming a bore by driving an open-ended steel casing using a percussive hammer from a drive pit. The soil may be removed from the casing by augering, jetting, or compressed air.

Pipe reaming: A variation of directional boring, pipe reaming can be used to replace existing clay, asbestos cement, non-reinforced concrete and PVC pipe. A reamer is pulled through the existing pipe, which cuts the pipe into small pieces. The pipe pieces are flushed out the borehole with the drilling fluid.

Pipe segment: A specific portion of the sewer or pipeline system; which usually runs between two structures (e.g., manhole, trap tanks, sumps); identified with unique sewer or pipe structure ID number.

Pipe splitting: Replacement method for breaking an existing pipe by longitudinal slitting. At the same time a new pipe of the same or larger diameter may be drawn in behind the splitting tool. See also pipe bursting.

Pipeline rehabilitation: See pipeline renewal.

Pipeline renewal: The in-situ renewal of an existing pipeline, which has become deteriorated. The selection of appropriate renewal method is dependent on type of application and characteristics and types of defects of the existing pipe. See Chap. 2 for method selection criteria.

Plastic: Any of a variety of thermoplastic and thermoset material used in pipeline construction and renewal (e.g., polypropylene, PVC, fiberglass reinforced plastics, polyester felt reinforced pipe, epoxy and polyester mortars, and so on).

Point CIPP: CIPP techniques entail impregnating fabric with a suitable resin, pulling this into place within the sewer around an inflatable packer or mandrel, and then filling the packer with water, steam, or air under pressure to press the patch against the existing sewer wall while the resin cures.

Point repairs: Repair works on an existing pipe, to an extent less than the run between two access points or manholes.

Point source repair: See localized repair.

Polyester: Resin formed by condensation of polybasic and monobasic acids with polyhydric alcohols.

Polyethylene: A ductile, durable, virtually inert thermoplastic composed by polymers of ethylene. It is normally a translucent, tough solid. In pipe grade resins, ethylene-hexene copolymers are usually specified with carbon black pigment for weatherability.

Polymer coating: It is a thermoset coating made up of inert plastic-like epoxies, urethanes and ureas, polyesters, which have a high resistance to corrosion; they are applied by trained professionals using special spraying equipments and as per the manufacturers specifications.

Polyolefin: A family of plastic material used to make pipes.

Polypropylene (PP): A type of plastic pipe from the polyolefin family.

Potholing: Digging a vertical hole to visually locate a utility.

Preparatory cleaning: Internal cleaning of pipelines, particularly sewers, prior to inspection, usually with water jetting and removal of material where appropriate.

Preventative maintenance: Routine maintenance designed to prevent pipeline system failures and resulting emergency repairs.

Product pipe: Permanent pipeline for operational use. Pipe for conveyance for water, gas, sewage, and other products.

Protruding: To be projecting outward.

Pull-back force: The tensile load applied to a drill string during the pull back process. Horizontal directional drilling rigs are generally rated by their maximum pull-back force.

PVC: Polyvinyl chloride; a form of thermoplastic pipe.

Quality assurance: includes developing inspection and testing methods to ensure products or services are designed and produced to meet or exceed owner requirements.

Quality control: includes providing evidence needed (test results) to establish confidence among all concerned, that quality-related activities are performed.

Radian: An arc of a circle equal in length to its radius; or the angle at the center measured by the arc.

Ramming: A percussive hammer is attached to an open-end casing, which is driven through the ground. See pipe ramming.

Receiving pit: (1) See exit pit. (2) An opening in the earth located at the expected exit of the cutting head or tunneling machine (TBM). (3) The pit that is dug at the end of the bore, opposite the jacking pit. Also target pit.

Receiving shaft: See reception shaft.

Reception or exit shaft or pit: Excavation into which trenchless technology equipment is driven and recovered following the installation of the product pipe, conduit, or cable. See receiving pit.

Registered professional engineer: means a person who is registered as a professional engineer in the state where the work is to be performed. However, a professional engineer, registered in any state is deemed to be a "registered professional engineer" when approving designs for "manufactured protective systems" or "tabulated data" to be used in interstate commerce.

Rehabilitation: See renewal.

Reinstatement: Method of backfilling, compaction, and resurfacing of any excavation to restore the original surface and underlying structures to enable it to perform its original function.

Remote-control system (microtunneling):   The remote-control system monitors and controls the MBTM, the automated transport system, and the guidance system from a location not in the MTBM.

Renewal: All aspects of rehabilitating, reconstructing, renovating, or upgrading with a *new design life* for the performance of existing pipeline systems.

Renovation: See renewal.

Repair: Reconstruction of short pipe lengths, but not the reconstruction of a whole pipeline. Therefore, *a new design life is not provided.* In contrast, in pipeline renewal, a new design life is provided to existing pipeline system.

Replacement: Replacing an old, existing, or deteriorated pipe with a new pipe by use of open-cut, inline replacement (pipe bursting), and/or other new installation methods (HDD, microtunneling, etc.). The new pipe may have a larger diameter and different pipe material from the existing or old pipe.

Resin impregnation (wet-out): A process used in cured-in-place pipe installation where a plastic coated fabric tube is uniformly saturated with a liquid thermosetting resin, while air is removed from the coated tube by means of vacuum suction.

Resins: An organic polymer, solid or liquid; usually thermoplastic or thermosetting.

Reverse: In horizontal auger boring, the counterclockwise rotation of the auger as viewed from the machine end.

Ring compression: The principal stress in a confined thin circular ring subjected to external pressure.

Robot: Remote-control device with closed-circuit television (CCTV) monitoring, used mainly in localized repair work, such as cutting away obstructions, reopening lateral connections, grinding and refilling defective areas, and injecting resin into cracks and cavities.

Robotic cutter: A device used to re-open house connections after the installation of a liner without excavation.

Rod pushing: Method of forming a pilot bore by driving a closed pipe head with rigid attachment from a launch pit into the soil, which is displaced. See thrust boring.

Roller cone bit or reamer: A bit or reamer in which the teeth rotate on separate, internal shafts that are usually aligned perpendicular to line. Used for boring rock.

Rollers: The rollers that control the exact amount of resin material that is required to properly saturate a felt tube.

Rotary rod machine: A machine used to drill earth horizontally by means of a cutting head attached to a rotating rod (not an auger). Such drilling may include fluid injected to the cutting head through a hollow rod.

Saddle: In horizontal auger boring, a vertical support mechanism to hold the casing in position while starting (collaring) the bore.

Saturated tube: See impregnated tube.

SBR (styrene butadiene): Type of rubber that has good abrasion resistance and excellent impact and cut-and-gouge resistance. Used as gasket material.

SDR: See standard dimension ratio

Segmental lining: See sliplining.

Segmental sliplining: See sliplining.

Self-cleansing: A consequence of good hydraulic design when the pipe invert is kept relatively free of sediments by ensuring adequate flow velocities.

Semi-structural liner: A liner that in its own entity does not have the required strength to withstand internal, external, or both types of loading from soil column, traffic, and groundwater pressure for the design life of the product, but will offer some level of structural support against internal pressure.

Separate system: A system that uses sanitary sewers to convey the wastewater and stormwater sewers to carry the stormwater.

Sewer: An underground pipe or conduit for transporting stormwater, or wastewater, or both.

Sewer cleaning: The use of mechanical or hydraulic equipment to dislodge, transport, and remove debris from sewer lines.

Sewer lateral: A building sewer (sometimes referred to as a sewer lateral or house lateral) is the pipeline between the public sanitary sewer line, which is usually located in the street, and the indoor plumbing.

Sewer pipe: A length of conduit, manufactured from various materials and in various lengths, that when joined together can be used to transport wastewaters from the points of origin to a treatment facility. Types of pipe are: acrylonitrile-butadiene-styrene (ABS); asbestos-cement (AC); brick pipe (BP); concrete pipe (CP); cast iron pipe (CIP); polyethylene (PE); polyvinyl chloride (PVC); and vitrified clay (VC).

Shield (shield system):   means a structure that is able to withstand the forces imposed on it by a cave-in and thereby protect employees within the structure. Shields can be permanent structures or can be designed to be portable and moved along as work progresses. Additionally, shields can be either pre-manufactured or job-built in accordance with 1926.652(c)(3) or (c)(4). Shields used in trenches are usually referred to as "trench boxes" or "trench shields."

Shoring (shoring system):   A structure such as a metal hydraulic, mechanical or timber shoring system that supports the sides of an excavation and is designed to prevent cave-ins.

Shotcrete: Spraying wet concrete (see also gunite).

Skin friction: Resistance to advancement caused by soil pressure around the pipe or casing.

Slipline: A renewal technique covering the insertion of one pipe inside an existing pipe.

Sliplining: (1) General term used to describe methods of lining with continuous pipes and lining with discrete pipes. (2) Insertion of a new pipe by pulling or pushing it into the existing pipe and grouting the annular space. The pipe used may be continuous or a string of discrete pipes. The latter is also referred to as segmental sliplining.

Slurry: A fluid, mainly water mixed with bentonite and sometimes polymers, used in a closed loop system for the removal of spoil and for the balance of groundwater pressure during tunneling and micro-tunneling operations.

Slurry chamber: Located behind the cutting head of a slurry micro-tunneling machine. Excavated material is mixed with slurry in the chamber for transport to the surface.

Slurry line: A series of hoses or pipes that transport tunnel muck and slurry from the face of a slurry microtunneling machine to the ground surface for separation.

Slurry separation: A process where excavated material is separated from the circulation slurry.

Slurry shield method: Method using a mechanical tunneling shield with closed face, which employs hydraulic means for removing the excavated material and balances the ground water pressure. See also earth pressure balance machine.

Social costs: Costs incurred by society as a result of underground pipeline construction and renewal. These include but not limited to traffic disruptions, environmental damages, safety hazards, inconvenience to general public, and business losses owing to road closures.

Soft lining: See cured-in-place pipe (CIPP).

Spiral lining: A technique in which a ribbed plastic strip is spirally wound by a winding machine to form a liner, which is inserted into a defective pipeline. The annular space may be grouted or the spiral liner expanded to reduce the annulus and form a close-fit liner. In larger diameters, the strips are sometimes formed into panels and installed by hand. Grouting the annular space after installation is recommended.

Spiral weld pipe (casing):   Pipe made from coils of steel plate by wrapping around a mandrail in such a manner that the welds are a spiral helix.

Spiral wound: In this process a new pipe is installed inside the existing pipe from the continuous strip of polyvinyl chloride (PVC). The strip has tongue and groove casting on its edges. It is fed to a special winding machine placed in a manhole, which creates a continuous helically wound liner that proceeds through the existing pipe. The continuous spiral joint is watertight. Upon completions of the annulus space between the lining and the existing pipe wall is usually required.

Spoil (muck):   Earth, rock, and other materials displaced by a tunnel, pipe or casing, and removed as the tunnel, pipe or casing is installed. In some cases, it is used to mean only the material that has no further use.

Spot repair: See localized repair.

Spray lining: A technique for applying a lining of cement mortar or resin by rotating a spray head, which is winched through the existing pipeline.

Springline: (1) An imaginary horizontal line across the pipe that passes between the points where the pipe has its greatest cross-sectional width. (2) Midpoint of a pipe cross section (equal vertical distance between the crown and the invert of the pipe).

Standard dimension ratio (SDR):   Standard dimension ratio is defined as the ratio of the outside pipe diameter to wall thickness. Same as DR.

Static mixer: A computerized device that provides fast, uniform resin and catalyst mixing.

Steering head: In horizontal auger boring, a moveable lead section of casing that can be adjusted to steer the bore.

Styrene: A component of polyester and many vinyl ester resin systems.

Subsidence: The settlement of the ground, pipeline, or other structure. The effects may not be evenly distributed and/or immediately noticeable. Differential settlement may occur.

Sump: A depression usually in the drive pit to allow the collection of water and the installation of a sump pump for water removal.

Swivel: In horizontal directional drilling, it is used to attach product pipe (to be pulled into drilled hole) to drill pipe to prevent it from rotating.

Target shaft or pit: See reception or exit shaft or pit.

TBM: See tunnel boring machine.

Teeth: See bits.

Televise: Process by which a sewer, pipeline or lateral is inspected with a closed-circuit television camera.

Thermocouple: A device used to measure the internal temperature of a resin-saturated felt tube during the installation process.

Thermoformed pipe: A type of renewal method that uses polyvinyl chloride (PVC) or polyethylene (PE) pipe that is expanded by thermo-forming to fit tightly to fit inside the existing, old, or host pipe.

Thermoplastic (TP):   A polymer material, such as polyethylene, that will repeatedly soften when heated and harden and reformed when cooled. TPs are generally much easier to recycle than their thermoset (see below) counterparts.

Thermoset (TS): A polymer material, such as rubber, that does not melt when reheated. TS polymers can be formed initially into almost any desired shape, but they cannot be reformed at a later time.

Thermoset resin: A material, such as epoxies, that will undergo or has undergone a chemical reaction by the action of heat, chemical catalyst, ultraviolet light, etc., leading to an infusible state.

Thrust: Force applied to a pipeline or drill string to propel it through the ground.

Thrust block: See backstop.

Thrust boring: A method of forming a pilot bore by driving a closed pipe or head from a thrust pit into the soil which is displaced. Some small-diameter models have steering capability achieved by a slanted pilot-head face and electronic monitoring. Back reaming may be used to enlarge the pilot bore. Also loosely applied to various trenchless installations methods. See rod pushing.

Thrust jacking method: Method in which a pipe is jacked through the ground without mechanical excavation of material from the front of the pipeline.

Thrust pit: See drive pit.

Thrust ring: A fabricated ring that is mounted on the face of the jacking frame. It is intended to transfer the jacking load from the jacking frame to the thrust bearing area of the pipe section being jacked.

Top Inversion: The CIPP tube is inverted from an inversion ring located at the top of the inversion platform.

Torque: The rotary force available at the drive chuck.

Track: A set of longitudinal rails mounted on cross members that support and guide a horizontal auger boring machine.

Trench (trench excavation): A narrow excavation (in relation to its length) made below the surface of the ground. In general, the depth is greater than the width, but the width of a trench (measured at the bottom) is not greater than 15 feet (4.6 m). If forms or other structures are installed or constructed in an excavation, so as to reduce the dimension measured from the forms or structure to the side of the excavation to 15 feet (4.6 m) or less (measured at the bottom of the excavation), the excavation is also considered to be a trench.

Trenching: See open-cut or conventional trenching.

Trenchless: A technology that is used for renewal of existing pipelines, typically without any excavation.

Trenchless methods: See trenchless technology.

Trenchless rehabilitation: See renewal.

Trenchless technology: Also NO-DIG, techniques for underground pipeline and utility construction and replacement, rehabilitation, renovation (collectively called renewal), repair, inspection, and leak detection, etc., with minimum or no excavation from the ground surface.

Tube: The fabric material tube used to carry and hold the thermoset resin materials in place against the existing pipe prior to curing.

Tuberculation: Localized corrosion at scattered locations resulting in knob like mounds.

Tunnel: An underground conduit, often deep and expensive to construct, which provides conveyance and/or storage volumes for wastewater, often involving minimal surface disruption.

Tunnel boring machine (TBM):   (1) A full-face circular mechanized shield machine, usually of worker-entry diameter, steerable, and with a rotary cutting head. For pipe jacking installation it leads a string of pipes. It may be controlled from within the shield or remotely such as in microtunneling. (2) A mechanical excavator used in a tunnel to excavate the front face of the tunnel (mole, tunneling head).

Ultra violet light cure: See light cure.

Uncased bore: Any bore without a lining or pipe inserted, that is, self-supporting, whether temporary or permanent. Not recommended except in special conditions.

Underground utility: Active or inactive services or utilities below ground level.

Upset: See machine upset.

Upsizing: Any method such as pipe replacement or pipe bursting that increases the cross-sectional area of an existing pipeline by replacing with a larger-diameter pipe.

Utility tunneling: It is general approach of constructing underground utility line by removing the excavated soil from the front of cutting face and installing liner segments to form continuous ground support structures. The product pipe is then transported and installed inside the tunnel. The annular space between the liner and the pipe is usually filled with grout.

VCP: Vitrified clay pipe.

VCT: Vitrified clay tile or vitrified clay tile pipe.

Velocity head: For water moving at a given velocity, the equivalent head through which it would have to fall by gravity to acquire the same velocity.

Vinyl-ester: Resin systems used for many industrial applications requiring a higher level of corrosion resistance.

Viscosity: That property of a fluid which determines the amount of its resistance to a shearing stress.

Voids: (1) Holes on the outside of the pipe in the surrounding soil or material. (2) A term generally applied to paints to describe holidays, holes, and skips in the film. Also used to describe shrinkage in castings or welds.

Walkover system: See locator.

Washover pipe: In horizontal directional drilling, sometimes (for drilling in rock) a rotating drill pipe of larger diameter than the pilot drill pipe is used and placed around it with its leading edge less far advanced. Its purpose is to provide stiffness to the drilling pipe to maintain steering control over long bores, to reduce friction between the drill string and the soil and to facilitate mud circulation.

Water jetting: (1) Method for the internal cleansing of pipelines using high-pressure water jets. (2) An obsolete and unauthorized method for cutting earth with water jetting.

Water table: (1) The depth of the ground water. (2) The upper limit of the portion of ground wholly saturated with water.

Weatherability: The properties of a plastic material that allows it to withstand natural weathering; hot and cold temperatures, wind, rain, and ultraviolet rays.

Wetout: The process of injecting resin into, and distributing it throughout, a hose or tube, which will then be installed into the pipeline and cured in place.

Winch: Mechanical device used to pull the CCTV cameras or cleaning tools through a pipe.

APPENDIX E

Conversion Table

From	To	Multiply by
Linear		
mil	in.	0.001
in.	mm	25.4
ft	m	0.3048
yards	m	0.9144
ft	in.	12
mile	km	1.609
mile	ft	5280
Area		
in.^{2}	mm^2	645.16
ft^2	m^2	0.0929
sq yards	m^2	0.8361
sq mi	km^2	2.5889
acres	km^2	4.0469×10^{-3}
ha	km^2	0.01
acres	ft^2	43,560
m^2	ha	10^4
Pressure		
psi	kN/m^2	6.895
psi	atm	0.0680
psi	kg/m^2	9.80665
psi	Pa	6894.757
psi	lb/ft^2	144
N/m^2	Pa	1
atm	psi	14.696
atm	Pa	101.325
bar	Pa	10^5
bar	psi	14.5

From	To	Multiply by
Specific Weight (weight/unit volume)		
lb/ft^3	n/m^3	157.1
lb/in^3	lb/ft^3	1728
Forces		
lb	kN	0.004448
tons	kN	9.96401
kN	kg(f)	102.0
Velocity		
mi/h	km/h	1.609
ft/s	m/s	0.3048
ft/min	m/s	0.00508
ft/min	m/min	0.305
Volume Flow Rate		
ft^3/s	gal/min	449
m^3/s	ft^3/s	35.3
m^3/s	gal/min	15,850
gal/min	L/min	3785
m^3/s	L/min	60,000
m^3/s	ft^3/min	2120
m^3/h	L/min	16.67
ft^3/s	m^3/h	101.9
Temperature Conversion		
°F to °C: deduct 32, multiply by 5, divide by 9		
°C to °F: multiply by 9, divide by 5, add 32		

Index

Note: Page numbers referencing figures are followed by an *"f"*; page numbers referencing tables are followed by a *"t"*.

AC. *See* asbestos cement pipes
access pits, 357–358
additional path curvature, 221–222
air scouring, 100*t*
alignment considerations, 286*t*, 328, 376–377
ambient cured resins, 288
American Society of Civil Engineers, 373
annular space, 305*t*
appurtenances, reconnecting, 105
arching effect of soils, 127–128, 128*f*
asbestos cement (AC) pipes
  advantages/limitations of, 149
  joint types in, 149, 149*f*
  manufacture of, 148–149
  overview of, 141
  usage of, 148
as-built drawings, 264

background assessment, 48–49
backstop, 310, 385
balling, 100*t*
bar-wrapped steel-cylinder concrete pipe, 143
beam effect, 136
bedding, 121*f*, 122*t*
bends, pipelines with, 278, 279*t*
berm, 385
bidding, 43, 43*t*
bits/reamer safety, 343
blisters, 103*t*, 301*t*
bore path entry/exit, 192, 192*f*
bore path layout/design
  accuracy/tolerance, 215
  depth of cover, 208–209, 208*f*
  depth/setback implications, 212
  horizontal distance to rise to surface, 209*f*, 212–213, 213*f*, 214*f*

bore path layout/design (*Cont.*):
  minimum depth at level, 211, 211*f*
  overall, 213–216
  path layout, 214–215
  planar trajectory (horizontal), 213
  profile/trajectory (vertical plane), 209, 209*f*
  safety and, 382
  separation from existing utilities, 216
  setback distance, 210–211, 211*f*
  vertical trajectory, 208–213
bore (product pipe) salvage, 27–28
borehole collapse, 269–271
branch connections, reinstatement of, 90
brick pipe defects, 285*t*
broken pipe, 284*f*
bubbles, 301*t*
buckling analysis for CIPP, 75–77, 79–80, 82
bulges, 302*t*
bulkhead, 360
bung holes, defective/leaking, 304*t*
buoyant weight, 222
buried pipe
  history, 1–3
  North America and, 1–2
  pipe-soil interaction and, 2–3
  Rome and, 1
bypass pumping, 256, 296, 296*f*

calculation
  CIPP deflection, 78–79
  CIPP design, 69, 75–80, 82–83
  CPJ force, 311–312
  HDD pipe load, 216–226
  load, 267–275, 273*t*
  MT force, 321, 321*f*–322*f*, 323
  pressure limited due to stress, 80, 83

441